THE ORANG UTAN
ITS BIOLOGY AND CONSERVATION

Perspectives in vertebrate science

Volume 2

Series Editor
EUGENE K. BALON

Dr W. JUNK PUBLISHERS   THE HAGUE — BOSTON — LONDON

The orang utan
Its biology and conservation

Edited by
LEOBERT E.M. DE BOER

Dr W. JUNK PUBLISHERS THE HAGUE — BOSTON — LONDON

Distributors:

for the United States and Canada

Kluwer Boston, Inc.
190 Old Derby Street
Hingham, MA 02043
USA

for all other countries

Kluwer Academic Publishers Group
Distribution Center
P.O. Box 322
3300 AH Dordrecht
The Netherlands

This volume mainly consists of the contributions, presented at the 'Workshop on the Conservation of the Orang Utan', held October 1979 at Rotterdam.

Library of Congress Cataloging in Publication Data

Workshop on the Conservation of the Orang Utan
     (1979 : Rotterdam, Netherlands)
     The orang utan, its biology and conservation.

     (Perspectives in vertebrate science ; v. 2)
     "This volume mainly consists of the contributions,
presented at the 'Workshop on the Conservation of
the Orang Utan,' held October 1979 at Rotterdam"--
T.p. verso.
     Includes index.
     1. Orangutan--Congresses.   2. Wildlife
conservation--Congresses.   I. Boer, Leobert E. M. de.
II. Title.   III. Series.
QL737.P96W67 1979   639.9'798842   82-7722
                                          AACR2

ISBN 90-6193-702-7 (this volume)
ISBN 90-6193-890-2 (series)

PRINTED IN THE NETHERLANDS

Preface

The orang utan or 'man of the forest', as one of the anthropoid apes, is one of man's closest evolutionary relatives, and as such has always had a place in our imagination. It was hunted by prehistoric man and its skull was used, as it is by some present-day tribes, as part of their religious rituals. It was regarded as a highly prized trophy both by the early colonizers on Borneo and Sumatra and, later, by the first naturalists collecting material for the world's great natural history museums. After a series of misunderstandings in the days of Linnaeus, the orang utan received its scientific name, *Pongo Pygmaeus*, and description, based on poor illustrations and material. The first live orang utans were brought to the Western world somewhat later, in the last decades of the 18th century, where they caused a great sensation in the 'menagerie' collections, the precursors of the zoological gardens. By the end of the 19th century and during the 20th century the orang utan attracted increasing attention from anthropologists and primatologists as a subject for the study of the relationship between man and the anthropoids. Nevertheless, this animal never obtained the attention that was given to the African apes, particularly the chimpanzee which, as man's closest evolutionary neighbour, has always been the favourite subject of scientific study. The highly arboreal, brachiating man of the South East Asian rain forests, with its specialized anatomy and apparent lack of complex social organization, seemed of much less importance in the search for the origin of the physical and social adaptations of the ground-dwelling, bipedal *Homo sapiens*.

During the early 1960s, however, the interest in the orang utan suddenly increased after it became apparent that the species was threatened with extinction. In the remote past its originally vast distribution area had become reduced to the islands of Borneo and Sumatra possibly due to human predation. During historic times, and with ever increasing speed, the ranges in Sumatra and Borneo were diminished due to expanding human settlements and increasing agricultural activities. More recently and more rapidly, its habitat has been destroyed by large-scale commercial logging. In addition, the numbers of orang utans in their last refuges have further decreased by hunting. Around 1960 it was first made clear to the world that the orang utan populations still in existence were so small that unless urgent measures were taken to rescue the species, total extinction would follow within the next few decades. Soon after these initial calls for help were made, large-scale research projects involving the orang utan in its natural habitat were undertaken and it became apparent that thus far almost nothing was known of the animal's social life, demography and ecology. The interest in the orang utan had never been in the 'man of the forest' as a biological species, but almost exclusively in the man of the forest's relationship with its greatest enemy, the 'man of wisdom'.

Thanks to the efforts of the early field workers and their successors and to the measures taken by the Indonesian government under whose care the orang utans live, these unique apes have survived up to the present day. This, however, does not mean that their prolonged existence is *a priori* ensured; they are still in need of strict protection without which their future is uncertain. Conservation, here as in many other instances, also means preservation under man-controlled circumstances. Reserves, however natural their faunal and floral composition may be, are of necessity smaller than the original range of the species to be conserved; their locations and boundaries are determined by man and natural exchanges between populations in different reserves are impossible unless induced by man. Populations thus conserved need at the very least to be monitored regularly and, if necessary, to be managed. This implies that proper conservation is impossible without sound knowledge of the species concerned, particularly knowledge of the structure of that species (intraspecific differences, population structure and dynamics, etc.), of its reproduction (birth rates, reproductive behaviour), social organization and ethology, and its interactions with its surroundings, food demands, migrations, etc. The aim of this volume is to provide information on the basic biology of the orang utan in the belief that this will be of importance for future conservation policy. As such, the present volume includes a number of contributions on various aspects of the biology and conservation of this ape; the emphasis is not, as is usual, on comparisons of the orang utan with other hominoids, but primarily on the species itself. The contributions are largely based on the papers presented at the 'Workshop on the Conservation of the Orang Utan', which was held in Rotterdam in October 1979, and which was jointly organized by the Royal Rotterdam Zoological and Botanical Gardens and the Erasmus University (Rotterdam). This meeting tried to bring together the three main categories of workers studying orang utans, viz. 1) fieldworkers who study orang utans in their natural habitat, 2) university researchers, especially those attached to the great primate centres where a large amount of basic theoretical and experimental knowledge has been compiled, and 3) zoologists affiliated with zoological gardens who are able to collect important information on breeding and maintenance of orang utans under captive conditions.

Although there is still some debate as to the role that institutions maintaining endangered species in captivity could play in their actual conservation, and although it is certainly not intended to overestimate this possible role, this volume does not restrict itself to information directly connected with practical conservation of the orang utan in its natural habitat. Many studies on this ape which may render basic data crucial to its conservation, are more readily undertaken on captive populations than on those in the wild. Rather than accentuating the debate on wild versus captive conservation, the aim of this volume is to bring together as much information as possible from different sources and to encourage discussions on topics of mutual interest.

It must be noted that this volume does not cover all aspects of the biology of the orang utan in equal detail; some aspects may be underrepresented or absent. It is hoped, however, that it may in its own way contribute to rescuing this precious animal from the fate that has already extinguished too many others.

Leobert E.M. de Boer

Acknowledgements

Since most of the contents of the present volume are based on the contributions presented at the 'Workshop on the Conservation of the Orang Utan', held October 1979 in Rotterdam (The Netherlands), it is appropriate to express our gratitude in the first place to all institutions and persons who made this meeting possible, viz. Professor and Mrs M. de Vlieger (Rotterdam), Professor M.W. van Hof (Rotterdam), Professor I. Steele Russell (London), Ir D. van Dam (Rotterdam), the Erasmus University, the Gerrit Jan Mulder Foundation, the Büttikofer Foundation, the H.M.A. Schadee Foundation, the Ngajji Foundation, the Koninklijke Volker Stevin Company, the Royal Rotterdam Zoological and Botanical Gardens and the City of Rotterdam. Mrs Barbara Harrisson (Leeuwarden) was kind enough to contribute to the meeting as a chairperson leading discussions on conservation, for which she is sincerely thanked.

Dr W. Junk Publishers is acknowledged for producing this publication. Mr G. Looten and Mr F.A.C.M. Hollink (both of Rotterdam Zoo) are thanked for designing or redesigning the majority of the drawn illustrations, while Dr A.R. Glatston (Biological Research Department, Rotterdam Zoo) contributed significantly by critically reading and correcting some of the manuscripts. The Royal Rotterdam Zoological and Botanical Gardens, finally, is acknowledged for discharging the editor from other duties during the editorial work for the present volume.

Contents

1. Distribution and evolution of the orang utan, *Pongo pygmaeus* (Hoppius)

Gustav H.R. von Koenigswald

New methods, new observations and new finds have helped us to determine 'Man's place in nature' more precisely than formerly possible. Absolute dating of fossils by the potassium/argon method has allowed us to develop a more exact time scale and to add a historical moment which is necessary for a better understanding of the relationship between modern and early man and also of man's relationships with the primates in general.

Our closest relatives among the living primates are the three great apes: gorilla, chimpanzee and orang utan. They are generally classified as Pongidae and they share quite a number of anatomical characteristics with man. Modern methods have led to a 'molecular anthropology', and according to many molecular anthropologists chimpanzee and gorilla are more closely related to man than is the orang utan (for a recent collection of reviews the reader is referred to Goodman and Tashian 1978). They even go so far as to suggest the inclusion of the first two species together with man in the Hominidae, leaving only the orang utan in the Pongidae.

All higher anthropoids show an astonishing variability, which in the past has led to the distinction of a great number of species, subspecies and races, often without sharp definitions. Their variety is as great as among domestic animals. If we follow Napier and Napier (1967) for the African anthropoid apes, one species and three subspecies of gorillas are presently distinguished:

Gorilla gorilla gorilla	Western lowland gorilla
Gorilla gorilla beringei	Eastern highland gorilla
Gorilla gorilla graueri	Eastern lowland gorilla

The highland gorilla, or mountain gorilla, living in the high mountains to the north and east of Lake Kivu, is often regarded as a good species, *Gorilla beringei*. It surpasses the lowland gorillas in size.

The chimpanzee includes two species and four subspecies:

Pan troglodytes troglodytes	Tschego
Pan troglodytes verus	Common chimpanzee
Pan troglodytes schweinfurthii	Eastern chimpanzee
Pan paniscus paniscus	Pygmy chimpanzee

It is of importance to note that the distribution areas of both chimpanzee species are separated by the Congo River.

For the Asiatic ape, the orang utan, we find one species with two subspecies:

The orang utan. Its biology and conservation, edited by L.E.M. de Boer
© *1982, Dr W. Junk Publishers, The Hague. ISBN 90 6193 702 7*

Pongo pygmaeus pygmaeus   Bornean orang utan
Pongo pygmaeus abelii   Sumatran orang utan

An old name for the orang utan, no longer in use, is *Simia satyrus*. In Sumatra orang utans occur only in the north of the island (Atjeh). In Borneo they are found in the north in Sabah and in the west in Sarawak. Napier and Napier (1967) state that the 'distribution on Indonesian Borneo (Kalimantan) is not known' (p. 267). The species, however, occurs in large parts of Kalimantan. Distribution maps are given by Reynolds (1968, Figs. 6–8). (See also Fig. 8 in this chapter, and Rijksen 1978, 1982.)

A classical study by Selenka (1898) on the intraspecific variations of the orang utan gives a good idea of the variability within the species. His work was based on limited material, but these were his own observations and field work, and it therefore is still valuable. He distinguished ten different races of the orang utan: two in Sumatra and eight in Borneo. Besides skin colour, he used the possession of marginal flanges and brain capacity. His list, in his own nomenclature, is given in Table 1. The borderlines between the various races in all cases are

Table 1. Races of Bornean and Sumatran orang utans distinguished by Selenka (1898)* with brain capacities in cm^3 (n = number of skulls measured by Selenka).

Borneo
   Males with cheek flanges:
      Brain 'megalencephal' (n = 38):
         males: 410–534 (av. 500)
         females: 400–490 (av. 430)   *Simia satyrus dadappensis*
      Brain 'mikrencephal' (n = 14):
         males: 380–460 (av. 430)
         females: 330–380 (av. 360)   *Simia satyrus batangtuensis*
      Brain 'mikrencephal' (n = 22):
         males: 410–440 (av. 430)
         females: 350–400 (av. 370)   *Simia satyrus landakkensis*
      Brain 'mikrencephal' (n = 6):
         females: 310–360   *Simia satyrus Wallacei*
   Males without cheek flanges:
      Brain 'megalencephal' (n = 89):
         males: 370–500 (av. 440)
         females: 300–450 (av. 370)   *Simia satyrus skalauensis*
      Brain 'megalencephal' (n = 6):
         females: 400–470   *Simia satyrus tuakensis*
      Brain 'mikrencephal' (n = 9):
         males: 420–430
         females: 320–360   (*Simia satyrus rantaiensis*)
                                       (small *S.s. tuakensis?*)
      Brain 'mikrencephal' (n = 22):
         males: 360–430 (av. 390)
         females: 350–410 (av. 370)   *Simia satyrus genepaiensis*
Sumatra
   Males with cheek flanges:
      Brain 'mikrencephal' (n = 6):
         males: 385–445
         female: 340   *Simia sumatranus deliensis*
   Males without cheek flanges:
      No information on brain size   *Simia sumatranus abongensis*

* Selenka's (1898) distribution map of the Bornean races is shown in Fig. 1.

formed by rivers; the areas between the rivers are like islands on which the populations are isolated (see Fig. 1 for Schlegel's distribution map of the eight Bornean races; Schlegel only studied material of orang utans from northwestern Borneo. He believed that in eastern Borneo comparable races were to be found. The Sumatran races both occur in northwestern Sumatra: the *deliensis* race in the left drainage-area of the Langkat river and in Dule, the *abongensis* race north of the Langkat drainage-area in the Mount Abong Abong region.) The distribution of the marginal flanges in male orang utans, generally regarded as typical, is irregular. North of the Ketungau river (a side river of the Kapuas) in Borneo we find sandwiched between the *tuakensis* (T) and the *genepaiensis* (G) race, both without marginal flanges, the *dadappensis* (D) race possessing marginal flanges (see Fig. 1). Also, since one of the Sumatran races has flanges while the other has not, we arrive at the conclusion that the occurrence of marginal flanges in male orang utans most probably is a rather recent acquisition.

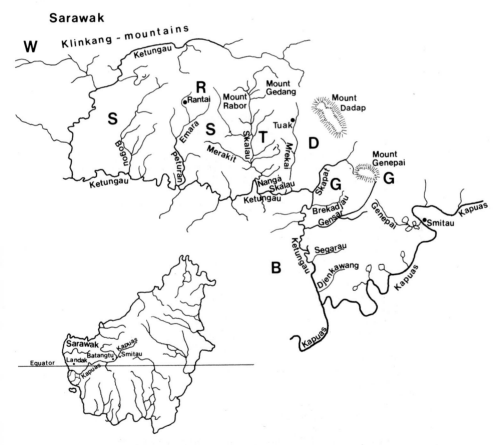

Fig. 1. The distribution of Schlegel's eight races of Bornean orang utans in northwestern Borneo. B = *Simia satyrus batangtuensis*; D = *S. s. dadappensis*; G = *S. s. genepaiensis*; R = *S. s. rantaiensis*; S = *S. s. skalauensis*; T = *S. s. tuakensis*; W = *S. s. Wallacei*; L = *S. s. landakkensis* (shown in the inset with name of the settlement).

3

In the following, we will limit ourself to skull and dentition of the orang utan, since these parts are most informative for tracing the species' palaeontological history.

The brain capacity of the orang utan is extremely variable. According to Schultz (1969) male brain capacity (n = 57) ranges between 334 and 502 cm^3 with an average of 416 cm^3, that of females ranges between 276 and 425 cm^3 with an average of 338 cm^3 (n = 52). He mentions six skulls with capacities of over 500 cm^3, four of which with 520 and one with 530 cm^3, which would be in the range for '*Homo habilis*'!

The skull of the orang utan shows an extreme posterior orientation of the foramen magnum. The breadth of the brain case (cranial index) varies from 70 to 99 'with a normal distribution of all intermediate variations' (Schlutz 1969). There is no supraorbital ridge as in the gorilla and chimpanzee. There is a forward position of the dental arch with respect to the neurocranium, and an elevation of the nuchal plate. The specific divergent skull topography is, according to Biegert (1963), due to the evolution of large laryngeal sacs, which in males extend far into the axillae.

There is a strong sexual dimorphism in orang utan skulls. Old males generally have a sagittal crest and a much broader face (Fig. 2). Male skulls are often asymmetrical, the face slightly turned to the right; Selenka (1898, Figs. 38–43) shows six such cases. All males have larger canines than females. Sagittal crests are never found in females.

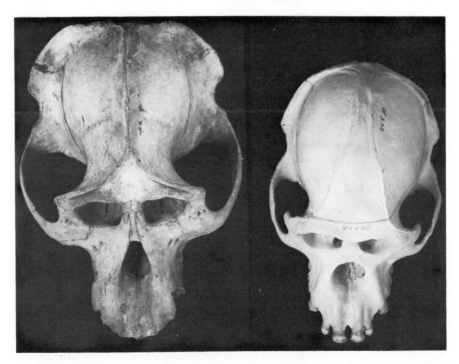

Fig. 2. The sexual dimorphism in the orang utan is well expressed in the skull. Full-grown skulls of male (left) and female (right). Females never show sagittal crests. Both sexes have no supraorbital tori as the African apes.

The dental arches are also variable. Among 158 mandibles Hellmann (1918) found the following types (Fig. 3):

1. U-shaped arch, in which the premolar/molar series are arranged in straight lines parallel to each other (n = 54);
2. Pyriform arch, in which the widest dimensions are in the first premolar region with the buccal teeth thereafter arranged in straight or nearly straight lines which converge posteriorly (n = 36);
3. Saddle-shaped arch, in which the premolar/molar series are arranged in curves with the convexity towards the tongue (n = 27);
4. Divergent arch, in which the premolar/molar series are arranged in straight or nearly straight lines diverging as they proceed backward (n = 24);
5. O-shaped or oval arch, in which the premolar/molar series are arranged in curved lines with the convexity towards the cheek (n = 17).

Among these types of arrangement of the teeth the diverging and the saddle-shaped types could easily be mistaken for typical hominid types, something which probably happened in the past to some fossils.

The teeth of the orang utan, like the animal itself, vary in size between those of gorilla and chimpanzee. They can be recognized by relatively low cusps in combination with a system of generally fine wrinkles, often irregular and bifurcated. Selenka mentions the presence of extra large teeth in his *dadappensis* and *batangtuensis* races (Schlegel 1898). Caries not only occurs in zoos, but also in wild and even in fossil animals (China).

Of the lower molars in *Pongo* (and also in *Gorilla*) the first is always the smallest (Remane 1921), which surely is a primitive condition. In male orang utans the third molar is the largest in 29.4% of the cases, in females only in 19.4%. This must be a more advanced trait, as in fossil anthropoids in nearly all cases the third molar is the biggest of the row (with the exception of *Ankarapithecus*).

All higher primates show a tendency to develop supernumerary molars, more often in the lower than in the upper jaw, and the orang utan more frequently than his African counterparts. Selenka (1898) found among 194 skulls 38 specimens with supernumerary molars, that is, in 20% of the cases (Fig. 4). He

Pyriform	U-shaped	Divergent	O-shaped	Saddle-shaped
36	54	24	17	27

Fig. 3. Lower jaws of the orang utan: various forms and frequencies of dental arches (after Hellmann 1918).

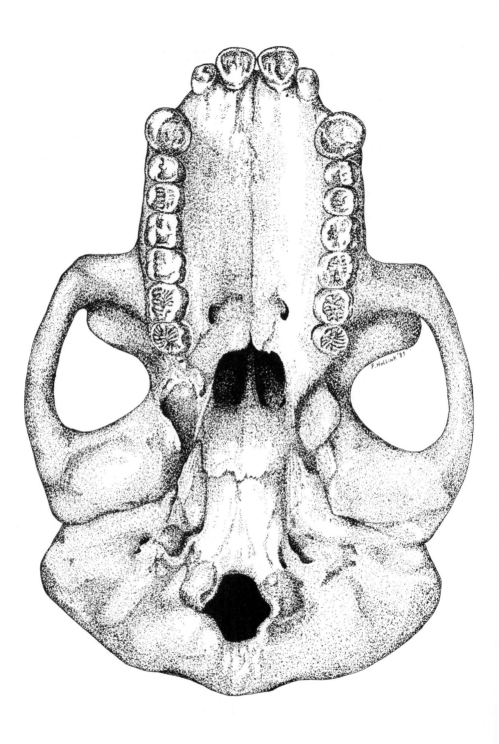

Fig. 4. Skull of a full-grown male orang utan with heavy superstructures. Note four molars on both sides. 'Batangtu' race (*Simia satyrus batangtuensis* Schlegel), Borneo, brain capacity 420 cm^3 (after Schlegel 1898).

Table 2. Length of fourth lower (P_4) and upper (P^4) premolar relative to length of first lower and upper molar in the anthropoid apes (after Remane 1962).

Species	Sex	P^4	P_4
Pan troglodytes	male	70.3	71.1
	female	73.1	71.9
Pan paniscus	male	72.6	78.0
	female	68.8	74.5
Gorilla gorilla	male	82.2	74.4
	female	73.6	75.0
Gorilla beringei	male	76.7	72.2
	female	75.1	68.8
Pongo pygmaeus	male	77.1	82.2
	female	79.3	82.1

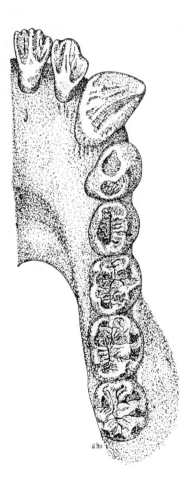

Fig. 5. Lower dentition of the orang utan. Note relative length of the last premolar.

even reports one case in which two additional molars were found on both sides of the upper jaw, together with one additional molar on both sides of the lower jaw, a total of six supernumerary molars. The Smithsonian Institution in Washington, D.C. is in possession of a mandible with two additional molars on the right side and one on the left side (Mus. no. 142199).

The most original pattern of the lower molars of the anthropoids is the so-called '*Dryopithecus*-pattern', originally described by Gregory and Hellmann (1926). A Y-shaped furrow isolating the hypoconid of the chimpanzee and especially of the gorilla shows this pattern perfectly. In the orang utan, however, this pattern is obscured, due to the wrinkles and low cusps, and in many cases a longitudinal valley has come into existence. We have not been able to make statistical analyses, but in the famous drawings of Selenka (1898) all molars show this condition.

Contrary to the African anthropoids, the dentition of the orang utan shows signs of molarisation, which means that the premolars more and more achieve the shape and pattern of the molars. This process is most common among perissodactyls (tapir, horse, rhinoceros) and some other groups, but never occurs in carnivorous mammals. In the orang utan premolar molarisation can be best illustrated by the increasing length of the fourth lower premolar relative to the length of the first molar (Table 2). In the orang utan the relative length of this element is clearly higher than in the other apes. In the upper jaw the predominance is somewhat less clear, the male gorilla by exception having an even relatively longer fourth upper premolar than the orang utan. In *Pongo* the shape of the last lower premolar is variable, but elongated specimens are frequent and can only be distinguished from molars by the absence of the hypoconid. A fine fossil specimen of an upper premolar from Sangiran (Java) on the hind of the trigon shows a minute hypocone, a clear illustration of a beginning 'molarisation'. This tendency of molarisation might also be expressed in rare bicuspid first lower premolars and in the bicuspid first deciduous molars.

The great complexity of the wrinkle system in the orang utan dentition is typical for this species and is most probably an adaptation to a pronounced frugivorous diet (see Fig. 7).

We have dealt in greater detail with the dentition of *Pongo* because teeth for the palaeontologist are often the only material for identification of a fossil species. To illustrate the hazards of fossilisation: more than 5,000 teeth of fossil and subfossil orang utans are known, and not a single jaw nor any parts of skulls or skeletons. This might be kept in mind to show how difficult it is to detect good specimens to reconstruct the evolutionary history of the higher primates in general and of the hominoids in particular. Nothing is known about the history of *Gorilla* and *Pan*; their remains are even absent from prehistoric human settlements. Due to ecological conditions they do not occur in sites with *Homo* and *Australopithecus* in Africa. Only Asia is different.

We will try here to sketch what is known of the history and former distribution of the orang utan. When E. Dubois arrived in the former Netherlands East Indies in 1889 to discover the 'missing link' between man and the apes as predicted by Charles Darwin, he began to excavate caves in

European fashion. In the Padang Highlands in Middle Sumatra he dug in three different caves (Lida Ajer, Sibrambang and Djambu), where he found 3,170 isolated teeth of orang utans. The few remains of other mammals all belonged to recent species (*Elephas, Tapirus, Rhinoceros, Helarctos* etc.), so the deposits are sub-recent. Of interest is that the orang utan here occurs outside its present distribution area. The astonishing number of teeth belong to full-grown individuals (there are only a few deciduous teeth).

A great number of orang utan remains (pieces of jaw and isolated teeth) have been found by Tom Harrisson during his excavation of the Niah Cave in Sarawak, Borneo. Here, the orang utan must have been the favourite game of prehistoric hunters. According to a preliminary report by Lord Medway, 'the orang utan occurs throughout the Niah Cave deposit in depth and area; it is second in abundance only to the pig'. The material has been described by Hooijer (1960). He mentions 111 specimens, 95 belonging to the permanent and 16 to the deciduous dentition, representing 14.4% of the total number of teeth recovered. This is more than in the larger samples from Central Sumatra and China, and certainly due to the activities of man. Among the canines there are twice as many belonging to females as to males. The deepest layers from Niah Cave are about 40,000 years old and contain apparently slightly larger specimens than the upper layers. There is no difference with the extant orang utan in the pattern and size of the teeth.

During our work in 1937 in the Gunung Kidul, the southern mountains of Central Java, we came across some fissure fillings near Punung and Patjitan. The fauna contains pigs, rhinoceros, a large tapir, bear, *Elephas* cf. *indicus* etc., and the teeth of primates, all isolated. There are many teeth of orang utans, a middle-sized form with large canines. This is the only site where we found lower molars of orang utans with a well-developed cingulum. This is also the only place where we found a fossil *Symphalangus syndactylus*, presently living only in Sumatra. Two lower teeth indicating a larger and a smaller form (probably representing sexual dimorphism) have been figured by Von Koenigswald (1940, Pl. 3, Figs. 9 and 10). The fauna has been described by Badoux (1959). The age of the fissure filling is most probably post-Trinil, probably Upper Middle Pleistocene, but certainly not sub-recent.

During his excavation in Trinil, where Dubois found the famous *Pithecanthropus* skull cap in 1891, he also discovered two large upper molars (Dubois 1924, Figs. 21–32) which he regarded as belonging to *Pithecanthropus*. There has been much discussion about them, and Weidenreich (1937) in his study of the dentition of *Sinanthropus* could show that this is impossible and that they might belong to the orang utan. The same opinion had been advocated by Miller (1923) and Von Koenigswald (1940) and Hooijer (1948) later came to the same conclusion.

Since then, we have collected a great number of teeth from the Pleistocene beds of Sangiran, Central Java. We have come to the conclusion, that the peculiar concave attrition of Dubois' second molar from Trinil is unlike anything we have found in the orang utan, but is typical for teeth we now refer to *Meganthropus* (Von Koenigswald 1967). Orang utan teeth from these beds occur in the Lower Pleistocene Djetis beds and the Middle Pleistocene Trinil beds in Sangiran (Von Koenigswald 1940, Pl. 2, Fig. 16). We only have isolated teeth, showing a great variety in size and pattern; they come from all parts of the section, especially the lower beds.

The upper jaw of *Pithecanthropus IV* (= *Homo modjokertensis*) shows the most primitive condition of any hominid specimen discovered thus far in Asia. The teeth rows are straight and divergent, the second molar is much larger than the first, and there is a 'simian gap'. A higher tribute could not be paid to this specimen than to attribute it to an anthropoid: only recently, Krantz (1975) described it as '*Pongo brevirostris*'! Nevertheless, there can be no doubt, that the orthognathous face with a typical human entrance into the nasal cavity with a nasal spine and the rest of the skull (capacity 900 cm^3) belong together and to a hominid.

In Indochina, Fromaget and Saurin (1936) excavated from the caves of Tam-Hang and Tam-Pá-Loi a number of orang utan teeth, supposed to be of Early Pleistocene age. Hooijer (1948), who had a chance to study the material, could not find specific differences between these and recent teeth, which are only a little larger. The same can be said of the material from Hang-Quit collected by Kahlke (1961).

The first fossil orang utan teeth from China were found in 1932 by C.C. Young, but were not recognised as such (Young 1932, refers to them as *?Ailuropodus*, p. 387, Fig. 4). His material consisted of a first lower premolar and a very worn third upper molar. The first recognisable material came from Chinese drugstores in Manilla and Hong Kong (Von Koenigswald 1935) and Nanning (Pei 1935). We found large quantities particularly in the apothecaries of Hong Kong and Canton; my own collection amounted to more than 1,500 specimens, and the collections in Peking are still larger.

Teeth from Yünnan have been illustrated in Weidenreich's (1937) extensive monograph on the dentition of Peking Man: upper molars, Figs. 282, 332 and 336, the latter with a 'Carabelli cusp'; lower molars, Figs. 179 and 243; and a lower canine, rather short and certainly belonging to a female, Figs. 61 and 258. For statistical reasons Hooijer (1948, p. 280) made this last specimen the holotype of *Pongo pygmaeus weidenreichi*. We believe this is rather premature, as within the Chinese material certainly several races if not species are represented.

The teeth occur mainly in the 'Yellow Deposit' quite common in rock fissures and caves in the south. The fauna is rich, but all bony parts have been gnawed by porcupines, leaving only the hard crown. The fauna contains *Elephas*, *Stegodon*, several species of deer and pig, *Rhinoceros sinensis* and bears; the giant panda (*Ailuropoda*) is not rare (therefore the name *Stegodon-Ailuropoda*-Fauna), while of the lesser panda a single tooth is mentioned; in addition the fauna contains *Capricornis*, *Macaca*, *Hylobates*, *Gigantopithecus*, and also Peking Man, recognisable by the very large first lower premolars (Von Koenigswald 1952). The age of the fauna is late Early to Early Middle Pleistocene. Faunal lists from this assemblage have been published by Von Koenigswald (1952), Kahlke (1961) and Aigner (1978). Remarks on the list of Aigner (1978, pp. 150–151): *Simia* is synonymous with *Pongo* and *Szechuanopithecus* is deciduous first molar of *Macaca* (see Von Koenigswald

1954). Aigner gives a list of eight localities with *Pongo*; this species is known from the provinces of Kwangsi, Kwantung, Szechuan Kiangsu and Yünnan, all in the south of China.

Kahlke (1961) not only gives a list of the distribution of *Stegodon-Ailuropoda* associations, but also the chronological positions. They range from the Upper Pleistocene (Würm; Mapa) through the whole of the Middle Pleistocene (Riss-Günz). The sites are spread over nine provinces.

The orang utan teeth from China show a great variability in size — the same, but not to the same extent, is also true for the fossils from Java — as is shown in Table 3 and Fig. 6. Also, the wrinkling of the surface shows great differences. This is not the place to go into detail, but some examples are shown in Fig. 7 to illustrate the conditions. There are also some teeth with no wrinkling at all; this might not be the result of attrition in all these cases. Some of the upper molars are so different from the typical orang utan, that we have described them as 'Hemanthropus'. During a visit to Peking, some years ago, I was not able to find additional material in the collections of the Geological Survey. Woo (1957) attributes the material selected by me to *Pongo*, but we are still convinced that this belongs to an independent species.

An isolated canine of 'Simia cf. *satyrus*' has been mentioned by Falconer (1868) from the Upper Siwaliks of India (Pinjor Beds of the Lower Pleistocene). The specimen is lost, and the determination based on such a specimen must be regarded as doubtful.

As we have seen above, there is much evidence that the fossil orang utan during the Pleistocene had a much wider distribution than the extant orang utan. The teeth (and nothing else is known) indicate a highly variable form, probably with lowland and (larger) mountain races or species. The oldest specimens come from the Lower Pleistocene of Sangiran, Java, and are about 2 million years old.

In Java the Pliocene beds are mostly marine, so no older specimens can be expected. In China the Pliocene 'Hipparion-Layers' do not contain the forerunners of the orang utan. However, intruders from the Indian Siwalik fauna, such as *Gigantopithecus*, *Ramapithecus* and *Sivapithecus*, indicate that the forerunner must be expected in that region.

There are indeed some signs from the Indian Siwaliks. In 1915, Pilgrim (Pl. 2, Fig. 9) described a wrinkled third upper molar from Chinji as *Palaeosimia rugosidens*, and already by the name has indicated that he regarded the new species as ancestral to the orang utan. This has been contradicted by Simons and Pilbeam (1965), who wanted to include this form in *Dryopithecus sivalensis*. We cannot agree with that opinion. I have seen unpublished lower molars from the Siwaliks with the same type of strong wrinkling and the development of a straight longitudinal valley as we have in the modern orang utan and I feel quite positive that here we are dealing with an ancestral form, about 10–12 million years old.

Andrews (1970) has described a palate from the Lower Miocene of Songhor, Kenya, East Africa, as possibly the earliest ancestor of the orang utan. The specimen, KNM-SO 700, shows highly wrinkled teeth, a certain molarisation of the premolars and great expansion of the maxillary sinus (also said to be typical

Table 3. Minimum and maximum length of third lower molars of recent anthropoid apes and fossil orang utans.

Recent (after Remane 1960):
Gorilla gorilla	13.9–20.3 mm
average male (n = 65)	17.2
average female (n = 29)	15.8
Gorilla beringei	16.6–23.0
average male (n = 20)	20.4
average female (n = 11)	17.9
Pongo pygmaeus	10.4–17.0
average male (n = 45)	14.2
average female (n = 28)	12.8
Fossil:	
Pongo	
from China	12.1–19.8 mm
from Sangiran, Java	17.5
from Sumatra*	9.5–16.8

*The lowest measurement from Sumatra is from one single exceptional specimen (no. 63), the next value is 10.9 mm, which is just within the variation for the living orang utan.

Fig. 6. Fossil orang utan from China: third lower molars, 'normal' and large form. Chinese drugstore, Hong Kong (Collection Von Koenigswald).

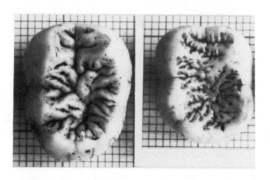

Fig. 7. Fossil orang utan from China: 'normal' wrinkling of lower premolar (left) and extreme fine wrinkling (right). The latter type is rare and perhaps belongs to a separate species. Chinese drugstore, Hong Kong (Collection Von Koenigswald).

for the orang utan). 'Parallelism is still a strong possibility; however, only the discovery of more complete Miocene and Pliocene material will show which explanation is the correct one' (Andrews 1970). Due to the strong internal cingulum on molars and premolars, typical for *Proconsul*, we favour parallelism. None of the Asiatic anthropoids shows a cingulum, which we regard as an over-specialisation rather than as a primitive character (Von Koenigswald 1973).

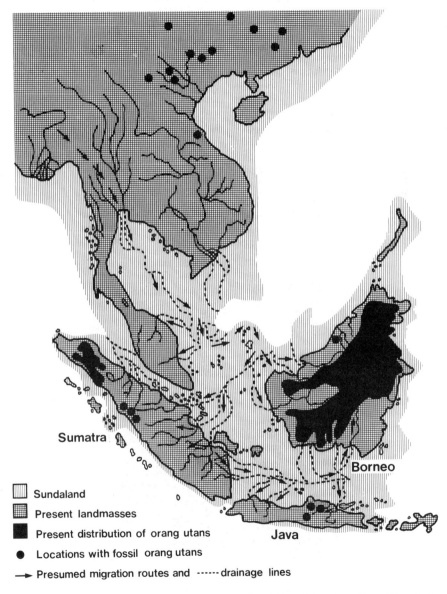

□ Sundaland
▥ Present landmasses
■ Present distribution of orang utans
● Locations with fossil orang utans
→ Presumed migration routes and ······ drainage lines

Fig. 8. Distribution of recent and fossil orang utan in South East Asia and southern China during the Pleistocene (after Kahlke 1967, and Rijksen 1978).

13

In the drugstores, orang utan teeth are frequently found together with the teeth of panda and *Capricornis sumatraensis*, the mountain goat. While in our time the orang utan is an inhabitant of the tropical rain forest, the panda and mountain goat are inhabitants of the high mountains. In Sumatra, *Capricornis* is only to be found at altitudes of 600 m and higher. Aigner (1978) gives three sites where panda and orang utan have been found together in situ.

Teeth of a very large orang utan are quite frequently found in the drugstores; they are larger than the average of the modern 'tropical orang utan'. In fact, the differences in size are comparable to those between the lowland gorilla (*Gorilla gorilla*) and the mountain gorilla (*G. beringei*). We therefore come to the conclusion, also based on the fauna composition, that in Pleistocene China a 'mountain orang utan' existed as a parallel form to the mountain gorilla in Africa.

The greatest area inhabited by the orang utan certainly existed in South East Asia during the Pleistocene, stretching from the Great Sunda Islands in the South to the southern part of China, bordering India in the West (Fig. 8). Since then, the species has disappeared from nearly the entire region, leaving only limited populations on Borneo and northern Sumatra. No reasons can be given for this. Climatic changes are certainly not responsible, as in the original area of distribution different types of environment were represented. These ranged from tropical conditions in the South (Java, Sumatra, Borneo, Vietnam) to more temperate regions in southern China, and from lowland to mountain ranges. The panda, a regular companion of the extinct orang utan, is still living, although its distribution now is also limited. That the ancient orang utan population has been diminished by man is also improbable. Where we have larger concentrations of orang utan teeth in Sumatra, Java, or from the rock fissures of Kwangsi and Kwantung, the number of immature individuals estimated from the number of deciduous teeth is small. If man had been involved as a hunter as he probably was in Niah (the only place with a higher percentage of deciduous elements) their number would have been much greater.

The extant orang utan occurs in small populations on islands. He must be regarded as an endemic species and we cannot say how much he deviates from the 'original orang utan' by isolation and inbreeding. Having seen the great variability, the Pleistocene forms might have been very different in appearance and perhaps in some respect closer to gorilla and chimpanzee.

References

Aigner, J.S. (1978). Pleistocene faunal and cultural stations in South China. In: F. Ikawa-Smith (ed.), Early Palaeolithic in South and East Asia. Mouton, The Hague.

Andrews, P. (1970). Two new fossil primates from the Lower Miocene of Kenya. Nature, 228: 537–540.

Badoux, D.M. (1959). Fossil mammals from two fissure deposits at Punung (Java). Ph.D. thesis, University of Utrecht.

Biegert, J. (1963). The evolution and characteristics of the skull, hands and feet for primate taxonomy. In: S.L. Washburn (ed.), Classification and human evolution. Viking Fund. Publ. 37: 116–145.

Dubois, E. (1924). Figures of the calvarium and endocranial cast, a fragment of the mandible and three teeth of *Pithecanthropus erectus*. Proc. Kon. Akad. Wetensch., Amsterdam, 37: 458–464.

Falconer, H. (1868). Palaeontological memoirs and notes of the late ... Compiled by Murchison. 2 vols. London.

Fromaget, J. and E. Saurin (1936). Note préliminaire sur les formations Cénozoignes et plus récentes de la chaine Annamitique septentrionale et du Haut Laos. Bull. Serv. Indochine, 12: 1–48.

Goodman, M., R.E. Tashian and J.H. Tashian (eds.) (1976). Molecular anthropology, genes and proteins in the evolutionary ascent of the primates. Plenum, New York.

Gregory, W.K. and M. Hellmann (1926). The dentition of *Dryopithecus* and the origin of man. Anthrop. Pap. Amer. Mus. Nat. Hist. 28: 1–123.

Hellmann, M. (1918). Observation on the form of the dental arch of the orang. Int. J. Orthodontia, 4: 3–15.

Hooijer, D.A. (1948). Prehistoric teeth of man and the orang-utan from Central Sumatra, with notes on the fossil orang-utan from Java and Southern China. Zool. Mededeel. Mus. Leiden, 29: 175–301.

Hooijer, D.A. (1960). The orang-utan in Niah Cave Pre-history. Sarawak Mus. J., 15–16: 408–421.

Kahlke, H.D. (1961). On the complex of the *Stegodon-Ailuropoda* Fauna of Southern China and the chronological position. Vertebrata Palasiatica, 6: 83–108.

Kahlke, H.D. (1967). Ausgrabungen auf vier Kontinenten. Uria Verlag, Berlin.

Koenigswald, G.H.R. von (1935). Eine fossile Säugetierfauna mit Simia aus Süd-China. Proc. Kon. Akad. Wetensch., Amsterdam, 38: 872–879. 27 + 1 figs.

Koenigswald, G.H.R. von (1940). Neue *Pithecanthropus*-Funde 1936–1938. Wetensch. Mededeel. Dienst Mijnbouw Ned. Indië (Batavia), 28: 1–232.

Koenigswald, G.H.R. von (1952). *Gigantopithecus blacki* v.K., a giant fossil hominoid from the Pleistocene of Southern China. Anthrop. Pap. Amer. Mus. Nat. Hist. New York, 43 (4): 295–325.

Koenigswald, G.H.R. von (1954). The status of *Szechuanopithecus* from the Pleistocene of China. Nature, 173: 643–644.

Koenigswald, G.H.R. von (1967). De *Pithecanthropus*-kiezen uit de collectie Dubois. Kon. Nederl. Akad. Wetensch., Versl. Verg. Afd. Natk., 76: 42–45.

Koenigswald, G.H.R. von (1973). The position of *Proconsul* among the pongids. IVth Congr. Primatology, Portland, vol. 3, Craniofacial biology of primates: pp. 148–153. Karger, Basel.

Krantz, G.S. (1975). An explanation for the diastema of the Javan *erectus* skull IV. In: R.H. Tuttle (ed.), Palaeoanthropology, morphology and palaeoecology. Mouton, The Hague.

Miller, G.S. (1923). Notes on the casts of the *Pithecanthropus* molars. Bull. Amer. Mus. Nat. Hist., 48: 527–530.

Napier, J.R. and P.H. Napier (1967). A handbook of living primates. Academic Press, London.

Pei, W.C. (1935). Fossil mammals from the Kwangsi Caves. Bull. Geol. Soc. China, 14: 413–425.

Pilgrim, G.E. (1915). New Siwalik primates and their bearing on the question of the evolution of man and the Anthropoidea. Proc. Geol. Surv. India, 45: 1–74.

Remane, A. (1921). Beiträge zur Morphologie des Anthropoidengebisses. Wiegmanns Arch. Naturgesch., A. 87: 1–179.

Remane, A. (1960). Zähne und Gebiss. In: H. Hofer, A.H. Schultz and D. Starck (eds.), Primatologia Vol. III. Karger, Basel.

Remane, A. (1962). Masse und Propertionen des Milchgebisses der Hominoidea. Biblioth. primat., 1: 229–238.

Reynolds, V. (1968). The apes. Cassell, London.

Rijksen, H.D. (1978). A fieldstudy on Sumatran orang utans (*Pongo pygmaeus abelii* Lesson 1827), ecology, behaviour and conservation. Mededel. Landbouwhogeschool Wageningen, 78–2: 1–420.

Rijksen, H.D. (1982). How to save the mysterious 'man of the rain forest'? In: L.E.M. de Boer (ed.), The orang utan. Its biology and conservation. Junk. The Hague.

Schultz, A.H. (1969). The life of primates. Weidenfeld, & Nicolson, London 1.

Selenka, E. (1898). Menschenaffen: Rassen. Schädel und Bezahnung des Orang-utan. Kreidel, Wiesbaden.

Simons, E.L. and D.R. Pilbeam (1968). Preliminary revision of the Dryopithecinae (Pongidae, Anthropoidea). Folia Primat., 3: 81–152.

Weidenreich, F. (1937). The dentition of *Sinanthropus pekinensis*. Palaeont. Sinica, 101: 1–180.

Woo, Ju-Kang (1957). *Dryopithecus* teeth from Keiyuan, Yunnan Province. Vertebrata Palasiatica, 1: 25–31.

Young, C.C. (1932). On some fossil mammals from Yünnan. Bull. Geol. Soc. China, 11: 383–393.

Author's address:
G.H.R. von Koenigswald
Senckenberg-Museum
Senckenberg-Anlage 25
D-6000 Frankfurt am Main
F.R.G.

2. The orang utan in captivity

Marvin L. Jones

Over 1,400 orang utans have been removed alive from Borneo and Sumatra since the early 18th century for the zoological gardens and research institutions of the world. In addition, more than 600 have been bred in captivity. This paper summarizes the captive history of the species based on a review of the available literature and previously unpublished information extracted from the archives of over one hundred collections, with special attention to breeding and longevity.

Introduction and methods

Barbara Harrisson (1963) mentioned that it was virtually impossible for her to secure accurate information on the hundreds of orang utans that had been held in captivity in zoos or placed into the trade by animal dealers. A search of the literature indeed offers very little as regards the majority of specimens that have been held in captivity. There are a few accounts of the first adults to arrive in Europe, of the large imports from Sumatra in 1927 and 1928, on breeding at the Nuremberg and Philadelphia zoos, and the work of Brandes at Dresden Zoo, but very little else.

Over the past 30 years the author has been given unlimited access to the archives of over one hundred major zoos around the world and has collected detailed information on over 3,000 species and subspecies of mammals that have been held in captivity. For this paper all of the information relative to the orang utan has been selected and compiled. The international studbook for the species, which the author keeps for the Zoological Society of San Diego, also serves as a reference point. The studbook appeared as a special publication of the Society (Jones 1980).

Early importations

The animal we call orang utan (and the name apparently is Malay rather than Bornean or Sumatran) had a rather confused early history and was not seen alive in captivity outside its native habitat until one arrived on the 29th of June,

The orang utan. Its biology and conservation, edited by L.E.M. de Boer
© *1982, Dr W. Junk Publishers, The Hague, ISBN 90 6193 702 7*

1776, at the private menagerie of the Prince of Orange in Holland. This young animal was fully described by Vosmaer (1778). Its death in January, 1777, and subsequent dissection were reported by Camper (1779). Reuther mentioned at a recent meeting of the International Union of Directors of Zoological Gardens that an orang utan came to America in 1789; however, I have been unable to confirm this report. The second known arrival outside the Indies was one that lived for five months at Empress Josephine's collection at Malmaison, France, in 1808. Two others arrived in London in 1816 and 1818, the latter brought back by Clarke Abel (for whom the Sumatran race, *Pongo pygmaeus abelii*, is named) (Abel 1825). Jeffries (1825) tells of one that came to Boston in that year, and it is known that two came to the menagerie of the Jardin des Plantes in Paris in 1831 and 1836. It is probable that one was seen at Lord Derby's Knowsley Hall collection before 1830. One brought in that year to the Bruton Street offices of the Zoological Society of London only survived for two days. The first orang utan exhibited in the London Zoo (then also called the Menagerie) arrived in 1838 and died in 1839. Fitzinger (1853) gives an excellent account of the history of the species until that point in time (but somehow overlooked completely the work of Hoppius [1763]) and mentions the live trade. A Viennese colleague named Kollar lived on Borneo and Sumatra from 1840 to 1848, and states that not less than 100 to 150 young animals arrived alive each year at the towns on the coast. The parents had been killed. Death due to malnutrition and dysentery took its toll and not more than 25 arrived in Singapore for shipment on to Europe, where probably only about five arrived alive. Add to this the vast numbers shot for the natural history museums of the world, and it is obvious that man in the early days was a major cause of the decline of the species.

From the mid-1800s through 1874, only a few animals reached the major animal collections of the day, only one or two each year, and these lived rather short lives. As far as is known all were infants only a few months of age.

In 1876 Hagenbeck brought to his then downtown Hamburg establishment what was called an adult male, as well as a younger female. They had been purchased in India, most likely in Calcutta. These were purchased immediately by Dr. Hermes for the Berlin Aquarium on Unter den Linden, and arrived there on the 2nd of March, 1876. It is not known just how long they survived, and while Friedel (1876) tells all about this arrival, it has somehow been ignored until now.

Brandes (1939) and Bungartz (1928) provide a detailed account of the arrival at the port of Antwerp of two large male orang utans in 1893. Like many imports of the time, they had been picked up by a ship captain who saw in the sale of the animals a lucrative side business. Captain Storm sold the animals to the Leipzig zoo director Pinkert, who took them at first to the private menagerie of Casten in Brussels, and then to the zoo run by the Société d'Acclimatisation in the Bois de Boulogne in Paris, which was the second major zoo of Paris at the time. They were an immediate sensation and it is said up to 35,000 people a day came to see them. Max was the younger animal, Moritz was estimated to be about 35 years of age. However, neither animal lived long enough to make the journey to Leipzig. Both died in January, 1894, of bronchial pneumonia, and their bodies were preserved.

In April, 1894, Captain Storm brought another adult male to Europe, this time to Hamburg, and it was also sold to Pinkert. Initially from April 3rd to 16th, it was shown at the then Hamburg Zoological Gardens (which closed in 1931) and was then taken on to Leipzig where it arrived on the 17th of April. It only survived until the 18th of May; cause of death was tuberculosis.

In May, 1895, Captain Storm again brought a large male to Germany, this time to his home port of Lübeck. Here it stayed for two weeks then also went to Leipzig, but eventually was sent to the Berlin Zoo in August, 1895. It lived a bit longer than the others, and died in 1899 of nephritis.

From 1776 to 1925 not less than 300 specimens are known to have arrived alive at the major collections of America, Australia, Europe and India, and a summary of these importations can be found at Tables 1 and 2.

Table 1. Wild-born orang utans exhibited outside the Dutch East Indies 1776–1925.

Year of import	Number*	Max longevity in months	Year of import	Number*	Max longevity in months
1776	0.1	6	1887	0.0.1	?
1789	0.0.1	?	1888	0.1.1	?
1808	0.0.1	?	1889	0.1.7	?
1816	0.0.1	?	1890	0.0.4	?
1818	0.0.1	?	1891	2.1.1	?
1825	0.0.1	?	1892	1.1.1	?
1830	1.0	<1	1893	4.0	48
1831	0.0.1	?	1894	1.3	6
1836	0.0.1	?	1895	1.2	48
1837	0.1	8	1896	2.0	37
1838	0.1	6	1897	1.2.1	?
1839	0.1.1	38	1898	0.1.5	106
1842	0.1	9	1899	0.3.2	29
1847	0.0.1	?	1900	1.2.2	108
1850	0.0.1	6	1901	5.0.3	?
1851	1.1	?	1902	3.2.1	56
1852	0.0.2	29	1903	4.0	240
1857	2.0	3	1904	2.3.1	28
1858	1.0	<1	1905	2.6.1	222
1860	1.1	?	1906	4.2	53
1861	0.1	3	1907	4.5.5	80
1863	0.0.2	2	1908	2.1.2	101
1864	0.1	?	1909	4.2.4	25
1866	1.1	?	1910	2.1	39
1867	0.0.1	?	1911	7.2.3	51
1868	0.0.2	?	1912	7.4	36
1872	0.0.2	5	1913	6.2.2	300
1873	0.0.1	?	1914	5.3.1	52
1874	0.2	?	1915	3.0.1	255
1875	3.2.2	24	1918	1.0	87
1876	1.0.1	?	1919	2.0	12
1877	1.1	36	1920	3.2.4	73
1878	3.2.1	21	1921	3.5.1	53
1879	0.1.4	39	1922	1.2.2	94
1880	1.0	<1	1923	5.2	41
1881	2.1.1	?	1924	5.4.4	102
1883	1.1	?	1925	7.3.10	116
1884	0.0.2	?			
1885	0.0.2	?	Totals	119.89.104	
1886	0.1.1	?		**312**	

* 1.0 = male; 0.1 = female; 0.0.1 = unknown sex.

Table 2. Collections outside the Dutch East Indies exhibiting the orang utan, 1776–1925.

Collection	First year of exhibit	Number exhibited
Adelaide Zoo, Australia	1889	7
Amsterdam Zoo, Holland[a]	1839	59
Antwerp Zoo, Belgium	1847	3
Basle Zoo, Switzerland	1900	4
Berlin Zoo, Germany	1872	14
Bronx Zoo, N.Y., USA	1899	28
Cairo Zoo, Egypt	1898	1
Calcutta Zoo, India	1875	4
Chicago (Lincoln Park Zoo), USA	1904	7
Cologne Zoo, Germany	1863	4
Dresden Zoo, Germany	1895	4
Dublin Zoo, Ireland	1913	1
Frankfurt (Main) Zoo, Germany	1903	5
Guadalajara Zoo, Mexico	1925	1
The Hague Zoo, Holland	1839	2
Hamburg Zoo, Germany	1873	22
Hannover Zoo, Germany	1924	2
Havana (Madame Abreu), Cuba[b]	1924	3
Leipzig Zoo, Germany	1894	2
London, collection unknown, England	1816	2
London (Zoological Society), England	1830	52
Marseilles Zoo, France	1892	1
Melbourne Zoo, Australia	1881	10
Paris (Jardin d'Acclimatation), France	1893	2
Paris (Jardin des Plantes), France	1808	6
Philadelphia Zoo, USA	1879	8
Menagerie of the Prince of Orange, Holland	1776	1
Rome Zoo, Italy	1911	2
Rotterdam Zoo, Holland	1875	38
St. Louis Zoo, USA	1918	6
San Diego Zoo, USA	1923	1
Stuttgart (Nihls Tiergarten), Germany	1899	1
Sydney Zoo, Australia	1898	10
USA, exact collection unknown	1789	2
Vienna Zoo, Austria[c]	1878	6+
Washington Zoo, USA	1908	2

[a] No exact data for the years 1873 to 1890.
[b] Three animals on exhibit in 1924, arrival dates unknown.
[c] In addition to the six shown, at least six others in the collection, only dates of death known, not arrival.

Importations 1926–1945

Starting in 1926, a number of collectors came back to America and Europe with adult specimens. These were principally from Sumatra, as for instance the pair that came to Berlin Zoo in May, 1926, and a few from Borneo, like the female Maggie who arrived at the Philadelphia Zoo, also in May, 1926. However, the major change took place in July, 1926 with the capture near Traemon in northern Sumatra of a large male by the Dutch collector Van Goens. He was called Jacob by the crew of the ship bringing both back to Europe. At the

Alfeld/Leine headquarters of the German dealer Ruhe, negotiations began to send the orang utan to Moscow. The arrangements broke down over problems associated with transport in the approaching winter of 1926 and he was sent instead to the Dresden Zoo, where director Dr. Gustav Brandes renamed him Goliath. Many pictures have been published of this huge male, and his arrival began a series of long-term behavior studies of the species at Dresden, which even today stand as classic examples.

Sent back to Sumatra by Ruhe for more animals, Van Goens returned to the port of Amsterdam in April, 1927, with the largest group of orang utans ever to reach Europe alive. There were 25 altogether, including six fully adult pairs each with a young, although it is not known if they were captured as family units. The group was taken for a few days of rest to the Palm House of the Amsterdam Zoo, and word flashed on to Alfeld of their safe arrival. Cables and telephone calls went out to the major zoo directors who began to assemble, along with writers such as Paul Eipper. The animals arrived at the new quarters in Alfeld at the end of April. The price was 25,000 marks a pair, and each of the directors left with as many as the budget permitted. Unfortunately only about half survived very long, as one might expect, but this only whetted the appetite of both dealer and customers for more. Supposedly Van Goens only lost one animal between Sumatra and Holland, and one was born on the 18th of April while the ship was in the Red Sea. The young male, taken by Brandes to Dresden and named Buschi, was left with its mother, Suma and was reared to adulthood. The book, *Buschi*, is based on Brandes' observations of his development and includes chapters on Goliath and other orang utans at Dresden (Brandes 1939). Brandes became the first to describe the development of the species in captivity. It has become a practice since then for German zoos to name orang utans Buschi and Suma, much as American zoos called their animals Jiggs and Maggie after the comic strip characters of the same name.

On the 29th of August, 1927, Van Goens returned with the second shipment, 33 animals this time, which were supposedly sold en bloc to the American circus impresario John Ringling who personally came to the port of Rotterdam to take delivery. Two of them died on the way to America, but just what became of the others remains somewhat of a mystery. Based on zoo records it appears some were actually sold in the U.S. German accounts indicate that the animals were never paid for by Ringling, and that supposedly two died en route to America. An account in *Billboard*, the American circus paper of the day, mentions that they arrived at the New Jersey port of Hoboken, but does not mention Ringling. It says, rather, that the orang utans were imported by the dealer, Bartels. The two adult pairs acquired in 1927 by Madame Rosalie Abreu for her large anthropoid colony at Quinta Palatina near Havana, Cuba, are believed to have come from this group, two of which, the famous Guas and Guarina, later went to the Philadelphia Zoo in 1931 with their youngster born in Cuba in 1929.

Following the revolution in Russia, an émigré named Georges Basilewsky purchased an estate at the town of Cros des Cagnes, between Nice and Cannes, on the French Rivièra, which he called the 'Centre d'Acclimatation de la Rivièra'. He was aided in this effort by the dealer, Ruhe and the Munich zoo director, Dr. Heinz Heck, and the Centre opened in 1923. This park afforded both the dealer and the zoo a warm weather place to acclimate new tropical animals, as the sun usually shone 300 days a year.

In the spring of 1928 Van Goens came back again with his third major Sumatran orang utan group. The relative ease of capturing so many adult animals in such a short period, gives some indication of the commonness of the species in those days. This time he had 44 animals, 30 were adult and of these there were 10 pairs, again each with a youngster. Rather than going to Alfeld, these were taken to the new Rivièra establishment of Basilewsky, and again were visited by prospective buyers and by Paul Eipper. After a period of rest in the sun of southern France the animals were sent to a number of collections, worldwide, and while some, as in the previous shipments, did not survive for long, many did, and in fact form the basis for the first captive breeding of the orang utan, in many collections.

This appears to have been the last large shipment allowed out by the Dutch authorities before World War Two intervened, although Van Goens continued to return to Europe with smaller groups. In fact, he was the first to bring adults out again following the end of hostilities in 1946.

Of the vast numbers brought out of Borneo and Sumatra before 1945, only a handful managed to survive the rigors of war, which not only created havoc in the bomb-damaged cities of Europe but also food shortages in other parts of the world. There were the breeding pair Guas and Guarina at Philadelphia, and scattered individuals at zoos in Adelaide, Antwerp, Brookfield, Edinburgh, London, Melbourne, Rotterdam, St. Louis, San Diego and Washington, often kept alone. Today none of these survive. Most died during the 1950s, although the Philadelphia pair lived until the late 1970s.

At Table 3 is a listing of known exports from 1926 to 1945.

Captive breeding 1928–1945

While a few individuals survived long enough to become sexually mature, prior to the arrival in 1926 of the adult pairs most had developed severe cases of rickets as youngsters and proved incapable of breeding. With the arrival of the new animals, already adult and in relatively good condition, it was anticipated that breeding would now take place.

The pair of Sumatran orang utans that arrived in Berlin in May, 1926, became the parents of the first captive-bred and born orang utan on the 12th of January, 1928. It was a female, which lived only until 18 February of the same year. One of the Ruhe shipment pairs, at Nuremberg, had a youngster in April, 1928, while at Philadelphia a male that came in 1927 bred with a 1926 female and this young was born 25 September, 1928.

From then until 1945 I can account for the birth of 31 individuals, one of which was not captive-conceived. Unfortunately, most did not live to maturity, due to malnutrition problems aggravated by the already poor wartime diets and often lack of adequate milk from the mother. Few were removed and hand-rearing attempted, although it became practice at Philadelphia to do this once the young was weaned, when it appeared the mother had not provided enough

Table 3. Wild-born orang utans exhibited outside the Dutch East Indies 1926–1945.

Year of import	Number[a]	Maximum longevity in months
1926	5.7.6	210
1927[b]	0.0.59	574
1928[c]	4.4.44	177
		176
1929	3.1.3	125
1930	6.2.3	294
1931	5.1	43
1932	3.1.1	120
1933	1.1.2	124
1934	4.2.1	73
1936	8.5.5	126
1937	3.2.2	263
1938	6.2	121
1939	4.2	173
1940	3.2	214
1941	0.2	200
1942	1.0	103
1943	1.0.1	?
Totals	57.34.127 **218**	

[a] 1.0 = male; 0.1 = female; 0.0.1 = unknown sex.
[b] 25 specimens arrived in Germany Apr. 1927, 33 in Aug. 1927.
[c] 44 specimens arrived in Europe Mar. 1928.

milk. Only one of these pre-1945-born youngsters ever bred. The female Ivy, born at Philadelphia in 1937, had several young sired by her father Guas, none of which survived for long. Usually the failure was due to a lack of milk and an inability to remove the young. This was before the invention of the dart gun which now renders such practice relatively easy.

At Table 4 is a list of all of the known young born from 1928 to 1945. Barbara Harrisson (1963) mentions that Moscow Zoo had one born in this period, but this cannot be confirmed.

Table 4. Captive-born orang utans 1928–1945 (S indicates the Sumatran subspecies, B the Bornean).

1. 12 Jan. 1928,   Sumatran, female, Berlin Zoo 1
     died   18 Feb. 1928
     Male   11 May 1926 Hassan x
     Female 11 May 1926 Cleo
2. 14 Apr. 1928,   Sumatran, male, Nuremberg 1, Bobby
     died   1931
     Male   May 1927 x
     Female   May 1927
     Was hand-reared and at six months of age weighed two kilos more than the mother-reared male Buschi at Dresden
3. 25 Sep. 1928,   Sumatran × Bornean, male, Philadelphia 1
     died   13 Sep. 1929
     Male   13 Sep. 1927 Chief Utan (S) x
     Female 12 May 1926 Maggie (B)

Table 4. (Continued)

4.	1929,	Sumatran, female, Abreu 1, Alberic, to Philadelphia Zoo 1 May 1931 with parents
		died 5 Mar. 1932
		Male 1927 Guas x
		Female 1927 Guarina
5.	5 May 1930,	Sumatran, male, Berlin Zoo 2
		died 16 Jan. 1933
		Male 19 Jun. 1929 Adam x
		Female 11 May 1926 Cleo
6.	11 Jul. 1930,	Sumatran × Bornean, Philadelphia 2, Norsuto
		died 14 Feb. 1943 (accident)
		Male 13 Sep. 1927 Chief Utan (S) x
		Female 12 May 1926 Maggie (B)
7.	14 Jul. 1930,	race unknown, Paris, Jardin des Plantes 1
		died 23 Oct. 1931
		Male 1928 x
		Female 1928
8.	9 Jan. 1931,	Sumatran, sex unknown, Düsseldorf 1
		died 9 Jan. 1931
		Male Apr. 1930 Mawas x
		Female 19 Sep. 1926 Julchen
9	18 Oct. 1931,	Sumatran, female, Düsseldorf 2
		date of death unknown
		Male Apr. 1930 Mawas x
		Female 19 Sep. 1926 Julchen
10.	11 Jun. 1932,	Sumatran × Bornean, Philadelphia 3
		died 11 Jun. 1932
		Male 1 May 1931 Guas (S) x
		Female 12 May 1926 Maggie (B)
11.	13 Aug. 1933,	race unknown, St. Louis 1, Patti Sue
		died 18 Oct. 1936
		Male 23 Oct. 1925 Bobby x
		Female 6 Feb. 1928 Bimbo
12.	13 Jan. 1934,	Sumatran, male, Berlin Zoo 3
		died 7 Jun. 1935
		Male 19 Jun. 1929 Adam x
		Female 11 May 1926 Cleo
13.	1 May 1934,	race unknown, Chicago (Lincoln Park) 1
		died 10 May 1934
		Male 4 Aug. 1928 Jiggs x
		Female 4 Aug. 1928 Nancy
14.	22 Jan. 1935,	Sumatran, female, Philadelphia 4, Cinderella, to Washington D.C. 8 Feb. 1944
		died 10 May 1947
		Male 1 May 1931 Guas x
		Female 1 May 1931 Guarina
15.	20 Oct. 1935,	race unknown, Chicago (Lincoln Park) 2
		died 31 Oct. 1935
		Male 4 Aug. 1928 Jiggs x
		Female 4 Aug. 1928 Nancy
16.	23 Apr. 1936,	Sumatran, sex unknown, Berlin Zoo 4, to Munich 28 Oct. 1938
		date of death unknown
		Male 19 Jun. 1929 Adam x
		Female 11 May 1926 Cleo
17.	3 Apr. 1937,	Sumatran, male, Munich 1, Heinz
		date of death unknown
		Male 10 Jul. 1936 Satan x
		Female arrival unknown, Joliette

Table 4. (Continued)

18.	13 Jun. 1937,	Sumatran, female, Philadelphia 5, Ivy
		died 24 Jul. 1972
		Male 1 May 1931 Guas x
		Female 1 May 1931 Guarina
19.	18 Aug. 1937,	race unknown, sex unknown, St. Louis 2
		died 18 Aug. 1937
		Male unknown, wild-bred
		Female 7 Aug. 1937 Maggie
20.	4 Apr. 1938,	Sumatran, female, Berlin Zoo 5, Musse, to Copenhagen 25 Nov. 1943
		died 14 Oct. 1946
		Male 19 Jun. 1929 Adam x
		Female 11 May 1926 Cleo
21.	17 Feb. 1939,	race unknown, female, St. Louis 3, Betsy
		died 8 Dec. 1939
		Male 7 Aug. 1937 Bob x
		Female 6 Feb. 1928 Bimbo
22.	6 Sep. 1940,	race unknown, male, St. Louis 4, George
		died 29 Oct. 1941
		Male 7 Aug. 1937 Bob x
		Female 7 Aug. 1937 Maggie
23.	7 Sep. 1940,	Sumatran, sex unknown
		date of death unknown
		Male 10 Jul. 1936 Satan
		Female unknown
24.	10 Oct. 1940,	race unknown, male, Paris, Jatdin des Plantes 2
		died 25 May 1941
		Male 23 Aug. 1938 x
		Female 16 Jan. 1939
25.	24 Nov. 1940,	Sumatran, male, Philadelphia 6, Rusty
		died 20 Feb. 1959
		Male 1 May 1931 Guas x
		Female 1 May 1931 Guarina
26.	24 Jun. 1941,	race unknown, male, St. Louis 5, Edwin
		died 21 Oct. 1941
		Male 7 Aug. 1937 Bob x
		Female 6 Feb. 1928 Bimbo
27.	12 Jul. 1942,	Sumatran, sex unknown, Edinburgh 1
		died Jul. 1942
		Male 18 Aug. 1937 Mickey x
		Female 18 Aug. 1937 Minnie
28.	11 Oct. 1942,	race unknown, male, St. Louis 6, Henry
		died 8 Jul. 1949
		Male 7 Aug. 1937 Bob x
		Female 6 Feb. 1928 Bimbo
29.	6 Oct. 1943,	Sumatran, female, Philadelphia 7
		died 20 Nov. 1944
		Male 1 May 1931 Guas x
		Female 1 May 1931 Guarina
30.	11 Jan. 1943,	race uncertain, female, San Diego 1
		died 13 Jan. 1943
		Male 22 Sep. 1934 Mike II (S) or
		Male 21 Aug. 1940 Kok-kok (B) x
		Female 20 Jun. 1936 Goola (S)
31.	1945,	race and sex unknown, Rome 1
		date of death unknown
		arrival date of parents unknown

Importations 1946–1978

The first studbook for the orang utan was compiled by Dr. Geoffrey Bourne and the staff of the Yerkes Regional Primate Research Center at Atlanta, Georgia. It listed animals held in captivity as of 31 December, 1969. When I assumed responsibility for the studbook in 1976 for the Zoological Society of San Diego, and following consultations with other studbook keepers, the decision was made to list all animals held in captivity (or bred in captivity) from 1 January, 1946. It was felt that sufficient records existed to make this practical, and it offered a good record of the development of the species in captivity after many of the problems of the pre-1945 period had been solved, by such invention as the dart gun, and the advent of new diets, advanced medication, and improved general health care.

During the period 1946 to 1978 I can account for the arrival at zoos and other collections, including Jakarta and Surabaya in Indonesia, of some 809 individuals (403 males, 399 females and 7 of undetermined sex). Of these, 180 males and 196 females are still alive today. It is probable that no more than an additional 50 to 100 unreported animals have been brought to the few collections not cooperating with the studbook. Contrary to the period from 1926 to 1945, few of the 809 individuals were captured as adults. As during the period from 1776 to 1925, the vast majority have been youngsters, and even with better attention, there have been serious losses. The Rotterdam Zoo, working with Van Goens, did bring in a number of adult animals in the early postwar years, which also bred. Frankfurt brought in the imposing male Moritz, and West Berlin the male Tuan and female Babu.

It has been assumed that the majority of the youngsters were captured after the adults were killed, hence a number of pleas for restraint on the part of humane groups. Today the back of the market in wild born orang utans has been broken, not only due to the various international regulations but also because of the longevity of captive specimens and the number born in captivity. In parts of the home range, it was common practice until just a few years ago to keep specimens as pets. This practice has now virtually halted, with the new re-introduction schemes as reported in this volume by Herman Rijksen and Rosalind Aveling.

The orang utan studbook (Jones 1980) offers detailed statistics on all 1946–1978 imported animals (copies of the studbook will be available on request from the author and sent to all owners as a matter of course).

Captive breeding 1946–1978

As previously mentioned the majority of post-1946 imported individuals were relatively young on arrival in captivity, and breeding thus did not really commence until they had reached sexual maturity. Preconceived ideas about age at sexual maturity sometimes prevented the introduction of animals until this threshold had long passed. Other animals were kept together from almost the day of arrival or birth, a practice not always conducive to good breeding results.

However, once sufficient numbers reached maturity, breeding commenced, and since 1946 some 508, many of which survive today, have been captive born in approximately 114 collections. Some of the young have in turn reproduced, although perhaps not as many captive-bred animals have been paired with other captive-bred mates to create the number of full second-generation young as many critics would like to expect.

While the practice of mating captive-bred young with parents, such as daughters to fathers, is deplored, genetically the mixing of captive-bred with unrelated wild-bred individuals is good for the captive gene pool. Sufficient numbers of captive-bred animals have in turn reproduced successfully to debunk the myth that such breeding is impossible. Quite to the contrary, several births in recent years have shown that the age at sexual maturity may be as low as six, and knowing that females can breed until 35, giving a possible 29 year breeding period, fears of a lack of captive-bred stock to perpetuate the species in captivity appear to be groundless.

At Table 5 is a list of births, by collection, for the period 1946 to 1978, and at Table 6 a list of captive-bred specimens that have reproduced, with data on young. Table 7 shows all of the twin captive-births to 31 December, 1978.

Table 5. Studbook registered captive-bred and born orang utans by collection, 1946–1978.

Collection	To wild-born parents	To one wild-born and one captive-born parent	Both parents captive-born
Aalborg	1.1.2*		
Al Ain	1.1		
Alberquerque	2.2		
Amsterdam	5.2		
Antwerp	2.2		
Arnhem	0.1		
Atlanta, Yerkes	25.17		
Basle	3.2	1.3	
Berlin Tierpark	5.5	2.0	
Berlin Zoo	7.5		
Birmingham			1.0.1
Bristol	3.2		
Bronx, N.Y.	1.1		
Brookfield	3.2		
Brownsville	6.4		
Buffalo	0.1		
Calgary	7.2		
Chester	4.1		
Chicago	1.4		
Cleveland	0.1		
Cologne	1.3		
Colombo	1.0		
Colorado Springs	12.11		
Columbus	1.0		
Copenhagen	1.1		
Dallas	5.3		
Denver		0.1	
Detroit	2.1		0.1
Dortmund			1.0
Dresden	4.6		

*1.0 = male; 0.1 = female; 0.0.1 – unknown sex.

27

Table 5. (Continued)

Collection	To wild-born parents	To one wild-born and one captive-born parent	Both parents captive-born
Dudley	1.1		
Duisburg	8.7		
Edinburgh	0.0.2		
Fort Worth	2.3		
Frankfurt	12.10	1.0	
Fresno	4.4		
Hamburg, Hagenbeck	3.1		
Hannover	0.1		1.0
Heidelberg	0.2		
Honolulu	0.1		
Houston	1.0		
Jakarta	0.5		
Jersey	2.2		
Kansas City	3.1		
Kobe	0.1		
Krefeld	1.1		
Kuala Lumpur	3.4		
Kyoto	0.1		
Leipzig	0.1	1.0	
London	7.6	2.2	
Los Angeles	2.4	1.1	
Madison	3.3		
Malton	1.0		
Manchester	2.1		
Maruyama	0.1		
Melbourne		1.1	
Memphis	4.0		
Miami, Monkey Jungle	2.1		
Milwaukee	1.1		
Monaco	1.0		
Monroe		3.0	
Munich	4.4	0.1	
Nagoya	1.2		
Natal	1.0		
Neunkirchen	0.1		
New Orleans	1.0		
New York, Central Park	1.0		
Nuremberg	2.6.1		
Oklahoma City	4.1	1.0	
Omaha	4.2	1.0	
Osnabrück	1.0		
Paris, Jardin des Plantes		0.1	
Perth	4.6	0.1	
Philadelphia	2.3	2.5.1	
Phoenix	2.1		
Pittsburgh	0.1		
Portland	2.1		
Prague	1.0		
Pretoria	1.1		
Rhenen	0.2		
Rio de Janeiro	0.1.3		
Rome	1.1		
Rostock	0.1		
Rotterdam	9.7	1.0	1.0
Sacramento	1.0		
St. Louis	11.3	1.1	
St. Paul	2.2		

Table 5. (Continued)

Collection	To wild-born parents	To one wild-born and one captive-born parent	Both parents captive-born
Salt Lake City	1.0		
Salzburg	1.0		
San Diego	2.4	5.4	0.1
San Francisco	0.1	0.2	
Seattle	1.1		
Shizuoka	0.0.1		
Singapore	4.0		
Stoneham	3.0		
Studen	0.1		
Stuttgart	2.11		
Surabaya	1.6		
Sydney	3.6	4.2	
Taiping	0.1		
Takamatsu	2.0		
Tarpon Springs	1.1		
Tel Aviv			0.0.2
Tokyo-Tama	0.4		
Tokyo-Ueno	2.1	1.3	
Toledo	5.2		
Toronto-Metro	1.1		
Toronto-Riverdale	1.0		
Toyohashi	0.0.1		
Twycross	2.5		
Vienna	0.2	0.0.2	
Washington	2.4		1.1
Wassenaar	2.2		
Zürich	0.1		
Totals	257.244.7 **508**	28.28.3 **59**	5.3.3 **11**
Plus born 1928–1945	11.11.8 **30**		
Grand totals	**538**	**59**	**11**

Table 6. Studbook registered captive-born orang utans that have reproduced, with data on young.

female, Ivy, Sumatran, Philadelphia 5 (13 Jun. 1937–24 Jun. 1972)
    all young sired by father, wild-born Sumatran:
    female   16 Aug. 1950–26 Oct. 1950
    male   27 Dec. 1953– 5 Jan. 1954
    female   3 Sep. 1955–15 Sep. 1955
    sex unknown 3 May 1961–16 May 1961
female, Blondie/Myrtle, Sumatran, Philadelphia 8 (20 Aug. 1946–to Birmingham 1 May 1958–8 Dec. 1974)
    all young sired by sibling, Philadelphia 11:
    aborted   4 Mar. 1963
    sex unknown 29 Oct. 1967 (stillborn)
    male   23 Oct. 1970 (stillborn)
male,   Ernst, Sumatran, Rotterdam 1 (20 Nov. 1951–13 Jan. 1965) to Rotterdam 2:
    male   27 Mar. 1963– 7 Jan. 1965
male,   Lucky/Jimmy, Sumatran, Philadelphia 11 (5 Sep. 1952–to Birmingham 1 May 1958–21 Jun. 1977)
    sired young to Philadelphia 8 as noted above

Table 6. (Continued)

female, Nina, Sumatran, Rotterdam 2 (11 Oct. 1953–8 Feb. 1965)
    young sired by Rotterdam 1 as noted above
female, Christine, Sumatran, Philadelphia 13 (25 Jan. 1955–living 31 Dec. 1978)
    all young sired by an unrelated wild-born male:
    female   17 Aug. 1950–living 31 Dec. 1978
    male   12 Mar. 1973–living at Kansas City 31 Dec. 1978
    female   30 Jul. 1975–living at Little Rock 31 Dec. 1978
    female   24 May 1977–living 31 Dec. 1978
female, Noell, Bornean, San Diego 3 (25 Dec. 1955–11 Oct. 1966)
    sired by wild-born male:
    female   6 Oct. 1966–living 31 Dec. 1978
male, Henri, Sumatran, Rotterdam 3 (18 Oct. 1956–to Paris 8 Oct. 1959–14 Sep. 1970)
    to wild-born female:
    female   24 Aug. 1964 (stillborn)
male, Peek, Sumatran, Rotterdam 4 (19 Jan. 1957–to Munich 19 Dec. 1974–4 Dec. 1977)
    to wild-born females:
    male   31 Oct. 1967–living at Dortmund 31 Dec. 1978
    female   20 Apr. 1978–25 Apr. 1978
female, Judy II, race unknown, Sydney 1 (26 Oct. 1957–living 31 Dec. 1978)
    all young sired by wild-born males:
    male   29 Dec. 1967– 4 Jan. 1968
    female   26 Mar. 1972–living 31 Dec. 1978
    male   15 Mar. 1975–living 31 Dec. 1978
    female   23 Aug. 1977–living 31 Dec. 1978
female, Roberta, Sumatran × Bornean, San Diego 4 (5 Feb. 1958–to Denver 22 Jan.
    1974–living 31 Dec. 1978)
    all young sired by unrelated wild-born males:
    male   4 Jan. 1969 (stillborn)
    female   25 Aug. 1971–living at Sao Paulo 31 Dec. 1978
    male   14 Feb. 1973–living 31 Dec. 1978
    female   30 Jul. 1978–living 31 Dec. 1978
female, Ellie II, race unknown, Vienna 1 (26 Dec. 1959–living 31 Dec. 1978)
    young sired by unrelated wild-born males:
    sex unknown 16 Jun. 1976 (stillborn)
    sex unknown 16 Aug. 1978 (stillborn)
female, Sali, Sumatran, Frankfurt 3 (11 Aug. 1960–12 May 1971)
    sired by wild-born father:
    male   9 May 1971 (stillborn)
female, Bulu, Bornean, London 1 (12 Mar. 1961–living 31 Dec. 1978)
    all young sired by wild-born males:
    male   21 Jan. 1970–21 Jan. 1970
    female   21 Jun. 1971–living 31 Dec. 1978
    male   30 May 1974–living 31 Dec. 1978
    aborted   15 Nov. 1976
    female   17 Sep. 1977–25 Sep. 1977
female, Wendy, race unknown, Sydney 2
    all young sired by wild-born males:
    male   3 Sep. 1973–15 Mar. 1974
    male   6 May 1975–living 31 Dec. 1978
female, Rosie, Sumatran, Fresno 1 (6 May 1961–to Monroe 11 Apr. 1968–17 Nov. 1973)
    all young sired by unrelated wild-born males:
    male   4 Jul. 1971–living 31 Dec. 1978
    twins, male 26 Oct. 1973–26 Dec. 1973
             male 26 Oct. 1973 (stillborn)
female, Hatsuko, race unknown, Toyoto-Uneo 1 (29 May 1961–living 31 Dec. 1978)
    some young sired by wild-born father, two by an unrelated wild-born male:
    female   5 Aug. 1971–living 31 Dec. 1978 (to father)
    female   2 Jan. 1975– 2 Jan. 1975
    female   21 May 1976–living at Sendai 31 Dec. 1978
    male   10 Jun. 1977–living 31 Dec. 1978 (to father)

Table 6. (Continued)

female, Maggie, Bornean, San Diego 7 (18 Jul. 1961–living 31 Dec. 1978)
    all young sired by unrelated wild-born males:
    male   13 Feb. 1971–living 31 Dec. 1978
    male   2 May 1973–living at Los Angeles 31 Dec. 1978)
    female   18 Dec. 1978–living 31 Dec. 1978
female, Kasih, Bornean × Sumatran, Basel 2 (19 Mar. 1962–living 31 Dec. 1978)
    all young sired by wild-born father:
    twins, male   8 Jul. 1973– 2 Aug. 1973
           male   8 Jul. 1973–living at Wuppertal 31 Dec. 1978
    aborted   5 May 1975
    female   10 Jun. 1976–16 Nov. 1976
    female   22 Apr. 1978–living 31 Dec. 1978
female, Tallulah, Sumatran × Bornean, Colorado Springs 4 (11 Jan. 1963–1 Feb. 1974)
    sired by unrelated male:
    aborted   1 Feb. 1974
male, Bobby, race unknown, Sydney 5 (7 Sep. 1964–to Perth 7 Apr. 1968–on loan by them to Melbourne 16 Mar. 1976–living 31 Dec. 1978)
    to wild-born female:
    female   9 Aug. 1973– 9 Aug. 1973
    twins, male   24 Jun. 1978–living 31 Dec. 1978
          fem.   24 Jun. 1978–living 31 Dec. 1978
male, Roger/Otis, Sumatran, Fresno 2 (6 Mar. 1965–to San Diego 16 Jul. 1966–living 31 Dec. 1978)
    to wild-born female:
    twins, male   25 Sep. 1977–living 31 Dec. 1978
          fem.   25 Sep. 1977–living 31 Dec. 1978
    to San Diego 8:
    female   7 Jun. 1978–living 31 Dec. 1978
female, Miri, Sumatran × Bornean, Colorado Springs 8 (2 May 1965–to Detroit 14 Jun. 1967–living 31 Dec. 1978)
    sired by Detroit 3:
    female   23 Jun. 1977–29 Jun. 1977
female, Sulong, Sumatran × Bornean, Los Angeles 1 (22 Apr. 1966–living 31 Dec. 1978)
    sired by unrelated wild-born male:
    female   27 Nov. 1976–living 31 Dec. 1978
male, Atjeh, Sumatran × Bornean, Washington 1 (2 Apr. 1966–living 31 Dec. 1978)
    all young born to female Yerkes 5:
    female   7 Jul. 1976–10 Jul. 1976
    male   14 Dec. 1977–living at Alberquerque 31 Dec. 1978
male, Samu, Bornean, Detroit 3 (9 Jul. 1966–living 31 Dec. 1978)
    to Colorado Springs 8:
    female   23 Jun. 1977–29 Jun. 1977
female, Carmen, Bornean, St. Louis 12 (17 Jul. 1966–living 31 Dec. 1978)
    all young sired by wild-born male (Sumatran):
    female   18 Jul. 1976–18 Jul. 1976
    male   28 Jan. 1978–16 Sep. 1978
male, Moro, Sumatran, Berlin Tierpark 1 (4 Aug. 1966–to Leipzig 9 Nov. 1973–returned to Berlin 13 Apr. 1976–living 31 Dec. 1978)
    to wild-born females:
    male   30 Jun. 1974–(stillborn)
    male   30 Jan. 1977–living 31 Dec. 1978
female, Robella, Bornean, San Diego 8 (6 Oct. 1966–living 31 Dec. 1978)
    sired by Fresno 2:
    female   7 Jun. 1978–living 31 Dec. 1978
male, Denny, Sumatran, Fresno 3 (18 May 1967–to San Francisco 12 Jul. 1968–living 31 Dec. 1978)
    all young born to wild-born female (Bornean):
    female   6 Feb. 1976– 6 Feb. 1976
    female   19 Nov. 1977–living 31 Dec. 1978

Table 6. (Continued)

female, Pensi, Sumatran, Yerkes 5 (19 Oct. 1967–to Washington 13 Sep. 1973–living 31 Dec.
1978)
all young sired by Washington 1:
female   7 Jul. 1976–10 Jul. 1976
male   14 Dec. 1977–living at Alberquerque 31 Dec. 1978
male,   Martijn, Sumatran, Rotterdam 10 (31 Oct. 1967–to Dortmund 12 Jul. 1976–living
31 Dec. 1978)
to Stuttgart 3:
male   9 Sep. 1978–living 31 Dec. 1978
female, Eloise, Bornean, Los Angeles 2 (10 Nov. 1968–living 31 Dec. 1978)
sired by unrelated wild-born male:
male   23 Nov. 1968–living 31 Dec. 1978
male,   Frank, Sumatran, Frankfurt 7 (23 Dec. 1968–to Tel Aviv 30 Sep. 1971–living 31 Dec.
1978)
to Nuremberg 6:
twins, sexes unknown, stillborn 30 Jun. 1978
female, Katja, Sumatran, Nuremberg 6 (6 Mar. 1969–to Tel Aviv 30 Sep. 1971–living 31 Dec.
1978)
young as above, sired by Frankfurt 7
female, Afra, Sumatran, Stuttgart 3 (6 Jun. 1969–to Dortmund 23 Aug. 1976–living 31 Dec.
1978)
to Rotterdam 10:
male   9 Sep. 1978–living 31 Dec. 1978
female, Tiye, Sumatran × Bornean, Omaha 3 (1 Mar. 1970–living 31 Dec. 1978)
sired by father (wild-born Sumatran):
male   4 May 1978–living 31 Dec. 1978
male,   Schorsch, Bornean × Sumatran, Frankfurt 11 (22 Mar. 1971–to Hannover 14 Dec.
1972–living 31 Dec. 1978)
to Hannover 1:
male   28 Aug. 1978–living 31 Dec. 1978
female, Uta, Sumatran, Hannover 1 (27 Jul. 1971–living 31 Dec. 1978)
Young sired by Frankfurt 11 as noted above

Table 7. Twin captive births in the orang utan, to 31 Dec. 1978.

male,	Towan, 19 Feb. 1968–living 31 Dec. 1978
female,	Chinta, 19 Feb. 1968–living 31 Dec. 1978
	to wild-born parents at Seattle
male,	Brunno, 18 Feb. 1969–living Dec. 1978
female,	Hella, 18 Feb. 1969–living Dec. 1978
	to wild-born parents at Munich
female,	Mawar, 27 Dec. 1971–28 Dec. 1972
female,	Melati, 27 Dec. 1971–loan to Seattle 25 Jun. 1974–living 31 Dec. 1978
	to wild-born parents at Washington
male,	Vedjar, 8 Jul. 1973–to Wuppertal 27 Jun. 1978–living 31 Dec. 1978
female,	no name, 8 Jul. 1973–2 Aug. 1973
	to wild-born male and captive-born female at Basle
male,	stillborn 26 Oct. 1973
male,	no name, 26 Oct. 1973–29 Dec. 1973
	to wild-born male and captive-born female at Monroe
male,	Trick, 31 Oct. 1974–to Milwaukee 23 Dec. 1974–living 31 Dec. 1978
female,	Treat, 31 Oct. 1974–to Milwaukee 23 Dec. 1974–living 31 Dec. 1978
	to wild-born parents at Madison (dam on loan from Milwaukee)
male,	Lock, 25 Sep. 1977–living 31 Dec. 1978
female,	Lisa, 25 Sep. 1977–living 31 Dec. 1978
	to captive-born male and wild-born female at San Diego
male,	Kember, 24 Jun. 1978–living 31 Dec. 1978

Table 7. (Continued)

female,	Kiani, 24 Jun. 1978–living 31 Dec. 1978
	to captive-born male and wild-born female at Melbourne
sex unknown, stillborn, 30 Jun. 1978	
sex unknown, stillborn, 30 Jun. 1978	
	to captive-born parents at Tel Aviv

Longevity

The majority of individuals that arrived in captivity from 1776 to 1925 lived only a few years at most. The average was under one year; however, there were a few that exceeded this norm. These were:

Male, Dresden Zoo, Sumatran, Peter: 2 Oct. 1898–2 Aug. 1907.
Female, Cologne Zoo, Sumatran: 1900 — to Frankfurt Zoo 23 Jul. 1903 — died 6 Apr. 1909.
Female, Melbourne Zoo, race unknown, Mollie: 1903–1923.
Male, London Zoo, Bornean, Sandy I: 7 Sep. 1905–10 Mar. 1924.
Male, London Zoo, Sumatran, Jacob: 5 Feb. 1908–29 Jul. 1916.
Male, Vienna Schönbrunn Zoo, Bornean, Emil I:1913–1938.
Male, London Zoo, Bornean, Gabong: 7 Aug. 1915 — to Bronx Zoo 28 Jun. 1916
  — died 20 Nov. 1936.
Male, Amsterdam Zoo, Sumatran, Sultan: 24 Sep. 1924–19 Apr. 1933.
Male, St. Louis Zoo, Sumatran, Bobby: 23 Oct. 1925–1 Jul. 1935.

Of the large group brought into captivity from 1926 to 1945, there are also a few outstanding individuals, who are:

Female, Berlin Zoo, Sumatran, Cleo: 11 May 1926–Nov. 1943.
Male, Abreu collection, Havana, Sumatran, Guas: 1927 — to Philadelphia Zoo 1 May.
  1931 — died 9 Feb. 1977.
Female, Abreu coll., Havana, Sumatran, Guarina: 1927 — to Philadelphia Zoo 1 May.
  1931 — died 16 Jan. 1976.
Female, London Zoo, Sumatran, Mary: 27 Jun. 1930–7 Jan. 1955.
Male, Edinburgh Zoo, Sumatran, Mickey: 18 Aug. 1937–17 Jul. 1959.
Male, Surabaya Zoo, Bornean, Katjeung: 1940 — to San Diego Zoo 21 Aug.1940, to Brookfield
  Zoo 28 Mar. 1941 — died 30 Jun. 1958 (may have been at Lorraburry Zoo since 1933).
Female, Melbourne Zoo, Sumatran, Noni: 23 Jan. 1941–Sep. 1957.

Of the over 800 orang utans that have come into captivity since 1946, only a handful were fully adult on arrival, that is to say not less than ten years of age. Mostly these adults were probably 15, 20 or older. The oldest surviving are two females at Rotterdam that came in 1950 and 1951 as adults (and which have not bred in many years) and other females who have been estimated as wild-bred from 1942 to 1947 and are now living in the zoos of Amsterdam, Omaha, Perth, Rome and San Diego, plus an old male at Colorado Springs.

Examination of carefully kept post-1946 records would indicate that from 30 to 40 is the average maximum lifespan for an orang utan, certainly the two at Philadelphia only lived longer due to strict diet control and medication. Males are reproductive until about 40, females until 35, although with so few records to go on, these may be extreme cases. Of the 180 males and 196 females still living that were wild-bred and brought into captivity since 1946, the majority date from 1962 onward. At Table 8 is a list of all of these individuals.

Table 8. Wild-born orang utans brought into captivity, 1946–1978.

Year of import	Number*	Maximum longevity in months or number living 31 Dec. 1978
1946	1.0	143
1947	5.4	0.1
1948	5.6.3	1.1
1949	9.9	322
1950	10.13	0.3
1951	4.4	0.1
1952	3.4	1.1
1953	8.5	1.2
1954	5.5	1.0
1955	20.19.1	3.2
1956	15.15.1	5.9
1957	9.5	2.1
1958	10.8	2.4
1959	18.18	7.8
1960	11.13	3.3
1961	10.10	2.6
1962	31.37.2	16.16
1963	33.30	14.12
1964	23.26	12.16
1965	19.18	6.5
1966	29.29	12.19
1967	19.15	12.10
1968	12.15	9.8
1969	24.12	16.8
1970	6.4	2.4
1971	14.18	12.10
1972	10.13	9.11
1973	9.14	6.9
1974	5.6	4.6
1975	9.6	7.5
1976	6.8	6.7
1977	9.8	7.7
1978	2.2	2.1
Totals	403.399.7	180.196
	809	**376**

* Numbers are animals brought to a studbook registered zoological collection; 1.0 = male; 0.1 = female; 0.0.1 = unknown sex.

References

Abel, C. (1825). Some accounts of an orang-outang of remarkable height found on the island of Sumatra. Asiatic Researches, 15: 489–398.

Aveling, R.J. (1982). Orang utan conservation in Sumatra, by habitat protection and conservation education. In: L.E.M. de Boer (ed.), The orang utan. Its biology and conservation. Junk, The Hague.

Bourne, G.H. (1969). Orang utan studbook as of December 31, 1969. Yerkes Regional Primate Research Center, Atlanta.

Brandes, G. (1929a). Der Durchbruch der Zähne beim Orang-Utan. Zoologischer Garten, N.F., 1: 25–28.

Brandes, G. (1929b). Zur Frage des Fettansatzes beim Orang-Utan. Zoologischer Garten, N.F., 1: 326–327.

Brandes, G. (1929c). Die Backenwülste des Orang-Mannes. Zoologischer Garten, N.F., 1: 365–368.

Brandes, G. (1929d). Der Tod unseres Riesenorangs 'Goliath'. Zoologischer Garten, N.F., 1: 396–400.

Brandes, G. (1930a). Zum Tode unseres Orang-Mannes 'Peter'. Zoologischer Garten, N.F., 3: 12–18.

Brandes, G. (1930b). Die Veränderungen des Orang-Kindes. Zoologischer Garten, N.F., 3: 286–289.

Brandes, G. (1931a). Wie alt wird der Orang-Utan? Zoologischer Garten, N.F., 4: 1–9.

Brandes, G. (1931b). Das Wachstum der Menschenaffen im Vergleich zu dem des Menschen in Kurven dargestellt. Zoologischer Garten, N.F., 4: 339–347.

Brandes, G. (1938). Die Stillzeiten des Orang. Zoologischer Garten N.F., 10: 139–141.

Brandes, G. (1939). Buschi — vom Orang-Säugling zum Backenwülster. Quelle und Meyer, Leipzich.

Bungartz, M.A.H. (1928). 'Manaburu' der Geist des Sumatranischen Urwaldes. Hamburger Zoo Zeitung, 1: no. 6.

Camper, P. (1779). Account of the organs of speech in the orang-outang. Philosophical Transactions of the Royal Society, 69: 139–159.

Fitzinger, L.J. (1853). Untersuchungen über die Existenz verschiedener Arten der Asiatischen Orang-Affen. Sitzbericht der Mathematischen und Naturwissenschaftlichen Klasse der Kaiserlichen Akademie der Wissenschaften, 11: 400–449.

Friedel, E. (1876). Die drei anthropomorphen des Berliner Aquariums. Zoologischer Garten, 17: 73–77.

Harrisson, B. (1963). Orang utan. Doubleday, Garden City.

Hoppius, C.E. (1763). Anthropomorpha. Section CV of C. Linnaeus, Amoenitas Academiae, 6: 63–76.

Jeffries, J. (1825). Some account of the dissection of a *Simia satyrus* — orang outang or wild man of the woods. Boston Journal of Philosophic Arts, 2: 570–586.

Jones, M.L. (1980). Studbook of the orang utan, *Pongo pygmaeus*. Zoological Society of San Diego, San Diego.

Rijksen, H.D. (1982). How to save the mysterious 'man of the rain forest'? In: L.E.M. de Boer (ed.), The orang utan. Its biology and conservation. Junk, The Hague.

Vosmaer, A. (1778). Beschryving van de zo zeldzaame als zonderlinge aap-soort genaamd orang-outang van het eiland Borneo. Pieter Meijer, Amsterdam.

Additional references

Abel, C. (1886). Account of an orang-outang. Miscellaneous Papers relating to Indo-China, Calcutta.

Aulmann, G. (1932). Geglückte Nachzucht eines Orang-Utan im Düsseldorfer Zoo. Zoologischer Garten, N.F., 5: 81–90.

Beeckman, D. (1714). A voyage to and from the island of Borneo. London.

Benchley, B. (1942). My friends, the apes. Little Brown, Boston.

Blyth (1886). On the different species of orang-utang. Miscellaneous Papers relating to Indo-China, Calcutta.

Bolau, D.H. and A. Pansch (1894). Der erste erwachsene Orang-Utan in Deutschland. Zoologischer Garten, 35: 97–102.

Bourdelle, E. and P. Rode (1932). Notes à propos d'un jeune orang *Pongo pygmaeus* Hopp. né à la menagerie du Jardin des Plantes. Bulletin Muséum d'Histoire Naturelle, 4: 472–477.

Brehm, A.E. (1873). Aus dem Leben des Schimpansen. Verhandlungen der Gesellschaft für Anthropologie.

Brooke, J. (1841). A letter to Mr. Waterhouse, dealing with the number of species of orang-outang and their distribution. Proceedings Zoological Society of London, 1841: 55–61.

Brown, C.E. (1936). Rearing wild animals in captivity and gestation period. Journal of Mammalogy, 17: 10–13.

Crandall, L.S. (1964). Management of wild mammals in captivity. University of Chicago Press, Chicago.

Cuvier, F. (1810). Description d'un orang outang. Annales du Muséum National d'Histoire Naturelle, 16: 46–48.

Delisle, F. (1893). Les orang-outangs au Jardin Zoologique d'Acclimatisation du Bois de Boulogne. L'Anthropologie, 4: 648–651.

Delisle, F. (1895). Notes sur l'ostéométrie et la cranologie des orang-outangs. Nouvelle Archieves du Muséum d'Histoire Naturelle, ser. 3, no. 7: 83–114.

Deniker, J. (1882). Sur les singes anthropoides de la menagerie. Bulletin Société Zoologique de France, 7: 301–304.

Eipper, P. (1932). In my zoo. Viking, New York.

Elliot, D.G. (1913). A review of the primates, III, Anthropoidea, *Miopithecus* to *Pan*. American Museum of Natural History, New York.

Felce, W. (1948). Apes. Chapman and Hall, London.

Fick, R. (1895a). Vergleichend anatomische Studien an einem erwachsenen Orang-Outang. Archiv für Anatomie und Physiologie, Anatomische Abteilung, 1895: 1–100.

Fick, R. (1895b). Beobachtungen an einem zweiten erwachsenen Orang-Outang und einem Schimpansen. Archiv für Anatomie und Physiologie, Anatomische Abteilung, 1895: 289–318.

Fick, R. (1929). Über die Körpermasze und den Kehlsack eines erwachsenen Orangs. Zeitschrift für Säugetierkunde, 4: 65–80.

Fox, H. (1929). The birth of two anthropoid apes. Journal of Mammalogy, 10: 37–51.

Grant, J. (1828). An account of the structure, manners and habits of an orang-outang from Borneo. Edinburgh Journal of Science, 9: 1–24.

Haeckel, E. (1909). Aus Insulinde, Malayische Reisebriefe. Leipzich (2. Auflage).

Hagenbeck, J. and V. Ottoman (1924). Sudasiatische Fahrten und Abenteuer. Dresden.

Heck, L. (1952). Animals, my adventure. Ullstein, Vienna.

Hermes, O. (1876). Die anthropomorphen Affen des Berliner Aquariums. Verhandlungen der Berliner Gesellschaft für Anthropologie, 1876: 88–94.

Hill, W.C.O. (1939). Observations on a giant Sumatran orang. American Journal of Physical Anthropology, 24: 449–510.

Hornaday, W.T. (1879). On the species of Bornean orangs with notes on their habits. Proceedings of the American Association for the Advancement of Science, 1879: 438–456.

Hornaday (1885). Two years in the jungle. New York.

Kerbert, C. (1914). Reuzen orang oetans. Natuur en Wetenschap, 11: 1–14.

Klös, H.G. (1969). Von der Menagerie zum Tierpark. Heude und Spenersche, Berlin.

Jantschke, F. (1972). Orang-Utans in zoologischen Gärten. R. Piper, Munich.

Lenz, H. (1885). Ein neuer groszer Orang-Outang. Zoologischer Garten, 36: 161–162.

Lucas, F.A. (1881). The species of orangs. Proceedings of Natural History, Boston, 21: 228–233.

MacKinnon, J. (1974). In search of the red ape. Holt, Reinhart and Winston, Chicago.

Milne-Edwards, A., J. Deniker, R. Boulart, E. de Pousargues and F. Delisle (1895). Observations sur deux orang-outans adultes morts à Paris. Nouvelles Archives du Muséum d'Histoire Naturelle, ser. 3, no. 7: 29–118.

Mobius, K. (1894). Lichtbilder alter Orang-Utans von Borneo. Zeitschrift für Ethnologie, 26: 382.

Morris, R. and D. Morris (1966). Men and apes. McGraw Hill, New York.

Napier, J.R. and P.H. Napier (1967). A handbook of living primates. Academic Press, New York.

Owen, R. (1835). On the osteology of the chimpanzee and orang-utan. Transactions of the Zoological Society of London, 1: 343–379.

Owen, R. (1836). On a new orang (*Simia morio*). Proceedings of the Zoological Society of London, 4: 91–96.

Portielje, A.F.J. (1939). Tierleben bezw. intelligente Auserungen beim Orang-Utan, *Pongo pygmaeus* Hoppius. Bijdragen tot de Dierkunde, 27: 61–114.

Rijksen, H.D. (1978). A field study on Sumatran orang utans (*Pongo pygmaeus abelii* Lesson 1827): ecology, behaviour and conservation. Mededelingen Landbouwhogeschool Wageningen. 78–2: 1–420.

Ruhe, H. (1960). Wilde Tiere, frei Haus. Copress, Munich.

Sanyal, K.B. (1876). Observations on the orang-utan in captivity. Journal of the Asiatic Society of Bengal, 71: 92–93.

Schlegel, H. and S. Müller (1839–1844). Bijdragen tot de natuurlijke historie van de orang oetan (*Simia satyrus*). Leiden.

Seitz, A. (1969). Einige Feststellungen zur Pflege und Aufzucht von Orang Utans, *Pongo pygmaeus*, Hoppius 1763. Zoologischer Garten, N.F., 36: 225–245.

Strauch, C. and G. Brandes (1927). Uber den Dresdener Orang 'Goliath'. Zeitschrift für Säugetierkunde, 2: 78–82.

Ulmer, F.A. (1957). Breeding of orang-utans. Zoologischer Garten, N.F., 23: 1–3.

Urbain, A., E. Dechambre and P. Rode (1941). Observations faites sur un jeune orang-outan né à la menagerie du Jardin des Plantes. Mammalia, 8: 82–85.

Wurmb, van (1780). Beschryving van de groote Borneoische orang-outang. Verhandelingen van het Bataviaasch Genootschap, II: 1–137.

Yerkes, R.M. (1925). Almost human. Century Co., New York.

Yerkes, R.M. and A.W. Yerkes (1929). The great apes. Yale University Press, New Haven.

Zuckerman, S. (1937). The duration and phases of the menstrual cycle in primates. Proceedings of the Zoological Society of London, 107: 315–329.

Author's address:
M.L. Jones
Zoological Society of San Diego
P.O. Box 551
San Diego, CA 92112
U.S.A.

3. Genetics and conservation of the orang utan

Leobert E.M. de Boer

Genetics and conservation

In this paper the term 'conservation' covers both the conservation of endangered species in the wild, in their natural habitat, and of captive populations. Although both modes of conservation do not necessarily have an
identical ultimate goal, genetics are of equal importance to both:

1. Genetic studies are pre-eminently suitable to detect the existence of
   intraspecific (subspecific, racial, geographical) differences. Knowledge of
   the possible existence of such differences is a prerequisite of a good
   conservation policy for wild populations and to proper management of
   captive stocks: we need to know which animals belong to (or originate
   from) the same natural population and which do not. It has been shown
   on many occasions that the division of species into subspecies or races on
   the basis of external characteristics may be inaccurate. In such cases the
   study of genetic markers often supplies useful information as to the exact
   delimitation of geographical forms. Sometimes differences are even
   discovered which indicate that what is believed to be a species, in fact
   represents two or more reproductively isolated populations.
   Identification of individuals on the basis of genetic markers therefore is
   highly preferable when external characteristics of two forms show
   overlapping variability ranges.

   It goes without saying that captive breeding programmes should aim at
   establishing populations which, at least genetically, resemble original,
   feral populations as closely as possible. Thus, whenever possible, subspecies or races should be maintained and bred separately and any
   available method of unambiguous identification should be applied. In
   cases where there are doubts as to the existence of different forms, genetic
   studies should be carried out in order to obtain conclusive evidence.

   Information on the existence of different geographic forms and their
   exact delimitation is of crucial importance to conservation in the wild
   since, ideally, reserves should be projected in the centres of the distribution areas of natural, randomly breeding populations. Location of
   reserves on the borders of two distinct forms is inadvisable, especially

The orang utan. Its biology and conservation, edited by L.E.M. de Boer
© *1982, Dr W. Junk Publishers, The Hague, ISBN 90 6193 702 7*

when there is a chance that these forms genetically differ more than their external characteristics suggest.

2. Conservation should take into consideration the genetic structure of the populations to be preserved. It should not primarily aim to conserve a certain number of specimens, but rather to establish healthy populations. These populations should resemble the original natural populations in so far as possible in terms of gene frequencies, not necessarily in numbers of animals. Only a population with a genetic structure resembling that of the original state will be fit to survive in the long term under natural circumstances, since that genetic population structure would be able to withstand the fluctuations in natural selective forces (by which, in fact, it would be shaped). Collection of genetic information at the population level such as the determination of gene frequencies of polymorphic loci and the detection of rare alleles, is a valuable contribution in assessing population structure. Once such information is available, populations can be monitored and possible ill developments can be detected.

3. Genetics are of importance with regard to the occurrence of hereditary diseases. This is mainly applicable to the maintenance of endangered species in captivity. In feral populations such diseases also occur, but in small captive populations absence of selective pressures and effects of genetic drift may increase the frequency of occurrence considerably. Early detection of affected animals, detection of carriers and information on modes of inheritance, are therefore crucial to prevent this.

4. Genetic information is indispensible with regard to the problems of inbreeding and maintenance of genetic variability in the captive situation. Considerable theoretical knowledge is available from population genetics which enables us to frame minimal inbreeding and maximum variability breeding schedules and more and more theoretical thought is given to the 'genetic management' of small captive populations.

Below, these aspects are elaborated with special reference to the conservation of the orang utan.

Genetics of the orang utan

Genetic studies of orang utans have mostly been undertaken as a means of tracing the phylogenetic relationships of this ape with other primates, particularly the other pongids and man. The conclusion drawn, almost unanimously, from such comparative genetic investigations has been that, of the extant great apes, the orang utan is the most distantly related to man. On several occasions this has led to the suggestion that the classification of hominids and pongids should be changed. Rather than placing *Pan*, *Gorilla* and *Pongo* in the family Pongidae and *Homo* in the Hominidae, it has been suggested to include the former two in the Hominidae (as a separate subfamily) and to consider *Pongo* as the only living representative of the Pongidae. As a consequence of the realization that the orang utan 'merely' represents a side branch of the human phylogenetic tree, the genetics of this species have been studied in far less detail than, for instance, those of the chimpanzee.

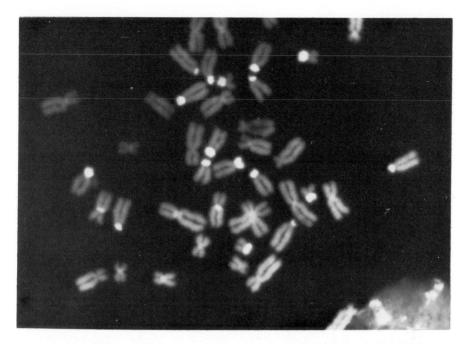

Fig. 1. Partial metaphase plate of a Bornean orang utan stained with distamycin/DAPI. Note the intensively fluorescing short arms of chromosomes 11–17 and 22–23 (courtesy of the Institute of Clinical Cytogenetics, Utrecht).

One of the results of this has been that geneticists have had almost no interest in the possible differences between Bornean and Sumatran orang utans. In spite of the fact that in the case of the orang utan there are better indications to expect intraspecific differences because of the strict geographic isolation of the two forms, much more attention has been given to the detection of genetic differences between local populations or races of the chimpanzee (e.g. Moor-Jankowski et. al. 1969). Apart from this lack of interest, identification problems may have added to the relative paucity of studies differentiating between *Pongo pygmaeus abelii* and *Pongo pygmaeus pygmaeus*. Most of the published work refers to the animals that have been studied only by the species name. Only in a few cases where the house names of the individual specimens are mentioned, is it possible to trace their geographic origins retrospectively with the aid of the Studbook of the Orang Utan (Jones 1980). Less than half a dozen papers indicate subspecific names themselves. This lack of specification of animal material studied also makes it impossible to combine data from different studies on animals from the same source. The genetics of the orang utans from the Yerkes Regional Primate Research Center, for instance, have been studied extensively, but since these studies were carried out in various laboratories and animal names or numbers were seldom given in the final publications, it is not possible to compare the results of each specimen individually.

It is currently difficult to define the term 'genetics' because of the overlap with fields such as molecular biology and chromosome cytology. For the sake of convenience we will limit ourselves below to basic information on the orang

41

utan genome and to a brief review of the intraspecific variation of genetic markers indicating subspecific differences or genetic polymorphisms.

The orang utan genome

The chromosome morphology of the orang utan has been extensively studied. During the past few years, in particular, much information on the structure of individual chromosomes has been obtained by the use of differential staining techniques. The chromosome banding patterns show a high degree of similarity to those of man and the other great apes (for a review see Seuánez 1979). Such detailed studies have also led to the discovery of a consistent difference in the second chromosome pair of Bornean and Sumatran orang utans (Seuánez et al. 1979) and the existence in both subspecies of a polymorphism involving a complex rearrangement in the 9th chromosome pair (see also De Boer and Seuánez 1982; this volume). In addition, variants have been found of chromosomes number 14 and 22, although to date not enough animals have been studied to facilitate assessment of their possible polymorphic occurrence.

Recently much work has also been done on the distribution of specific types of DNA in the orang utan genome, such as centromeric heterochromatin, nucleolus organizing regions (NORs), rDNA genes and satellite DNAs. Table 1 gives a summary of the results of these studies. No specific studies have so far been undertaken to trace the possible existence of polymorphisms with regard to the distribution of such specific DNA-types. Nevertheless, there seem to be indications that polymorphisms may be found in the short arms of chromosome pairs 11–17 and 22–23. The short arms of these 9 chromosomes contain rDNA (as well as satellite DNAs I, II and III), the distribution of which in *Homo* and *Pan* is confined to five chromosomes and in *Gorilla* even to two (see Table 2). Variability was found in at least some of these short arms in a *Pongo pygmaeus* studied by Anderle et al. (1979) and Schweitzer et al. (1979) after the staining of chromosome preparations with Dystamicin/DAPI and chromomycin (see also Fig. 1). Detailed studies of these regions on larger groups of orang utans may bring to light interesting and useful polymorphisms.

Sequences of unique orang utan DNA have been studied by DNA hybridization experiments (e.g. Hoyer et al. 1972; Kohne et al. 1972). This type of work, however, mainly involved interspecific comparisons and thus is of little interest in the present context.

Also primarily comparative, but of more importance because of studies on genetic markers in the orang utan, is the work on gene mapping by means of somatic cell hybridization. Part of the interspecific chromosome homologies proposed on the basis of comparison of banding patterns of *Pongo* and the other anthropoids and man have been confirmed by the localization of homologous genes on homologous chromosomes. The 1979 report of the Committee for Comparative Gene Mapping (Pearson et al. 1979) mentions 26 genes to be localized in the orang utan, distributed over 14 linkage groups (exclusive of satellite DNA and rDNA genes; see Table 3). The number of genes localized in *Pongo* is only slightly lower than in chimpanzee (36) and gorilla (38), but still lies far behind that in man (345 in 1979).

Table 1. Distribution of specific types of DNA in the chromosomes of the orang utan (references are given at the bottom of the table).

Chromosome number	Centromeric heterochromatin (1)	Nucleolus organizing regions, Ag staining (2, 3)	Chromosome associations (4)	rDNA genes (4, 5, 6)	Satellite DNA I (7)	Satellite DNA II (7)	Satellite DNA III (8)	Satellite DNA IV (7)	Distamicyn/DAPI positive regions (9)	Chromomycin positive regions (9)
PPY1	+									
PPY2	+									
PPY3	+									
PPY4	+									
PPY5	+									
PPY6	+									
PPY7	+					+				
PPY8	+				+				+	
PPY9					+			+		
PPY10	+		?							
PPY11	+	+	+	+	+	+	+	+	+	+
PPY12	+	?	+	+	+	+	+			
PPY13	+	+	+	+	+	+	+		+	
PPY14	+	+	+	+	+	+	+		+	+
PPY15	+	+	+	+	+	+	+		+	
PPY16	+	+	+	+	+	+	+	+	+	
PPY17	+	+	+	+	+	+	+		+	+
PPY18	+									
PPY19	+									
PPY20	+									
PPY21	+									
PPY22	+	+	+	+	+	+	+		+	+
PPY23	+	+	+	+	+	+	+		+	
PPYX	+									
PPYY	+				+	+	+	+		

References: 1: De Boer and Seuánez (1982); 2: Tantravahi et al. (1976); 3: Anderle et al. (1979); 4: Henderson et al. (1979); 5: Henderson et al. (1976); 6: Gosden et al. (1978); 7: Gosden et al. (1977); 8: Mitchell et al. (1977); 9: Schweitzer et al. (1979).

Table 2. Distribution of rDNA genes in the great apes and man (data according to Tantravahi et al. 1976, and Henderson et al. 1976, 1979).

Chromosome number (human numbering system)	rDNA genes			
	Homo	Gorilla	Pan	Pongo
2q	−	−	−	+(PPY11)
2p	−	−	−	+(PPY12)
9	−	−	−	+(PPY13)
13	+	−	+(PTR14)	+(PPY14)
14	+	−	+(PTR15)	+(PPY15)
15	+	−	−	+(PPY16)
18	−	−	+(PTR17)	+(PPY17)
21	+	+(GGO22)	+(PTR22)	+(PPY22)
22	+	+(GGO23)	+(PTR23)	+(PPY23)

43

Table 3. Genes localized in the orang utan genome (exclusive of satellite DNAs and rDNA genes) (data from Pearson et al. 1979).

Chromosome number	Symbol	Gene locus
PPY1	ENO1	Enolase-1
	PGD	Phosphogluconate dehydrogenase
	PGM1	Phosphoglucomutase-1
	FH	Fumarate hydratase
	PEPC	Peptidase-C
	RN5S	5S RNA
PPY2	ACP1	Acid phosphatase-1
	MDH1	Malate dehydrogenase-1
PPY3	–	–
PPY4	HEXB	Hexosaminidase-B
PPY5	SOD2	Superoxide dismutase-2
PPY6	GSR	Glutathione resuctase
PPY7	GOT1	Glutamic-oxaloacetic transaminase-1
PPY8	LDHA	Lactate dehydrogenase-A
PPY9	LDHB	Lactate dehydrogenase-B
	GAPD	Glyceraldehyde-3-phosphate dehydrogenase
	PEPB	Peptidase-B
	TPI	Triosephosphate isomerase
PPY10	H1-H4	Histone genes 1–4
PPY11	IDH1	Isocitrate dehydrogenase (soluble)
PPY12	–	–
PPY13	–	–
PPY14	–	–
PPY15	–	–
PPY16	MPI	Mannosephosphate isomerase
	HEXA	Hexosaminase-A
	PKM2	Pyruvate kinase (M2)
PPY17	–	–
PPY18	–	–
PPY19	–	–
PPY20	GPI	Glucose phosphate isomerase
PPY21	ITPA	Inosine triphosphatase-A
PPY22	–	–
PPY23	–	–
PPYX	G6PD	Glucose-6-phosphate dehydrogenase
	GLA	α-Galactosidase
PPYY	–	–

Amino acid sequencing of proteins provides a direct reflection of the DNA base sequences of the genes producing these proteins. To date four types of orang utan proteins have been completely or partly sequenced, fibrinopeptide (see Goodman 1971) myoglobin (Romero-Herrera et al. 1976), haemoglobin α- and β-chain (Maita et al. 1978) and carbonic anhydrase-1 (Tashian and Stroup 1970; Tashian et al. 1976). The orang utan myoglobin sequence causes some confusion with regard to the phylogeny of the hominoids, but is of little interest for intraspecific studies on the orang utan. Carbonic anhydrase sequence is interesting because of the occurrence of three CA1 alleles in the orang utan (see Meera Khan et al. 1982, this volume). The haemoglobin α- and β-chain sequences are of importance since the orang utan exhibits extensive haemoglobin polymorphism (see De Boer and Meera Khan 1982a and b, this volume).

Apart from the chromosomal difference between the Bornean and Sumatran orang utans and the confirmed or possible chromosomal polymorphisms mentioned above and described in detail by De Boer and Seuánez (1982, this volume), studies on a number of genetic systems and markers point to the existence of more such differences and polymorphisms. However, because of the incomplete specifications in the original articles of the study animals it is often impossible to distinguish between true subspecific differences on the one hand and polymorphisms on the other, or to assess possible differences in allelic frequencies between the two subspecies of *Pongo pygmaeus*. Some of the most important systems exhibiting variation of any kind are listed below.

Blood group systems

Blood group systems of primates have been studied rather extensively during the 1960s and early 1970s. Nevertheless, data on the orang utan blood groups are rather poor, with the exception of the ABO system.

In the ABO system of the orang utan the O-allele apparently is lacking (in contrast to the situation in the chimpanzee, which lacks the B-allele, and the gorilla which shows exclusively a B-like blood group factor). Schmitt (1968) gives a list of 13 references including a total of 44 individuals studied, none of which exhibit the O blood group, and he believes only two alleles to be acting in the orang utan, the A-allele (analogous to the human Al-allele) and the B-allele. This results in three possible genotypes, AA, AB and BB. However, in the AB heterozygotes the B-allele apparently suppresses the A-allele, resulting in a much weaker reaction of AB blood with A-antibodies than seen in AA blood.

Table 4 lists the ABO findings in a total of 74 orang utans from various sources. It is obvious from the totals that for a population in Hardy-Weinberg equilibrium there is a relative excess of B animals and a deficiency of AB

Table 4. ABO blood group factors in the orang utan.

| Reference | Number studied | Phenotypes | | | |
		A	B	AB	O
Schmitt et al. 1963	10	5	–	5	–
Wiener 1963	10	9	–	1	–
Wiener and Moor-Jankowski 1971[a]	16	13	1	2	–
Tippit and Green (pers. comm.)	14	7	1	6	–
Others[b]	24	7	11	6	–
Totals	74	41	13	20	–
Expected according to Hardy-Weinberg equilibrium (A = 0.69, B = 0.31)	74	35	7	32	–
Bornean orang utans[c]	16	7	1	8	–
Sumatran orang utans[d]	4	4	–	–	–

[a] Only the number of animals that were not already studied by Wiener (1963) is given.
[b] Full references are given by Schmitt et al. (1963).
[c] 5 identifiable animals of Schmitt et al. (1963) plus 11 of Tippit and Green (pers. comm.).
[d] 3 identifiable specimens of Schmitt et al. (1963) plus 1 of Tippit and Green (pers. comm.).

specimens, a conclusion that Schmitt et al. (1963) had already drawn on the basis of 34 specimens. However, since there are two orang utan populations the Hardy-Weinberg law is not applicable here. So far, no attempts have been made to investigate possible differences in AB(O) frequencies between Bornean and Sumatran orang utans. Such differences may be expected to exist, since they are well known in human races and in geographical populations of the chimpanzee (Moor-Jankowski et al. 1966). With the aid of the house names available for the animals studied by Schmitt et al. (1963) and by Tippit and Green (pers. comm.) it has been possible to identify the subspecific status of 20 specimens tested for AB(O) (see Table 4). It is notable that all of the four Sumatran animals in this group have blood group A, while in the 16 Bornean animals A, B and AB are represented. The numbers, however, are much too low to allow any conclusion.

Lewis blood group antigens have only been incompletely studied in the orang utan. Both Le(a) and Le(b) blood group substances have been demonstrated, the former in variable amounts (Schmitt 1968; Wiener et al. 1963, 1964a). Variation may well exist in the ABH secretor status. Moor-Jankowski and Wiener (1964) found a single non-secreting orang utan among 200 apes, which might indicate that in this species both secretors and non-secretors exist, in contrast to *Pan* and *Gorilla* which are exclusively secretors.

In the MN bloodgroup system N-like substances are probably consistently absent in the orang utan. A total of 52 specimens tested by Wiener et al. (1964c), Moor-Jankowski et al. (1964), Wiener and Moor-Jankowski (1971), Schmitt (1968) and Tippitt and Green (pers. comm.) failed to react with anti-N reagents, particularly *Vicia graminea* lectin. Polymorphism, however, is likely to exist in the M-like blood group substances. Twelve of the 24 specimens studied by Wiener and Moor-Jankowski (1971) did not react with anti-M sera; Schmitt (1968) obtained variable results with *Bowhinia pupurea* phytagglutinin in 14 animals, while all 14 orang utans tested by Tippit and Green (pers. comm.) reacted with *Phaseolus lunatus*, *P. limensis* and *Molucella laevis* lectins. Orang utan M-like substances are probably not identical to those in man, and further investigations are necessary to assess the genetics of this polymorphism.

Rh blood groups have been studied by Wiener (1963), Wiener et al. (1964b), Schmitt et al. (1963) and Schmitt (1968), among others. The results are not very clear. D-like and C-like receptors seem to be present polymorphically and E-like receptors seem to be absent. Recently, a case of Rh-antagonism was reported in an orang utan baby of Rhenen Zoo (pers. comm., Van Foreest) so that further investigation of this blood group system is desirable for captive breeding.

In spite of the indications of the existence of polymorphism with regard to several blood group systems, nothing is known as to possible differences in gene frequencies between Bornean and Sumatran orang utans.

Serum protein systems
An interesting polymorphism is known to occur in the 'serum group specific component' (Gc-system) of the orang utan. In man two alleles occur on the Gc-locus, resulting in three possible phenotypes: Gc 1-1, Gc 2-1 and Gc 2-2. No such polymorphism has been found in large samples of *Pan troglodytes* and *Macaca mulatta*, nor in smaller groups of *Gorilla gorilla*, *Symphalangus*

syndactylus and *Pan paniscus* (Kitchin and Bearn 1965; Schmitt et al. 1965). The orang utan, however, shows polymorphism identical to that in man. Among 20 specimens Kitchen and Bearn (1965) and Schmitt et al. (1965) found four individuals to be homozygous Gc 1-1, eight individuals homozygous 2-2 and eight individuals heterozygous Gc 2-1. No information is available on geographic origin of the animals studied.

Orang utans are also polymorphic with regard to the Gm-factors of gamma immunoglobulins. The only other primates exhibiting this type of polymorphism are man and the chimpanzee (Van Loghem et al. 1968; Van Loghem and Litwin 1972; Schmitt 1968). Seventeen of 25 orang utans studied by Van Loghem et al. possessed the combination of factors $s + c^5 + b^0 + b^1 - b^3 + b^4 -$, while the remaining eight had $s + c^5 + b^0 - b^1 - b^3 - b^4 -$ (γ2c-locus) (the latter animals are also exceptional among subhuman primates since, having s and c^5, they lack the b^0 factor). In man the frequencies of occurrence of the various combinations of Gm-factors are clearly geographically determined, certain combinations occurring only in certain races. Thus, it is possible that in the case of the orang utan we are not dealing with a polymorphism, but with a difference between the Bornean and Sumatran populations.

Unlike man, but in common with the other great apes and some lesser primates, the orang utan shows polymorphism in the serum transferrin system. In man, except for some local polymorphisms, only rare variants are found (apart from the usual C-allele). Twenty-nine orang utans studied by Nute and Buettner-Janusch (1968) and 12 studied by Lange and Schmitt (1963) have revealed the existence of two transferrin alleles and three phenotypes: AA ($= D_0D_0$) (7 of 41 animals), BB ($= B_3B_3$) (15 of 41) and AB ($= B_3D_0$) (19 of 41). The geographic origins of 10 of the 12 animals of Lange and Schmitt (1963) could be traced by their house names: five Bornean orang utans, including one AB and four BB individuals, and five Sumatran orang utans, all AB individuals. Thus, the two alleles apparently occur polymorphically in both populations, but so far nothing can be said on possible differences in allelic frequencies.

Like man and the other pongids, *Pongo* is polymorphic for the occurrence of the Lp(a) lipoprotein factor. Eight of 11 individuals tested for this factor by Schmitt (1968) reacted to anti-Lp(a) [Lp(a +)], three did not [Lp(a −)].

No polymorphisms were found in the haptoglobins (e.g. Lange and Schmitt 1963), albumin, alkaline phosphatase, ceroeloplasmin, esterases and leucine aminopeptidase (Bruce and Ayala 1979, who tested a sample of eight Sumatran and three Bornean orang utans).

Leucocyte antigens
Leucocyte antigens of the HLA-system have been studied in 32 orang utans by Seigler et al. (1972) and Metzgar et al. (1972) and in eight orang utans by Schmitt (1970). Some of the antigens occurring in man and the chimpanzee are also found in the orang utan. The latter species, however, has not been nearly as extensively studied as the chimpanzee, and it is difficult to make this highly specialized type of research applicable to the question of subspecific differences and population structure.

Red cell enzymes

These are treated separately by Meera Khan et al. (1982) and Wijnen et al. (1982) (both this volume). Polymorphisms occur in various isozymes and there are indications that some alleles are typical for either Bornean or Sumatran orang utans.

Haemoglobins

Polymorphism of the haemoglobin α-chain has been found in both Bornean and Sumatran orang utans with clearly different allelic frequencies in the two populations. Polymorphism of the β-chain possibly only occurs in Sumatra. Orang utan haemoglobins are reviewed separately by De Boer and Meera Khan (1982a and b, this volume).

Conclusions

The most reliable genetic characteristic which can be used to distinguish orang utans so far would seem to be the structure of the second chromosome pair (see De Boer and Seuánez 1982, this volume). There are, however, indications that there may be more differences in the occurrence of variants of several genetic markers and in allelic frequencies of polymorphic systems. Further study of the genetics of the orang utan would certainly provide useful information on the exact genetic distance between the two forms. It may well be that this distance proves to be as great or even greater than that between the common and the pygmy chimpanzee which are recognized as distinct species (*Pan troglodytes* and *Pan paniscus*).

No attempts have been made to date to gain more insight into the possible existence of regional differences between (sub)-populations in Borneo and Sumatra. The Bornean orang utans in particular need such attention since in the past up to eight local races have been recognized on the island (Selenka 1898). It would be of great interest to conservation strategy to assess the distribution of local races and the extent of racial diversification. The genetic data presently available seem to indicate that there are enough starting points to ensure a chance of success for this type of research. Attention should also be given to the orang utans of Sumatra. Formerly two forms were recognized on the island, (e.g. Selenka 1898) and recently Rijksen (1978) again pointed to the existence of two phenotypically clearly different, though interbreeding forms.

The genetic data on orang utans briefly reviewed above clearly indicate that many genetic polymorphisms exist in this species (see Table 5). Even several systems that are monomorphic in the other great apes appear to be polymorphic in *Pongo* (e.g. serum group specific component, haemoglobin α- and β-chain, chromosome number 9 structure). As such, orang utans do not give the impression of belonging to relict populations which often show seriously decreased genetic variability. On the contrary, genetically the species appears to be still rather variable, which probably is a good starting point for conservation success. In addition, the high rate of polymorphism provides a good tool for monitoring both captive and feral populations over generations. Here, again,

Table 5. Genetic systems of the orang utan exhibiting intraspecific variation either of polymorphic nature or potentially indicating subspecific differences.

Blood group systems:	AB(O) system
	Lewis system
	Secretor status
	M(N) system
	Rhesus system
Serum protein systems:	Serum group specific component
	Gm-factors of γ-immunoglobulins
	Transferrin system
	Lipoprotein system
Leucocyte antigens:	HL-A system
Red cell enzyme systems:	Adenosine deaminase
	Adenylate kinase-1
	Carbonic anhydrase-1
	Carbonic anhydrase-2
	Enolase-1
	Fumarate hydratase
	Glucose-6-phosphate dehydrogenase activity
	Guanylate kinase-1
	Guanylate kinase-3
	Nucleoside phosphorylase
	Peptidase-C
	Phosphoglucomutase-1
	Phosphoglucomutase-2
	Diaphorase-1*
Haemoglobins:	Haemoglobin α-chain
	Haemoglobin β-chain
Chromosomes:	Chromosome number 2 structure
	Chromosome number 9 structure
	Short arms of some of the chromosomes 11–17 and 22–23

* Fisher, pers. comm.; for other references see text.

the performance of more detailed studies on larger numbers of specimens is a prerequisite to the assessment of accurate gene frequencies.

Genetic diseases

However important the knowledge of genetically determined disorders and their modes of inheritance may be for conservation and in particular for the captive breeding of endangered species, little can be said on this topic with regard to the orang utan. No more than two clear cases are known for the species.

The first is a case of alkaptonuria, a classic metabolic hereditary disease in man, which was found in an orang utan by Keeling et al. (1973). Alkaptonuria is caused by a deficiency of the enzyme homogentisic acid oxidase which prevents the conversion of dietary phenylalanine and tyrosine to acetoacetic acid. This causes the excretion of homogentisic acid in the urine. The colour of the urine of such patients gradually turns dark to black when left standing, due to the oxidation of homogentisic acid to a melanin-like product (La Du 1966; Henry 1964). Humans with this disease usually show no pathologic signs during the first decade of life. Later, however, they may develop ochronosis because of accumulation of dark pigments (polymers of homogentisic acid), while deposition of the pigments in cartilaginous tissue may lead to arthritis. In man, alkaptonuria is usually transmitted as a Mendelian recessive trait. Since the trait is rare, consanguinity is often involved in cases of occurrence.

Keeling et al. (1973) discovered homogentisic acid in the urine of a 2-year-old orang utan of the Yerkes Regional Primate Research Center. At the age of 5 years this animal did not exhibit other abnormalities. At the time of their study only half sibs of the propositus were available, none of which showed signs of the disease. No follow-up data were published. Alkaptonuria is also known to occur in the chimpanzee (Watkins et al. 1970).

The second case of genetic disease concerns a female orang utan in Vienna Zoo with a trisomy of the 22nd chromosome pair, reported by Anderle et al. (1979). The 22nd pair of chromosomes in the orang utan is believed to be homologous to the 21st pair in man (see Seuánez 1979). Trisomy of this pair results in the clinical symptoms of Down's syndrome (mongoloidism). Trisomy 21 occurs with higher frequency in children of older parents and slightly more frequently when the parents are very young. Carriers of a chromosome 21 translocated to one of the D-group chromosomes, however, have a higher risk at all ages of having children with Down's syndrome.

The trisomic orang utan showed the normal trisomy, not involving a translocation. Both parents (at Stuttgart Zoo) were relatively young when the propositus was born, but had already produced three young (Fig. 2). The chromosomes of two of the sibs of the propositus were studied by De Boer and Seuánez (1982, this volume) and proved to be normal. Thus, apparently we are dealing here with an incidental case of trisomy.

The symptoms of the trisomic orang utan were not as distinct as in Down's syndrome in man, probably the reason why the animal did not attract special attention in the zoos where it lived during the first years of life (it was karyotyped at the age of 7). According to Anderle et al. (1979) it weighed 6.8 kg at the age of 1 year (which is about normal according to Seitz 1969) and 23.4 kg at the age of 7 (which is somewhat below normal). The clinical features included: more pronounced mongoloid position of the eyelid axis, malformations of the ear (imperfect convolution), permanently slightly open mouth, anomalies of the teeth, abnormal position of the right foot (probably caused by

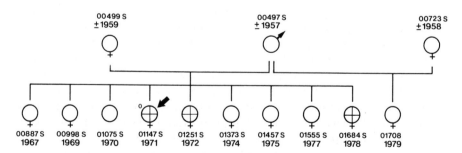

Fig. 2. Pedigree of the orang utan with trisomy of the 22nd chromosome pair described by Anderle et al. (1979). The propositus is indicated by an arrow. For each animal the studbook number is given as well as the year of birth. The chromosomes of the animals 01075S, 01251S and 01684S have been studied. The latter two did not show abnormal karyotypes.

functional retraction of the second and fifth toes) and hypotonia of all skeletal muscles. The animal's behaviour was characterized by hypomimy, lack of concentration, reduced motor activity and general inactivity, while the animal was said to be rather withdrawn when in the company of other orang utans (Anderle et al. 1979).

A single case of trisomy of the 22nd chromosome pair has been reported in the chimpanzee (McClure et al. 1969, 1970, 1971). The symptoms in that case, however, were more pronounced and more clearly resembled those of Down's syndrome in man than those in the above-described orang utan. A gorilla with an additional chromosome 22 was not noticed to show clinical or behavioural abnormalities at all (Turleau et al. 1972). Thus, the expression of Down's syndrome seems to be rather variable from species to species.

Genetic management of the captive orang utan populations

From the point of view of population genetics captive populations of endangered species are unnatural in several ways.
1. Generally, captive populations are rather small compared to natural populations.
2. There is no random breeding in captive populations, since they are mostly subdivided into various numbers (up to over 100) of small breeding units, often spread over zoological collections in different continents. Even within such breeding units there is no random breeding, since often the keepers rather than the animals determine which couples should mate.
3. The animals constituting the captive 'population' often do not all originate from the same natural population; sometimes they even originate from populations that are reproductively isolated in the wild.
If in spite of these unnatural circumstances it is intended to maintain a captive population over a considerable number of generations without changing its basic characteristics, genetic management of the population with the aim of eliminating the ill effects of the unnatural situation as much as possible, is necessary.

There are two dangerous aspects of the unnatural captive populations — the high risk of inbreeding and the risk of loss of genetic variability. Particularly in zoological gardens the existence of an inbreeding problem has often been denied. Two arguments against the possible ill effects of inbreeding which are often heard are: (1) that (captive) populations of some species were founded by only one or a few couples, and (2) that inbreeding may lead to 'higher quality' animals as well as to 'inferior' animals. If generalized, the first argument denies the existence of deleterious genes in wild animals (genes that could possibly have negative effects when present in the homozygous state). The second argument in fact proves, rather than disproves, the risks of inbreeding; just as inbreeding may cause the homozygous expression of 'favourable' genes, it may cause the homozygous expression of 'bad' genes. It may be stated that diploidy and the sexual mode of propagation were 'invented' by nature to reduce the risks of homozygosity and to profit by the advantages of genetic diversity. This implies

51

that high rates of inbreeding are generally unnatural since inbreeding has exactly the opposite effects. Diploidy and the resulting possibility of the existence of a large genetic variability render a population fit to withstand the changes in environmental selective factors because they facilitate the presence of genes (in the heterozygous state and often in low frequencies) that are not advantageous at a given time, but may be so in the future (or have been in the past). Although disadvantageous genes are most probably present in any feral population, this certainly does not imply that any given animal of a population necessarily carries such genes. If a breeding group is set up with animals that, by chance, do not carry deleterious genes, there is no risk of inbreeding at all. Such examples, however, may not be taken to disprove inbreeding risks in general. If deleterious genes exist in a given population, the probability of which is very high, the risk of their expression is greatly increased by inbreeding.

Unlike the situation in the zoo populations of several other endangered animal species, such as the Przewalski horse (Bouwman 1977, 1979), the okapi (Foose 1978), the dorcas gazelle (Ralls 1980), the leopard (Shoemaker and Wharton 1981) and the Siberian tiger (Seal 1980), in which there are clear indications of inbreeding depression, there is virtually no inbreeding problem in the captive orang utan. The reason for this simply is that breeding of orang utans on a larger scale only started a few decades ago, while their generation time is rather long compared to that of other species. Although over 600 orang utans have been born in captivity (see Jones 1982; this volume), the vast majority of births have occurred to wild-caught parents, which, by definition means that the offspring have inbreeding coefficients of zero. As of 31 December, 1979, 418 captive-born orang utans were living, of which 370 were born to two wild parents. Thirty eight were born to one wild parent and one captive-born parent and only nine of these were inbred. Five animals were born to two captive-born parents and one of these was inbred (see Table 6 for a summary of birth and inbreeding data).

It may be hoped and expected, however, that the number of second, third and fourth generation births in zoological gardens and primate centres will rapidly increase in the next decade. Simultaneously, the risk of higher inbreeding coefficients will increase if no proper breeding policy is followed. To prevent this risk, old-fashioned views and practices such as mating daughters to their fathers and sons to their mothers or selling of full sibs as breeding pairs should be eliminated. Good breeding policy, however, not only involves the avoidance of breeding animals with high inbreeding coefficients for the time being, it also involves the avoidance of inbreeding in the long run. Thus, breeders should be careful in continuing breeding with the same pair for too many years, and in using one breeding male for too many females, while other males are excluded from propagation (a practice unfortunately often met with in zoological gardens). Such practices result in large numbers of closely related animals which, although having inbreeding coefficients of zero, have high coefficients of relationship resulting in an increased risk of inbreeding in the next generations.

As important as the deleterious effects of inbreeding (which in fact causes a decrease of the heterozygozity in individual animals, and thus an increased homozygous expression of recessive genes including deleterious ones) is the risk

Table 6. Population sizes of captive orang utans specified as to subspecies and origin. Numbers (with numbers of males, females and animals of unknown sex in brackets) are those of living animals per 31 December 1979. Data collected from the Studbook of the Orang Utan (Jones 1980).

Subspecies	Population size	Origin	Generation	Inbreeding coefficients*
Bornean	363 (176.187)	wild-caught 217 (110.107)		$F = 0.000$
		captive-born 146 (66.80)	wild × wild 136 (61.75)	$F = 0.000$
			wild × capt.-b. 9 (5.4)	$F = 0.000$
			capt.-b. × capt.-b. 1 (0.1)	$F = 0.250$
Sumatran	242 (106.134.2)	wild-caught 106 (45.61)		$F = 0.000$
		captive-born 136 (61.73.2)	wild × wild 126 (53.71.2)	$F = 0.000$
			wild × capt.-b. 9 (7.2)	$F = 0.000(7), F = 0.250(2)$
			capt.-b. × capt.-b. 1 (1.0)	$F = 0.000$
Unknown	93 (46.45.2)	wild-caught 58 (33.25)		$F = 0.000$
		captive-born 35 (13.20.2)	wild × wild 24 (9.13.2)	$F = 0.000(7), F = 0.250(4)$
			wild × capt.-b. 11 (4.7)	$F = 0.000$
			capt.-b. × capt.-b. 3 (2.1)	$F = 0.000$
Hybrids	96 (56.40)	captive-born 96 (56.40)	wild × wild 84 (52.32)	$F = 0.000(6), F = 0.250(3)$
			wild × capt.-b. 9 (2.7)	$F = 0.000$
			capt.-b. × capt.-b. 3 (2.1)	$F = 0.000$
Totals	794 (384.406.4)	wild-caught 381 (188.193)		$F = 0.000$
		captive-born 413 (196.213.4)	wild × wild 370 (175.191.4)	$F = 0.000$
			wild × capt.-b. 38 (18.20)	$F = 0.000(29), F = 0.250(9)$
			capt.-b. × capt.-b. 5 (3.2)	$F = 0.000(4), F = 0.250(1)$

* When two inbreeding coefficients are given, the numbers in brackets indicate the corresponding numbers of animals.

of loss of genetic variability in small captive populations. Genetic heterogeneity, the existence of more than one allele over a considerable proportion of the total number of gene loci, is currently considered to be an important characteristic of wild populations. In fact, the gene pool, the total of all alleles present in the population, each with its own frequency, forms the genetic basis of the population (i.e. the species) and genetic management of captive populations should aim at retaining this gene pool as well as possible without considerably changing allelic frequencies or losing alleles entirely. Otherwise, features of domestication may arise rapidly and the population may lose characteristics that are of crucial importance to life under wild or semi-wild conditions. In recent years population geneticists have given much thought to the question of how to manage small populations to ensure optimal preservation of gene pools. For general work in this context the reader is referred to Foose (1977, 1980), Flessness (1977), Seal et al. (1977) and Fowler and Smith (1973), while specific populations have been analysed by Foose (1978, okapi), Foose (1981, Père David deer), and Seal (1980, Siberian tiger).

It is intended that the maintenance of orang utans in captivity should not have any negative effect on the conservation of the species in the wild. This means among other things that there can be no refreshing of captive blood lines with additional animals from the wild. The application of the genetic models to preserve gene pools, therefore, is certainly desirable. A particular problem in the orang utan is that, because of the difference between Bornean and Sumatran orang utans, there are in fact two gene pools to be conserved strictly separately. The following considerations are of importance when speaking of preserving gene pools in general and those of the captive orang utans in particular:

1. The founder population (the wild-caught specimens that form the basis of the captive population) represents only a small part of the original wild population, and consequently does not include all the genetic variants characteristic of the species. Recent work, however, has made it clear that a relatively small number of specimens (provided that these are unrelated), represent a relatively large proportion of the total genetic variability of the entire population (Foose 1977, 1980). As few as 10 unrelated animals would carry as much as 95% of the genetic variation of the total population, while in 50 animals this percentage would be very close to 100 (see Fig. 3). The numbers of Bornean and Sumatran orang utans living in captivity as of 31 December, 1979 were 363 and 242, respectively. Of these, 146 and 136 animals, respectively, were born in captivity, the remainder being wild-caught (numbers from the Studbook of the Orang Utan, Jones 1980; see also Table 6). The captive-born animals originated from approximately 140 Bornean and 105 Sumatran wild-caught parents. However, the exact size of the founder population cannot yet be estimated. On the one hand, the number of founders may still increase since many of the wild-caught animals have not bred yet, but are still of breeding age, while animals of unknown subspecific identity (96 as of 31 Dec., 1979) may be identified and added to the Bornean and Sumatran stocks. On the other hand, the founder population may decrease if first generation captive-born orang utans (carrying genetic

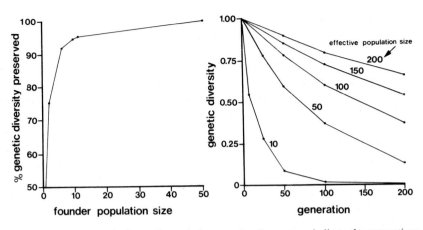

Fig. 3. Left: genetic diversity in small population samples. Percentages indicate the proportions of the total genetic variability of a large population which are present in small samples of unrelated animals. Right: loss of genetic variability over the generations in stable populations of various sizes. The number given with each curve indicates the effective population size which is generally lower than the actual number of animals (from Flesness 1977 and Foose 1980).

material of the founders) fail to breed. Nevertheless, it can be safely stated that the founder populations of both Bornean and Sumatran orang utans will be considerably larger than those of several other zoo populations that have been bred in captivity for many generations (e.g. 13 for the Przewalski horse, 28 for the okapi and 10 for the gaur [Bouwman 1979; Foose 1978, 1980]). Accordingly, the initial amount of genetic variability in the captive orang utan populations certainly will be more favourable.

2. The size of the captive population determines the influence of genetic drift, which reduces the genetic variability. The smaller the population, the smaller the probability that all genes present in one generation will reappear in the next. In this respect it is not the size of the founder population that is of importance but that of the actual population in captivity. Thus, to keep effects of genetic drift at a minimum, the founder population should be increased as rapidly as possible until the maximum population size is reached. This is determined by the maximum number of animals that can be kept in captivity. Currently the total number of orang utans in captivity is approximately 800. Since the great apes are among the most expensive zoo animals (at least when they are to be maintained under appropriate conditions in traditional zoos), it cannot be anticipated that this number could be increased substantially in the near future. In economic terms zoos will (and should) aim at the creation of better facilities for the present stock rather than at increasing the number of 'cages'. Since this total capacity is used for two types of orang utans, the maximum population size for Bornean and Sumatran orang utans may not be much higher than 400 to 500 specimens each. However, at present approximately 12% of the total capacity is occupied by hybrid animals that are of no use to either population. From the point of view of

55

optimal genetic management, it would therefore be advisable to reduce the number of hybrids to a minimum (however, it would seem advisable to maintain a small hybrid population with which fertility experiments can be performed; see also De Boer and Seuánez 1982; this volume). Another possibility to increase the carrying capacity might be to investigate the possibilities of less expensive and less intensive ways of maintaining larger groups of orang utans in wild-parks.

3. The sex ratio is of importance in maintaining maximum genetic variability. The more the sex ratio deviates from the 1:1 ratio, the more rapidly genetic variation is lost from the population. As can be seen from Table 6, in the captive orang utans the sex ratio in this respect is almost optimal. However, when we look at the number of breeding animals, the number of females is somewhat in excess to that of the males. This will be discussed in the following point.

4. Maintenance of maximum variability requires equal family size. The more the family size varies, the more rapidly variability is lost. In this respect the situation in the captive orang utans is far from optimal. The captive born stock of Bornean orang utans originated from 66 Bornean males and 76 Bornean females, the captive-born Sumatran stock from 53 Sumatran males and 60 Sumatran females. This means that the majority of wild-caught animals have not (yet) bred. In addition, a relatively small number of the breeding animals have produced a relatively large part of the captive-born stock (see Table 7). Thus, family size is largely unequal. Admittedly, it is too early to present definite figures, because in the orang utan populations the breeding results of separate generations cannot yet be analysed as the wild-caught generation still breeds while the captive born generations have hardly started breeding. Nevertheless, a clear tendency is apparent, a tendency of breeding more frequently with animals that have proved to breed easily in captivity. Apart from the fact that this practice considerably reduces the genetic variability in the population, this easily leads to unwanted selection in favour of animals breeding easily under captive conditions. This inevitably leads to features of domestication. However difficult it may be, one should strive to breed all of the available animals in the population equally.

Nevertheless, even if sex ratio, number of reproducing animals and equality of family size are kept at an optimum, the genetic variability will decrease with the number of generations the population lives in captivity because of its limited size. As can be seen from Fig. 3 in orang utan populations numbering several hundreds of animals, this decrease would be very slow if all the requirements of genetic management were met. However, in practice there are severe problems. In the first place, the captive orang utans are distributed over no less than 160 institutions all over the world. This means that the average group of orang utans consists of some five individuals, but since many of the institutions keep the Bornean as well as the Sumatran subspecies (sometimes together with hybrids), the averages for each of the two subspecies are even lower. Add to this the fact that adult breeding males in particular seem to be very sensitive to relocation from one colony to another, and it will be clear that it is extremely

Table 7. Family sizes of the parents of 146 captive-born Bornean and 136 captive-born Sumatran orang utans living per 31 December 1979 (data collected from the Studbook of the Orang Utan; Jones 1980).

| | Number of offspring per parent | | | | | | | | | | |
	1	2	3	4	5	6	7	8	9	10	Totals
Bornean orang utans											
Number of breeding males	26	22	8	5	1	1	2	1	–	–	66 breeding males
Total number of offspring	26	44	24	20	5	6	14	8	–	–	146 offspring
Number of breeding females	35	22	10	5	1	2	–	–	–	–	76 breeding females
Total number of offspring	35	44	30	20	5	12	–	–	–	–	146 offspring
Sumatran orang utans											
Number of breeding males	21	11	6	10	2	1	–	–	1	1	53 breeding males
Total number of offspring	21	22	18	40	10	6	–	–	9	10	136 offspring
Number of breeding females	27	11	13	6	3	–	–	–	1	–	60 breeding females
Total number of offspring	27	22	39	24	15	–	–	–	9	–	136 offspring

difficult to manage such collections of orang utans as populations in which there is some semblance of random breeding and equal family size. Only very well coordinated programmes of exchanging young orang utans may be effective in this respect. Another practical problem in larger colonies might be the suppression of the ability to breed when two or more males are kept in close contact (see Kingsley 1982; this volume). This phenomenon increases the practice of breeding with one male and several females.

It might be argued that the genetic management measures advocated above do not seem to reflect the natural reproductive characteristics of the orang utan. In the wild polygyny certainly occurs (see Schürmann 1982; this volume), and possibly there is a certain degree of consanguineous breeding in family groups. The effects of inbreeding, and the loss of genetic variability, however, caused by this in the wild are largely made up by the large size of the population. Natural populations of orang utans will consist of at least several thousands of individuals. The unnaturally small population size in captivity, on the other hand, requires unnatural modes of reproduction to retain optimal genetic structure.

References

Anderle, M., W. Fielder, A. Rett, P. Ambros and D. Schweitzer (1979). A case of trisomy 22 in *Pongo pygmaeus*. Cytogen. Cell Gen., 24: 1–6.

Boer, L.E.M. de and P. Meera Khan (1982a). Orang utan haemoglobins: a short review. In: L.E.M. de Boer (ed.), The orang utan. Its biology and conservation. Junk, The Hague.

Boer, L.E.M. de and P. Meera Khan (1982b). Haemoglobin polymorphisms in Bornean and Sumatran orang utans. In: L.E.M. de Boer (ed.), The orang utan. Its biology and conservation. Junk, The Hague.

Boer, L.E.M. de and H.N. Seuánez (1982). The chromosomes of the orang utan and their relevance to the conservation of the species. In: L.E.M. de Boer (ed.), The orang utan. Its biology and conservation. Junk, The Hague.

Bouman, J. (1977). The future of Przewalski horses in captivity. Int. Zoo Yearb., 17: 62–68.

Bouman, J.G. and H. Bos (1979). Two symptoms of inbreeding depression in Przewalski horses living in captivity. In: L.E.M. de Boer, J. Bouman and I. Bouman (eds.), Genetics and hereditary diseases of the Przewalski horse. Foundation for the Preservation and Protection of the Przewalski Horse, Rotterdam.

Bruce, E.J. and F.J. Ayala (1979). Phylogenetic relationships between man and the apes: electrophoretic evidence. Evolution, 33: 1040–1056.

Foose, T. (1977). Demographic models for management of captive populations. Int. Zoo Yearb., 17: 70–76.

Foose, T. (1978). Demographic and genetic models for the management of the okapi (*Okapia johnstoni*) in captivity. Acta Zool. Pathol. Antverpiensa, 73: 119–195.

Foose, T. (1980). Demographic and genetic management of endangered species in captivity. Int. Zoo Yearb., 20: 154–166.

Foose, T. (1981). Demographic and genetic status and management of Pere David deer (*Elaphurus davidianus*) in captivity. In: Pere David deer: Biology and captive management (in press).

Flesness, N. (1977). Gene pool conservation and computer analysis. Int. Zoo Yearb., 17: 77–81.

Fowler, C.W. and T. Smith (1973). Characterizing stable populations: an application to the African elephant population. J. Wildlife Management, 37: 513–523.

Goodman, M., A.L. Koen, J. Barnabas and G.W. Moore (1971). Evolving primate genes and proteins. In: A.B. Chiarelli (ed.), Comparative genetics in monkeys, apes and man. Academic Press, London.

Gosden, J.R., A.R. Mitchell, H.N. Seuánez and C.M. Gosden (1977). The distribution of sequences complementary to human satellite DNAs I, II and IV in the chromosomes of chimpanzee (*Pan troglodytes*), gorilla (*Gorilla gorilla*) and orang utan (*Pongo pygmaeus*). Chromosoma (Berl.), 63: 253–271.

Gosden, J., S. Lawrie and H. Seuánez (1978). Ribosomal and human-homologous repeated DNA distribution in the orangutan (*Pongo pygmaeus*). Comparison with the distribution of these DNAs in the other species of the Hominidae. Cytogenet. Cell Genet., 21: 1–10.

Henderson, A.S., K.C. Atwood and D. Warburton (1976). Chromosomal distribution of rDNA in *Pan paniscus*, *Gorilla gorilla beringei* and *Symphalangus syndactylus*: comparison to related primates. Chromosoma (Berl.), 59: 147–155.

Henderson, A.S., D. Warburton, S. Megraw-Ripley and K.C. Atwood (1979). The chromosomal location of rDNA in the Sumatran orang utan, *Pongo pygmaeus albei*. Cytogenet. Cell Genet., 23: 213–216.

Henry, R.J. (1964). Clinical chemistry: principles and technics. Harper and Row, New York.

Hoyer, B.H., N.W. van de Velde, M. Goodman and R.B. Roberts (1972). Examination of hominid evolution by DNA sequence homology. J. Hum. Evol., 1: 645–649.

Jones, M.L. (1980). Studbook of the orang utan. Zoological Society of San Diego, San Diego.

Jones, M.L. (1982). The orang utan in captivity. In: L.E.M. de Boer (ed.), The orang utan. Its biology and conservation. Junk, The Hague.

Keeling, M.E., H.M. McClure and R.F. Kibler (1973). Alkaptonuria in an orangutan (*Pongo pygmaeus*). Am. J. Phys. Anthrop., 38: 435–438.

Kingsley, S. (1982). Causes of non-breeding and the development of the secondary sexual characteristics in the male orang utan — a hormonal study. In: L.E.M. de Boer (ed.), The orang utan. Its biology and conservation. Junk, The Hague.

Kitchin, F.D. and A.G. Bearn (1965). The serum group specific component in non-human primates. Am. J. Hum. Genet., 17: 42–50.

Kohne, D.E., J.A. Chiscon and B.H. Hoyer (1972). Evolution of primate DNA sequences. J. Hum. Evol., 1: 627–644.

La Du, B.N. (1966). The metabolic basis of inherited disease. McGraw-Hill, New York.

Lange, V. and J. Schmitt (1963). Das Serumeiweissbild der Primaten unter besonderer Berücksichtigung der Haptoglobine und Transferrine. Folia Primat., 1: 208–250.

Loghem, E. van and S.D. Litwin (1972). Antigenic determinants on immunoglobulins of nonhuman primates. In: H. Balner and J.J. van Rood (eds.), Transplantation genetics of primates. Transplantation Proceedings IV (1).

Loghem, E. van, J. Shuster and H.H. Fudenberg (1968). Gm-factors in non-human primates. Vox Sang., 14: 81–94.

Maita, T., A. Araya, M. Goodman and G. Matsuda (1978). The amino acid sequences of the two main components of adult hemoglobin from orangutan (*Pongo pygmaeus*). Hoppe-Seyler's Z. Physiol. Chem., 359: 129–132.

McClure, H.M., K.H. Belden, W.A. Pieper and C.B. Jacobson (1969). Autosomal trisomy in a chimpanzee: resemblance to Down's syndrome. Science, 165: 1010–1012.

McClure, H.M., K.H. Belden, W.A. Pieper, C.B. Jacobson and D. Picciano (1970). Cytogenetic studies and observations in the Yerkes great ape colony. Med. Primat. (Proc. 2nd Conf. Exp. Med. Surg. Primates, New York 1969): 281–296.

McClure, H.M., W.A. Pieper, M.E. Keeling and C.B. Jacobson (1971). Mongoloid-like condition in a chimpanzee. Proc. 3rd Int. Congr. Primat., Zürich 1970, vol. 2: 110–115. Karger, Basel.

Meera Khan, P., H. Rijken, J. Wijnen, Lucie M.M. Wijnen and L.E.M. de Boer (1982). Red cell enzyme variation in the orang utan; electrophoretic characterization of 45 enzyme systems in Cellogel. In: L.E.M. de Boer (ed.), The orang utan. Its biology and conservation. Junk, The Hague.

Metzgar, R.S., H.F. Seigler, F.E. Ward, E.D. Hill and T. Mohanakumar (1972). Characterization of chimpanzee leucocyte alloantisera. In: H. Balner and J.J. van Rood (eds.), Transplantation genetics of primates. Transplantation Proceedings IV (1).

Mitchell, A.R. and J.R. Gosden (1968). Evolutionary relationships between man and the great apes. Sci. Prog., Oxf., 65: 273–293.

Mitchell, A.R., H.N. Seuánez, S.S. Lawrie, D.E. Martin and J.R. Gosden (1977). The location of DNA homologous to human satellite III DNA in the chromosomes of chimpanzee (*Pan troglodytes*), gorilla (*Gorilla gorilla*) and orang utan (*Pongo pygmaeus*). Chromosoma (Berl.), 61: 345–358.

Moor-Jankowski, J. and A.S. Wiener (1964). Blood groups and serum specificities of apes and monkeys. Lab. Primate Newsletter, 3: 1–14.

Moor-Jankowski, J., A.S. Wiener, C.H. Kratochvil and J. Fineg (1966). Chimpanzee blood groups; blood group distribution as racial characteristics. Science, 152: 219.

Moor-Jankowski, J., A.S. Wiener and E.B. Gordon (1964). Bloodgroups of apes and monkeys. III. The M-N blood factors of apes. Folia Primat., 2: 129–148.

Nute, P.E. and J. Buettner-Janusch (1968). Serum transferrins of primates: *Pan*, *Pongo* and *Hylobates*. Folia Primat., 8: 282–289.

Pearson, P.L., T.H. Roderick, M.T. Davisson, J.J. Garver, D. Warburton, P.A. Lalley and S.J. O'Brien (1979). Report of the committee on comparative mapping. Cytogenet. Cell Genet., 25: 82–95.

Ralls, K., K. Brugger and A. Glick (1980). Deleterious effects of inbreeding in a herd of dorcas gazelle (*Gazella dorcas*). Int. Zoo Yearb., 20: 137–146.

Rijksen, H.D. (1978). A field study on Sumatran orang utans (*Pongo pygmaeus abelii* Lesson 1827), ecology, behaviour and conservation. Meded. Landbouwhogeschool Wageningen, 78–2: 1–420.

Romero-Herrera, A.E., H. Lehmann, O. Castillo, K.A. Joysey and A.E. Friday (1976). Myoglobin of the orang utan as a phylogenetic enigma. Nature (London), 261: 162–164.

Schmitt, J. (1968). Immunbiologische Untersuchungen bei Primaten. Bibliotheca Primatologica, 8: 1–146. Karger, Basel.

Schmitt, J. (1970). Die Blut- und Serumgruppen der Primaten. Humangenetik, 8: 261–279.

Schmitt, J., W. Spielmann and M. Weber (1963). Serologische Befunde beim Orang-Utan (*Pongo pygmaeus* Linnaeus 1760). Z. Säugetierk., 27: 93–102.

Schmitt, J., A. Ensgraber, M. Krüpe and H. Cleve (1965). Die Gc-Erbmerkmale im Serum der anthropoiden Affen. Humangenetik, 1: 289–295.

Schürmann, C. (1982). Mating behaviour of wild orang utans. In: L.E.M. de Boer (ed.), The orang utan. Its biology and conservation. Junk, The Hague.

Schweitzer, D., P. Ambros, M. Anderle, A. Rett and W. Fiedler (1979). Demonstration of specific heterochromatin segments in the orangutan (*Pongo pygmaeus*) by a distamycin/DAPI double staining technique. Cytogenet. Cell Genet., 24: 7–14.

Seal, U.S. (1980). Genetics and demography of studbook Siberian tigers in North American zoos with evidence for inbreeding depression. (Unpublished).

Seal, U.S., D.G. Malcey, D. Bridgewater, L. Simmons and L. Murtfeldt (1977). ISIS: a computerized record system for the management of wild animals in captivity. Int. Zoo Yearb., 17: 68–70.

59

Seigler, H.F., R.S. Metzgar, F.E. Ward and D.M. Reid (1972). Reactions of human HL-A sera with orang utan and gorilla lymphocytes. In: H. Balner and J.J. van Rood (eds.), Transplantation genetics of primates. Transplantation Proceedings, IV-1.

Selenka, E. (1898). Menschenaffen: Rassen, Schädel und Bezahnung des Orang-utan. Kreidel, Wiesbaden.

Seitz, A. (1969). Notes on the body weights of new-born and young orang-utans, *Pongo pygmaeus*. Int. Zoo Yearb., 9: 81–94.

Seuánez, H.N. (1979). The phylogeny of human chromosomes. Springer, Heidelberg.

Seuánez, H., H.J. Evans, D.E. Martin and J. Fletcher (1979). An inversion in chromosome 2 that distinguishes between Bornean and Sumatran orangutans. Cytogenet. Cell Genet., 23: 137–140.

Shoemaker, A. and D. Wharton (1981). Analysis of inbreeding within leopards in captivity. Zool. Garten (in press).

Tantravahi, R., D.A. Miller, V.G. Dev and O.J. Miller (1976). Detection of nucleolus organizer regions in chromosomes of human, chimpanzee, gorilla, orangutan and gibbon. Chromosoma (Berl.), 56: 15–27.

Tashian, R.E., M. Goodman, R.E. Ferell and R.J. Tanis (1976). Evolution of carbonic anhydrase in primates and other mammals. In: M. Goodman and R.E. Tashian (eds.), Molecular anthropology. Plenum, New York.

Turleau, C., J. de Grouchy and M. Klein (1972). Phylogénie chromosomique de l'homme et des primates hominiens (*Pan troglodytes, Gorilla gorilla* et *Pongo pygmaeus*). Essai de réconstitution de l'ancêtre comun. Ann. Génét., 15: 225–240.

Watkins, S.P., H. Benley and N.R. Shulman (1970). Alkaptonuria in a chimpanzee. In: Medical Primatology 1970. Karger, Basel.

Wiener, A.S. (1963). Blood groups in anthropoid apes and baboons. Science, 142: 67–69.

Wiener, A.S. and J. Moor-Jankowski (1971). Bloodgroups of non-human primates and their relationship to the bloodgroups of man. In: A.B. Chiarelli (ed.), Comparative genetics of monkeys, apes and man. Academic Press, London.

Wiener, A.S., J. Moor-Jankowski and E.B. Gordon (1963). Bloodgroups of apes and monkeys. II. The A-B-O bloodgroups, secretor and Lewis types in apes. Am. J. Phys. Anthrop., 21: 271–281.

Wiener, A.S., E.B. Gordon and J. Moor-Jankowski (1964a). The Lewis bloodgroups in man. A review with supporting data on non-human primates. J. Forensic Med., S. Africa, 11: 67.

Wiener, A.S., J. Moor-Jankowski and E.B. Gordon (1964b). Bloodgroups of apes and monkeys. IV. The Rh-Hr blood types of anthropoid apes and monkeys. Am. J. Human Genet., 16: 246.

Wiener, A.S., J. Moor-Jankowski and E.B. Gordon (1964c). Bloodgroups and cross-reacting antibodies in primates, including man. II. Studies on the M-N types of orang utans. J. Immunol., 93: 101–105.

Author's address:
L.E.M. de Boer
Biological Research Department
Royal Rotterdam Zoological and Botanical Gardens
Rotterdam
The Netherlands

4. Red cell enzyme variation in the orang utan: electrophoretic characterization of 45 enzyme systems in Cellogel

P. Meera Khan, Herbert Rijken, Juul Th. Wijnen, Lucie M.M. Wijnen and Leobert E.M. de Boer

Introduction

Even though eight local races of orang utan were identified in Borneo and two in Sumatra as early as 1898 by Selenka no attempts have been made so far to search for genetic differences between these populations except the recently reported chromosomal studies (see De Boer and Seuánez 1982, this volume, for a review).

Availability of genetic markers in the orang utan should, in principle, facilitate the estimate of the genetic distances between different populations of present day orang utans, the construction of a microevolutionary tree (Cavalli-Sforza and Bodmer 1971) and the correlation of the inferences thus derived with the available geological evidence for geographical isolation of the populations during the 'recent' past. Apart from the inherent academic interests, it appears important to investigate these populations to establish the degree of genetic divergence between them — a knowledge useful in developing effective breeding programmes to conserve this endangered species in its natural as well as captive habitats (De Boer 1982, this volume).

In this preliminary search for genetic markers we screened 45 red cell enzyme systems in 12 Bornean, two Sumatran, one full and one partial hybrid (3/4 Sumatran and 1/4 Bornean) orang utans for electrophoretic variants. Such variants are presumed to be determined by different alleles at distinct genetic loci scattered in their genomes, by analogy with the human variants (Harris 1980). The aims of this paper are to describe the procedures of enzyme electrophoresis and stainings employed, to present the zymogram patterns observed, and to report the intraspecific and intersubspecific enzyme variations found in the present material.

Material

Peripheral venous blood erythrocytes were obtained from eight male and eight female orang utans including animals of Bornean, Sumatran and hybrid origin (Table 1). Pedigree charts depicting their relationships to each other, if any, and origin have been described in other articles appearing in this volume (Wijnen

The orang utan. Its biology and conservation, edited by L.E.M. de Boer
© *1982. Dr W. Junk Publishers, The Hague, ISBN 90 6193 702 7*

et al. 1982; De Boer and Seuánez 1982). The animals were maintained at various zoos in Germany, the Netherlands and Belgium at the time of bleeding during 1979.

Methods

The procedures of collection and processing of the blood samples and the preparation and storage of haemolysates have been described elsewhere (De Boer and Meera Khan 1982, this volume). The samples (Table 1) were screened for all the enzyme systems listed in Table 2 following the general procedure of enzyme electrophoresis in Cellogel described by Meera Khan (1971).

The various procedures described in this section have been routinely employed by us in screening the human and other primate materials including the primate-rodent somatic cell hybrids for identifying the expression of species specific enzyme molecules both variant and invariant (Meera Khan and Rattazzi 1968; Meera Khan et al. 1971; Meera Khan et al. 1972; Meera Khan and Balner 1972; Ebeli-Struyk et al. 1976; Meera Khan and Doppert 1976; Meera Khan et al. 1976; Estop et al. 1978; Garver et al. 1978; Pearson et al. 1978). The original descriptions of most of the staining procedures and the principles of reactions involved in the identification of specific gene products through zymograms have been summarized elsewhere (Harris and Hopkinson 1976; Meera Khan 1971; van Someren et al. 1974). Therefore, only a comprehensive account of the special features of methodology relevant to individual systems are indicated below.

The abbreviations of various reagents and methodological terms are listed and explained in Table 3. A few of the examples are: XOD = xanthine oxidase; PEP = phosphoenolpyruvate; **BS** = buffer system; **QSPA** = quantity of sample per application. Unless otherwise mentioned, the abbreviations of the respective biological reagents used in the following text mean not only their expanded name, but also the concentration of the reagents in the solution, whenever the reagents are added as solutions to the reaction mixtures. The concentrations are also indicated in Table 3. For example, 0.2 ml MTT in the text means 0.2 ml of an aqueous solution of MTT (2 mg/ml).

Acid phosphatase-1 (ACP1)

Electrophoresis. BS: 1.5 g Na_2-α-glycerophosphate, 1.5 g Na_2-β-glycerophosphate per litre, pH adjusted to 6.4 with 2 M citric acid; 0.4 ml/l 1 M $MgCl_2$ added before use. **QSPA: 4** μl of undiluted haemolysate. **LSA:** 1/3 C. **TE:** 4°C (cold room). **IV:** 250 V. **DE:** 4 h.

Staining. RM-1: 5 mg 2-naphthylphosphate dissolved in 1 ml 0.05 M trisodium citrate-citric acid buffer pH 6.0. **IG-1:** 1 h at 37°C in a moist incubation chamber. **RM-2:** 1 mg Fast Blue RR (Gurr) in 1 ml 0.36 M Tris–HCl buffer pH 8.0. **IG-2:** a few minutes at RT.

Result: scarlet red bands against a white background indicate the ACP1 activity. The patterns are seen in Fig. 1.

Table 1. The orang utans included in the present study.

Number	Name	Sex	Studbook number	Number in De Boer and Seuánez[a]	Origin	Born	Age[b]	Zoo
OU1	Geert	♂	01294B M	BI16	Borneo	wild	12 y	Arnhem
OU2	Joop	♂	01172S M	SI16	Sumatra	wild	15 y	Rotterdam
OU3	Piku	♂	00379B M	BI38	Borneo	wild	21 y	Antwerp
OU4	Timo[c]	♂	01664X M	HI12	hybrid[c]	Hannover	11 m	Hannover
OU5	Petra	♀	01093B F	BI41	Borneo	wild	14 y	Rotterdam
OU6	Connie[d]	♀	01660X F	HI21	hybrid[d]	Rotterdam	12 m	Hannover
OU7	Fred	♂	00738B M	BI20	Borneo	wild	20 y	Rotterdam
OU8	Joke	♀	00828B F	BI17	Borneo	wild	24 y	Arnhem
OU9	Sarita	♀	01461B F	BI19	Borneo	wild	7 y	Arnhem
OU10	Siljo	♂	01460B M	BI18	Borneo	wild	8 y	Arnhem
OU11	Anak[e]	♀	01503B F	BI16	Borneo[e]	Rhenen	3 y	Rhenen
OU12	Bernadine[f]	♀	01521B F	BI7	Borneo	Rhenen	$2\frac{1}{2}$ y	Rhenen
OU13	Bongo	♂	01255B M	BI14	Borneo	wild	10 y	Rhenen
OU14	Sjaan	♀	01256B F	BI15	Borneo	wild	10 y	Rhenen
OU15	Ori	♂	00650B M	BI12	Borneo	wild	18 y	Rhenen
OU16	Djambi	♀	01684S F	SI138	Sumatra	Stuttgart	12 m	Wuppertal

[a] The subspecific status of all orang utans was identified by chromosome analysis; for further information and pedigrees see De Boer and Seuánez 1982; and Wijnen et al. 1982 (both this volume).

[b] The age of the wild-caught animals is estimated; y = years, m = months.

[c] The father of this animal was a captive-born hybrid, the mother a captive-born pure-bred Sumatran animal.

[d] OU6 is the hybrid daughter of Sumatran male OU2 and Bornean female OU5.

[e] OU11 is the daughter of OU15 (father).

[f] OU12 is the daughter of OU13 (father) and OU 14 (mother).

Table 2. List of enzyme markers.

Locus symbol[a]	Chromosome assignment[b] HSA	PPY	Enzyme name	E.C. number	Animals tested (OU numbers)	Figure number
ACP1	2	12	Acid phosphatase-1	3.1.3.2	1–12	1
ADA	20		Adenosine deaminase	3.5.4.4	4, 11–16	2
AK1	9		Adenylate kinase-1	2.7.4.3	1–16	3
ALAD			δ-Aminolevulinate dehydrase	4.2.1.24	1–16	4
ALD			Aldolase	4.1.2.13	1–16	5
BLVR	7		Biliverdin reductase	1.3.1.24	1–16	6
CA1			Carbonic anhydrase-1	4.2.1.1	1–16	7
CA2			Carbonic anhydrase-2	4.2.1.1	1–16	8
DIA1	22		Diaphorase (NADH)	1.6.2.2	1–12	9
DIA2			Diaphorase (NADPH)	1.6.*.*	1–12	10
ENO1	1		Enolase-1	4.2.1.11	1–12	11
ESD	13	1	Esterase-D	3.1.1.1	1–12	12
FH	1		Fumarate hydratase (Fumarase)	4.2.1.2	1–5, 7–16	13
FK			Fructokinase	2.7.1.4	1–16	14
GLO1	6		Glyoxalase-1	4.4.1.5	1–12	15
GOT1	10	7	Glutamic oxaloacetic transaminase-1	2.6.1.1	1–12	16
GPI	19	20	Glucosephosphate isomerase	5.3.1.9	1–12	17
GPT1	16		Glutamic pyruvic transaminase-1	2.6.1.2	1–16	18
GPX1	3		Glutathione peroxidase-1	1.11.1.9	1–12	19
G6PD[c]	X	X	Glucose-6-phosphate dehydrogenase	1.1.1.49	1–16	20
GSR	8	6	Glutathione reductase	1.6.4.2	1–12	21
GUK1	1		Guanylate kinase-1	2.7.4.8	1–16	22
GUK3			Guanylate kinase-3	2.7.4.8	1–16	22

Symbol	HSA	PPY	Enzyme	EC number		
ITPA	20		Inosine triphosphatase	3.6.1.19	1–12	23
IDH1	2	11	Isocitrate dehydrogenase-1	1.1.1.42	1–12	24
LDHA	11	8	Lactate dehydrogenase-A	1.1.1.27	1–12	25
LDHB	12	9	Lactate dehydrogenase-B	1.1.1.27	1–12	25
MDH1	2	12	Malate dehydrogenase-1	1.1.1.37	1–12	26
MPI	15	16	Mannosephosphate isomerase	5.3.1.8	1–16	27
NP	14		Nucleoside phosphorylase	2.4.2.1	1–16	28
PEPA	18		Peptidase-A	3.4.11.*	1–12	29
PEPB	12	9	Peptidase-B	3.4.11.*	1–12	30
PEPC	1	1	Peptidase-C	3.4.11.*	1–16	31
PEPD	19		Peptidase-D	3.4.13.9	1–12	32
PFK			Phosphofructokinase	2.7.1.56	1–16	33
PGD	1	1	Phosphogluconate dehydrogenase	1.1.1.44	1–12	34
PGK1	X		Phosphoglycerate kinase-1	2.7.2.3	1–12	35
PGM1	1	1	Phosphoglucomutase-1	2.7.5.1	1–16	36
PGM2 (PPM)	4	3	Phosphoglucomutase-2	2.7.5.6	1–16	37–39
PGP	16		Phosphoglycolate phosphatase	3.1.3.18	1–12	40
PK1	15		Pyruvate kinase-1	2.7.1.40	1–12	41
PP	10		Pyrophosphatase (inorganic)	3.6.1.1	1–12	42
SOD1	21		Superoxide dismutase	5.3.1.1	1–12	43
TPI	12	9	Triosephosphate isomerase	5.3.1.1	1–12	44
UMPK	1		Uridine monophosphate kinase	2.7.4.*	1–12	45

[a] Symbols as per the International System for Human Gene Nomenclature (1979).
[b] HSA = human chromosome number; PPY = orang utan chromosome number; see Pearson et al. 1978, 1979; Estop et al. 1978; Garver et al. 1978.
[c] Data on G6PD have been presented by Wijnen et al. 1982 (this volume).

Table 3. List of abbreviations.

Abbreviation	Expansion	Stock solution	Supplier
A. Biochemical reagents			
ADP	Adenosine-5'-diphosphate		Boehringer
ALD	Aldolase	20 mg/2 ml	Boehringer
AMP	Adenosine-5'-monophosphate	10 mg/ml	Boehringer
ATP	Adenosine-5'-triphosphate	35 mg/ml	Boehringer
DCIP	2,6-Dichloroindophenol	2 mg/ml	Merck
Fl,6diP	Fructose-1,6-diphosphate	20 mg/ml	Boehringer
F6P	Fructose-6-phosphate	10 mg/ml	Boehringer
GAPD	Glyceraldehyde-3-phosphate dehydrogenase	10 mg/ml	Boehringer
G1,6diP	Glucose-1,6-diphosphate	1 mg/6 ml	Boehringer
GMP	Guanosine-3'-monophosphate	10 mg/ml	Boehringer
G1P$^+$	Glucose-1-phosphoric acid, containing 1% Glucose-1,6-diphosphoric acid	16 mg/ml	BDH
G3P	D,L-Glycerinaldehyde-3-phosphate Ba-salt	12 mg/ml	Boehringer
G6P	Glucose-6-phosphate	20 mg/ml	Boehringer
G6PD	Glucose-6-phosphate dehydrogenase	1 mg/ml	Boehringer
GPD	Glycerol-3-phosphate dehydrogenase	2 mg/ml	Boehringer
GSH	Glutathione, reduced form		Boehringer
GSR	Glutathione reductase	1 mg/ml	Boehringer
GSSG	Glutathione, oxidised form	—	Boehringer
ITP	Inosine-5'-triphosphate	—	Boehringer
LDH	Lactate dehydrogenase (pig heart)	10 mg/ml	Boehringer
MDH	Malate dehydrogenase	5 mg/ml	Boehringer
M6P	Mannose-6-phosphate	10 mg/ml	Boehringer
MTT	Tetrazolium salt solution	2 mg/ml	Sigma
NAD	β-Nicotinamide adenide dinucleotide	10 mg/ml	Boehringer
NADH	β-Nicotinamide adenine dinucleotide reduced	10 mg/ml	Boehringer
NADP	β-Nicotinamide adenine dinucleotide phosphate	4 mg/ml	Boehringer
NADPH	β-Nicotinamide adenine dinucleotide phosphate reduced	10 mg/ml	Boehringer
NP	Purine nucleoside phosphorylase	1 mg/ml	Boehringer
PEP	Phosphoenolpyruvate	25 mg/ml	Boehringer
6PG	Gluconate-6-phosphate	10 mg/ml	Boehringer
2PGA	Glycerate-2-phosphate	—	Boehringer
3PGA	Glycerate-3-phosphate	10 mg/ml	Boehringer
PGI	Phosphoglucose isomerase	2 mg/ml	Boehringer
PK	Pyruvate kinase	2 mg/ml	Boehringer
PMS	Phenazine methosulphate	0.4 mg/ml	Sigma
POD	Horse radish peroxidase	5 mg/ml	Boehringer
R1P	Ribose-1-phosphate	13.8 mg/ml	Boehringer
TBH	t-Butyl hydroperoxide	80%	Fluka
TPI	Triosephosphate isomerase	10 mg/ml	Boehringer
UMP	Uridine-5'-monophosphate	10 mg/ml	Boehringer
Venom	Snake venom (No. V-4500)	5 mg/ml	Sigma
XOD	Xanthine oxidase	20 mg/ml	Koch-Light
B. Methodological terms			
BS	Buffer system		
QSPA	Quantity of sample per application		
LSA	Line of sample application on the gel C : cathodal; 1/3 C : one third on the cathodal side M : middle; 1/3 A : one third on the anodal side		
TE	Temperature of the gel environment during electrophoresis		
IV	Initial voltage		
DE	Duration of electrophoresis		
RT	Room temperature		
RM	Reaction mixture (RM-1, RM-2, first and second RM)		
IG	Incubation of the gel (IG-1, IG-2, first and second IG)		

Adenosine deaminase (ADA) (van Someren et al. 1974)

Electrophoresis. BS: 0.02 M Tris, pH adjusted to 7.5 with 2 M citric acid. **QSPA:** 1 μl. **LSA:** C. **TE:** RT. **IV:** 200 V. **DE:** 1 h and 30 min.

Staining. RM: 2.5 mg adenosine is dissolved in 2 ml 0.025 M sodium phosphate buffer, pH 7.5 and 10 μl XOD, 10 μl NP, 0.2 ml MTT and 0.2 ml PMS are added. **IG:** 15–30 min at 37°C. If the bands of activity do not appear satisfactorily within 30 minutes, treat the gel once again with freshly prepared total reaction mixture.

Result: blue bands against a white background (Fig. 2).

Adenylate kinase-1 (AK1) (van Someren et al. 1974)

Electrophoresis. BS: 0.02 M NaH_2PO_4 buffer, adjusted to pH 6.5 with concentrated NaOH; 0.4 ml of 1 M $MgCl_2$ solution per litre is added before use. **QSPA:** 1 μl. **LSA:** 1/3 A. **TE:** RT. **IV:** 200 V. **DE:** 3 h and 30 min.

Staining: RM-1: 0.4 ml 1 M Tris–HCl–Na_2EDTA (4 mM) (pH 8.6), 0.2 ml PEP, 0.2 ml AMP, 0.2 ml ATP, 0.2 ml freshly prepared NADH, 0.2 ml freshly prepared 1 M $MgCl_2$, 5 μl PK, 5 μl LDH. The **RM-1** is mixed thoroughly and left at room temperature for about 5 min prior to use. **IG-1:** 5–15 min at 37°C. Appearance of bands monitored under long wave UV light. Quenched areas in the fluorescent background indicate the AK1 activity. **RM-2:** 1 ml 1 M Tris–HCl–Na_2EDTA (4 mM) (pH 8.6), 0.5 ml MTT and 0.5 ml PMS. **IG-2:** 5 min at RT.

Result: white bands against a blue background (Fig. 3).

δ-Aminolevulinate dehydrase (ALAD)

Electrophoresis. BS: 0.02 M Tris, 0.02 M maleic acid, 0.002 M $MgCl_2$, pH adjusted to 7.8 with concentrated NaOH; 0.4 ml 1 M β-mercaptoethanol is added per litre of buffer before use. **QSPA:** 10 μl. **LSA:** C. **TE:** RT. **IV:** 200 V. **DE:** 3 h.

Staining: RM-1: 2 mg δ-aminolevulinic acid (Sigma), 1 mg dithioerythrit (Serva, Heidelberg), 1.0 ml 0.2 M Na_2HPO_4–HCl, pH 7.0. **IG-1:** 1 h at 37°C. **RM-2:** 16 mg p-dimethyl-amino benzaldehyde in 0.6 ml acetic acid; 13 mg $HgCl_2$ in 0.8 ml 70% perchloric acid; these two solutions are mixed together and diluted 1:1 with distilled water. **IG-2:** 20 min. After appearance of the bands some more of RM-2 is dropped at the zones of activity.

Result: pink bands against a white background (Fig. 4).

Aldolase (ALD)

Electrophoresis. BS: 0.02 M NaH_2PO_4, pH adjusted to 7.0 with concentrated NaOH.

QSPA: 4 μl. **LSA:** C. **TE:** RT. **IV:** 200 V. **DE:** 2 h.

Staining. RM-1: 0.7 ml 1 M Tris–HCl–Na_2EDTA (4 mM) (pH 8.6), 2 mg F-1,6-diP, 2 mg NAD, 0.1 ml Na_3arsenate (35 mg/ml), 5 μl GAPD. **IG-1:** 30 min to 1 h at 37°C. Appearance of bands monitored under long wave UV light. Fluorescent bands in a dark background indicate the ALD activity. **RM-2:** 1.0 ml 1 M Tris–HCl–Na_2EDTA (4 mM) (pH 8.6), 0.5 ml MTT, 0.5 ml PMS. **IG-2:** about 5 min at RT.

Result: blue bands against a white background (Fig. 5).

Biliverdin reductase (BLVR)
Electrophoresis. BS: 0.02 M Tris, 0.02 M citric acid, 0.002 M Na₂EDTA, 0.002 M MgCl₂, pH adjusted to 5.3 with concentrated NaOH. **QSPA:** 8 µl. **LSA:** 1/3 C. **TE:** RT. **IV:** 200 V. **DE:** 2 h.
Staining. RM: 1 mg biliverdin is dissolved in 1.0 ml 0.36 M Tris–HCl (pH 8.0); 2 mg NADPH is dissolved in 0.5 ml 0.36 M Tris–HCl (pH 8.0) and 0.2 ml 1 M MgCl₂; the solutions are then mixed. **IG:** 1–4 h at 37°C.
Result: yellow bands against a light green background (Fig. 6).

Carbonic anhydrase-1 (CA1)
Electrophoresis. BS: 0.02 M barbituric acid, 0.02 M Tris, pH 8.0, to which 0.4 ml/l 1 M MgCl₂ and 0.2 ml/l 1 M β-mercaptoethanol are added. **QSPA:** 5 µl. **LSA:** A. **TE:** RT. **IV:** 200 V. **DE:** 3 h.
Staining: RM: 2 mg 4-methylumbelliferylacetate, 1 ml 70% ethanol. **IG:** 5–10 min at 37°C. Appearance of bands is monitored under long wave UV light.
Result: fluorescent bands against a dark background under long-wave UV light indicate CA1 activity (Fig. 7).

Carbonic anhydrase-2 (CA2)
Electrophoresis. BS: 'Smithies' (0.9 M Tris, 0.5 M boric acid, 0.02 M Na₂EDTA) diluted 1:40 with distilled water, pH 8.6. **QSPA:** 5 µl. **LSA:** M. **TE:** RT. **IV:** 200 V. **DE:** 2 h.
Staining. RM: 2 mg fluorescein diacetate in a few drops of acetone just to dissolve the substrate; then add 1 ml of 0.1 M NaH₂PO₄–NaOH, pH 6.5. **IG:** 30–60 min at 37°C. Appearance of bands is monitored under long wave UV light.
Result: fluorescent bands against a dark background indicate CA2 activity (Fig. 8).

Diaphorase (NADH) (DIA1) (Los and Meera Khan 1977)
Electrophoresis. BS: 0.02 M Tris, 0.002 M citric acid, 0.004 M Na₂EDTA; this buffer is diluted 7:3 with distilled water, pH 7.0. **QSPA:** 4 µl. **LSA:** C. **TE:** RT. **IV:** 200 V. **DE:** 3 h.
Staining. RM: 1.0 ml 1 M Tris–HCl–Na₂EDTA (4 mM) (pH 8.6), 0.4 ml 0.4 M sodium lactate, 0.2 ml NAD, 5 µl LDH, 0.25 ml DCIP, 0.3 ml MTT. **IG:** 1 h at 37°C. White bands on a pink background will appear. Treat the gel with 10% formalin till the bands turn red. Then wash it in water till the red bands turn dark blue. Finally destain the background DCIP with 10% Na₂S₂O₅ solution.
Result: blue bands against a white background (Fig. 9).

Diaphorase (NADPH) (DIA2) (Los and Meera Khan 1977)
Electrophoresis. BS: 0.061 M Tris, 0.014 M citric acid, 0.004 EDTA, pH 7.0. **QSPA:** 2 µl. **LSA:** C. **TE:** RT. **IV:** 200 V. **DE:** 3 h.
Staining. RM: 1 ml 1 M Tris–HCl–Na₂EDTA (4 mM) (pH 8.6), 0.2 ml G6P, 0.2 ml NADP, 5 µl G6PD, 0.2 ml methylene blue (0.4 mg/ml), 0.2 ml MTT. **IG:** 15–30 min at 37°C.
Result: blue zones against a white background indicate DIA2 activity (Fig. 10).

Enolase-1 (ENO1) (van Someren et al. 1974)
Electrophoresis. BS: 0.008 M Na_2HPO_4-citric acid, pH 8.5, to which 0.2 ml/l 1 M β-mercaptoethanol and 0.4 ml/l 1 M $MgCl_2$ are added. **QSPA:** 2 µl. **LSA:** 1/3 C. **TE:** RT. **IV:** 200 V. **DE:** 5 h.
Staining. RM-1: 1 ml 0.2 M Tris–HCl (pH 7.5), 3 mg 2 PGA, 0.2 ml 1 M $MgCl_2$, 4 mg ADP, 3 mg NADH, 5 µl PK, 5 µl LDH. This reaction mixture is left to stand at room temperature for 5 min prior to use. **IG-1:** 15–30 min at 37°C. Appearance of bands is monitored under long wave UV light. Quenched areas in the fluorescent background indicate the ENO activity. **RM-2:** 0.5 ml 1 M Tris–HCl–Na_2EDTA (4 mM) (pH 8.6), 0.5 ml MTT, 0.5 ml PMS. **IG-2:** about 5 min at RT.
Result: white bands against a blue background (Fig. 11).

Esterase-D (ESD) (Ebeli-Struijk et al. 1976)
Electrophoresis. BS: solution of 15 g Tris, 8 g citric acid, 2.2 g boric acid and 0.08 g LiOH in 10 l of distilled water (pH 7.2). **QSPA:** 3 µl. **LSA:** C. **TE:** RT. **IV:** 200 V. **DE:** 2 h.
Staining. RM: 2 mg 4-methylumbelliferylacetate in 1 ml 70% ethanol. **IG:** 5–15 min at 37°C. Appearance of bands is monitored under long wave UV light.
Result: fluorescent zones against a dark background indicate the ESD activity (Fig. 12).

Fumarase = Fumarate hydratase (FH) (van Someren et al. 1974)
Electrophoresis. BS: 0.01 M Na_2HPO_4, pH adjusted to 7.0 with citric acid. **QSPA:** 5 µl. **LSA:** C. **TE:** 4°C (cold room). **IV:** 200 V. **DE:** 5 h.
Staining. RM: 0.5 ml Na_2HPO_4–HCl buffer pH 7.5, 0.1 ml NAD, 0.2 ml MTT, 0.2 ml PMS, 0.2 ml 1 M fumaric acid (pH 7.0), 5 µl MDH. **IG:** 1 h at 37°C in the dark (inside an aluminium foil).
Result: blue bands on a white background (Fig. 13).

Fructokinase (FK)
Electrophoresis. BS: 0.01 M Tris, 0.01 M maleic acid, 0.001 M $MgCl_2$, 0.001 M Na_2EDTA, adjusted to pH 7.4 with NaOH. **QSPA:** 10 µl. **LSA:** C. **TE:** 4°C (cold room). **IV:** 200 V. **DE:** 1 h and 30 min.
Staining. RM: 0.5 ml 0.002 M cysteine–HCl (pH 8.0), 0.2 ml 0.054 M Na_2EDTA (pH 7.0), 0.5 ml 1 M Tris–HCl–EDTA (pH 8.6), 0.1 ml 1 M $MgCl_2$, 0.2 ml NADP, 7 mg ATP, 10 mg fructose, 5 µl PGI, 0.2 ml MTT, 0.2 ml PMS. **IG:** 30–60 min at 37°C.
Result: blue bands against a white background (Fig. 14).

Glyoxalase-1 (GLO1) (Meera Khan and Doppert 1976)
Electrophoresis. BS: 0.03 M barbituric acid, 0.03 M Tris, pH 8.0, 0.2 ml/l 1 M β-mercaptoethanol and 0.4 ml/l 1 M $MgCl_2$. **QSPA:** 2 µl. **LSA:** C. **TE:** RT. **IV:** 200 V. **DE:** 2 h and 15 min.
Staining. RM-1: 1.6 ml 0.1 M NaH_2PO_4–NaOH (pH 6.5), 100 µl methylglyoxal

(40%, Sigma), 12 mg GSH, 0.4 ml MTT. The gel is treated with RM-1 and RM-2 in succession with zero interval. **RM-2:** 1.8 ml 0.1 M Tris–HCl (pH 7.8), 0.2 ml DCIP. **IG-2:** RT. Initially the total gel appears pink. The pink of the background disappears after 15–30 min, while the pink of the zones due to GLO1 activity sustains for some hours.
Result: pink bands against a white background (Fig. 15).

Glutamic oxaloacetic transaminase-1 (GOT1) (van Someren et al. 1974)
Electrophoresis. BS: 0.01 M Tris adjusted to pH 7.5 with 2 M citric acid. **QSPA:** 1 µl. **LSA:** C. **TE:** RT. **IV:** 200 V. **DE:** 3 h.
Staining. RM-1: 2 ml 0.3 M Na_2HPO_4–HCl buffer (pH 7.5), 60 mg aspartic acid, 4 mg α-ketoglutaric acid, 2 mg NADH. One drop of concentrated NaOH is added before the addition of 5 µl MDH. **IG-1:** 10 min at 37°C. Appearance of bands is monitored under long wave UV light. Quenched areas in a fluorescent background indicate the GOT1 activity. **RM-2:** 0.5 ml 1 M Tris–HCl–Na_2EDTA 0.004 M (pH 8.6), 0.5 ml MTT, 0.5 ml PMS. **IG-2:** about 5 min at RT.
Result: white bands against a blue background (Fig. 16).

Glucosephosphate isomerase (GPI) (van Someren et al. 1974)
Electrophoresis. BS: 0.02 M Tris, 0.002 M citric acid, 0.004 M Na_2EDTA, pH 8.0. **QSPA:** 2 µl. **LSA:** M. **TE:** RT. **IV:** 200 V. **DE:** 3 h.
Staining. RM-1: 1.0 ml 1 M Tris–HCl–Na_2EDTA (4 mM) (pH 8.6), 0.2 ml F6P, 5 µl G6PD, 0.2 ml MTT, 0.2 ml PMS, 0.2 ml 0.25 M $CoCl_2$. **IG:** 5 min at 37°C.
Result: blue bands appear on a white background indicating GPI activity (Fig. 17).

Glutamic pyruvic transaminase-1 (GPT1)
Electrophoresis. BS: 0.01 M Tris–citric acid, pH 7.5. **QSPA:** 40 µl haemolysate which has been incubated with 10 µl 1 M β-mercaptoethanol for 1 h. **LSA:** C. **TE:** RT. **IV:** 200 V. **DE:** 3 h.
Staining. RM-1: 2.0 ml 0.3 M Na_2HPO_4–HCl (pH 7.5), 60 mg alanine, 4 mg α-ketoglutaric acid, 2 mg NADH, 5 µl LDH. **IG-1:** 1–2 h at 37°C. Appearance of bands monitored under long wave UV light. Dark areas in a fluorescent background indicate the GPT activity. **RM-2:** 1.0 ml 1 M Tris–HCl–Na_2EDTA (4 mM) (pH 8.6), 0.5 ml MTT, 0.5 ml PMS. **IG-2:** about 5 min at RT.
Result: white bands against a blue background (Fig. 18).

Glutathione peroxidase-1 (GPX1) (Wijnen et al. 1978)
Electrophoresis. BS: 0.02 M NaH_2PO_4, pH adjusted to 7.0 with concentrated NaOH. **QSPA:** 1 µl. **LSA:** C. **TE:** RT. **IV:** 200 V. **DE:** 3 h.
Staining. RM-1: 1 ml 0.1 M KH_2PO_4–KOH (pH 7.0), 3 mg GSH, 5 µl GSR, 1.5 mg NADPH, 0.2 ml 0.0054 M Na_2EDTA (pH 7.0) and 5 µl TBH. TBH should be added to the mixture just before use. **IG-1:** 5–15 min at 37°C. Appearance of bands monitored under long wave UV light. Quenched areas in the fluorescent background indicate the GPX1 activity. **RM-2:** 0.5 ml 1 M

Tris–HCl–Na$_2$EDTA (4 mM) (pH 8.6), 0.5 ml MTT, 0.5 ml PMS. **IG-2:** about 5 min at RT.
Result: white bands against a blue background (Fig. 19).

Glucose-6-phosphate dehydrogenase (G6PD) (Rattazzi et al. 1967)
Electrophoresis. BS: 0.06 M Tris, 0.014 M citric acid, 0.004 M Na$_2$EDTA, pH 7.5. **QSPA:** 2 µl. **LSA:** C. **TE:** RT. **IV:** 200 V. **DE:** 3 h.
Staining. RM: 0.5 ml 1 M Tris–HCl–Na$_2$EDTA (4 mM) (pH 8.6), 0.2 ml NADP, 0.2 ml G6P, 0.2 ml MTT, 0.2 ml PMS, 0.2 ml 0.25 M CoCl$_2$. **IG:** 5–15 min at 37°C.
Result: blue bands against a white background indicate G6PD activity (Fig. 20).

Glutathione reductase (GSR) (van Someren et al. 1974)
Electrophoresis. BS: 0.02 M Tris, 0.002 M citric acid, 0.004 M Na$_2$EDTA, pH 7.5. **QSPA:** 2 µl. **LSA:** C. **TE:** RT. **IV:** 200 V. **DE:** 4 h and 30 min.
Staining. RM-1: 1.0 ml 0.36 M Tris–HCl (pH 8.0), 2 mg NADPH, 8 mg GSSG. **IG-1:** 5–30 min at 37°C. Appearance of bands monitored under long wave UV light. Quenched areas in the fluorescent background indicate the GSR activity. **RM-2:** 0.5 ml 1 M Tris–HCl–Na$_2$EDTA (4 mM) (pH 8.6), 0.5 ml MTT, 0.5 ml PMS. **IG-2:** about 5 min at RT.
Result: white bands against a blue background (Fig. 21).

Guanylate kinase-1 and -3 (GUK1 and GUK3) (Meera Khan et al. 1974)
Electrophoresis. BS: 0.02 M Tris–citric acid (pH 7.4), added with 0.2 ml/l 1 M β-mercaptoethanol and 0.4 ml/l 1 M MgCl$_2$. **QSPA:** 6 µl. **LSA:** C. **TE:** RT. **IV:** 200 V. **DE:** 2 h.
Staining. RM-1: 1.0 ml 0.36 M Tris–HCl (pH 8.0), 2 mg GMP, 10 mg ATP, 5 mg PEP, 0.2 ml 1 M MgCl$_2$, 2 mg NADH, 5 µl PK, 5 µl LDH. **IG-1:** 30 min to 1 h at 37°C. Appearance of bands monitored under long wave UV light. Quenched areas in the fluorescent background indicate the GUK activity. **RM-2:** 0.5 ml 1 M Tris–HCl–Na$_2$EDTA (4 mM) (pH 8.6), 0.5 ml MTT, 0.5 ml PMS. **IG-2:** about 5 min at RT.
Result: white bands against a blue background (Fig. 22).

Inosine triphosphatase (ITPA) (Herbschleb-Voogt et al. 1981)
Electrophoresis. BS: 0.01 M Tris, 0.01 M maleic acid, 0.001 M MgCl$_2$, 0.001 M Na$_2$EDTA, pH adjusted to 7.4 with concentrated NaOH. **QSPA:** 4 µl. **LSA:** 1/3 C. **TE:** RT. **IV:** 200 V. **DE:** 1 h.
Staining. RM-1: 1.0 ml 0.2 M Tris–HCl (pH 7.5), 1.0 ml 1 M MgCl$_2$, 2 mg ITP. **IG-1:** 30 min at 37°C. **RM-2:** 50 mg ascorbic acid in 1.0 ml 1.25% ammonium-heptamolybdate in 2 N H$_2$SO$_4$. **IG-2:** about 5 min at RT.
Result: blue bands appear indicating the ITPA activity (Fig. 23). Note: avoid all possible sources of contamination by inorganic phosphate during the procedure, since the system is specific for its detection.

Isocitrate dehydrogenase-1 (IDH1) (Meera Khan 1971)
Electrophoresis. BS: 0.01 M Na$_2$HPO$_4$, adjusted to pH 6.0 with citric acid.

QSPA: 2 µl. **LSA:** C. **TE:** RT. **IV:** 200 V. **DE:** 3 h.
Staining. RM-1: 1.0 ml 0.36 M Tris–HCl (pH 8.0), 0.2 ml NADP, 0.2 ml trisodium isocitrate (85 mg/ml), 0.2 ml 0.0125 M $MnCl_2$, 0.2 ml MTT, 0.2 ml PMS. **IG:** 30 min at 37°C.
Result: blue bands in a white background (Fig. 24).

Lactate dehydrogenase-A and -B (LDHA and LDHB) (Herbschleb-Voogt and Meera Khan 1981)
Electrophoresis. BS: 0.02 M Tris, 0.002 M citric acid, 0.004 M EDTA, pH 8.0; add 0.4 ml/l 1 M $MgCl_2$ and 0.2 ml/l 1 M β-mercaptoethanol before use.
QSPA: 1 µl. **LSA:** C. **TE:** RT. **IV:** 200 V. **DE:** 3 h.
Staining. RM: 0.5 ml 1 M Tris–HCl–Na_2EDTA (4 mM) (pH 8.6), 0.5 ml 0.4 M sodium lactate, 0.2 ml NAD, 0.2 ml MTT, 0.2 ml PMS. **IG:** 2–10 min at 37°C.
Result: blue bands in a white background (Fig. 25).

Malate dehydrogenase-1 (MDH1) (Meera Khan 1971)
Electrophoresis. BS: 0.245 M NaH_2PO_4, 0.15 M citric acid. This buffer is diluted 1:40 with distilled water and pH is adjusted to 5.9 with concentrated NaOH. **QSPA:** 2 µl. **LSA:** 1/3 C. **TE:** RT. **IV:** 200 V. **DE:** 2 h and 30 min.
Staining. RM: 1.0 ml 0.36 M Tris–HCl (pH 8.0), 0.2 ml 2 M *l*-malic acid (neutralized with NaOH), 0.2 ml NAD, 0.2 ml 0.0125 M $MnCl_2$, 0.2 ml 0.003 M KCN, 0.3 ml MTT, 0.2 ml PMS. **IG:** 15–45 min at RT.
Result: blue bands in a white background (Fig. 26).

Mannosephosphate isomerase (MPI) (van Someren et al. 1974)
Electrophoresis. BS: 0.01 M Na_2-β-glycerophosphate, adjusted to pH 6.4 with citric acid; add 0.4 ml/l 1 M $MgCl_2$ and 0.2 ml/l β-mercaptoethanol before use.
QSPA: 2 µl. **LSA:** C. **TE:** RT. **IV:** 200 V. **DE:** 3 h.
Staining. RM-1: 0.6 ml 0.36 M Tris–HCl (pH 8.0), 0.2 ml M6P, 0.2 ml NADP, 5 µl G6PD, 5 µl PGI. **IG-1:** 30 min to 1 h at 37°C. Appearance of bands monitored under long wave UV light. Fluorescent bands against dark background indicate the MPI activity. **RM-2:** 1 ml 1 M Tris–HCl–Na_2EDTA (4 mM) (pH 8.6), 0.5 ml MTT, 0.5 ml PMS. **IG-2:** about 5 min at RT.
Result: blue bands aginst a white background (Fig. 27).

Nucleoside phosphorylase (NP) (van Someren et al. 1974)
Electrophoresis. BS: 0.02 M Na_2HPO_4, 0.02 M NaH_2PO_4, pH 7.0. **QSPA:** 1 µl (haemolysate diluted 1:25). **LSA:** C. **TE:** 4°C (cold room). **DE:** 3 h.
Staining. RM-1: 0.7 ml 0.055 M Tris–HCl (pH 7.4) with 0.008 M $MgSO_4$, 0.1 ml inosine (5 mg/ml), 5 µl XOD. **IG-1:** 10 min at 37°C. **RM-2:** as RM-1 but adding 0.1 ml MTT and 0.1 ml PMS. **IG-2:** 5–10 min in a dark moist chamber at 37°C.
Result: blue bands against a white background (Fig. 28).

Peptidase-A (PEPA) (van Someren et al. 1974)
Electrophoresis. BS: 0.03 M Tris 0.03 M maleic acid, pH adjusted to 7.5 with NaOH, to which is added 0.4 ml/l 1 M $MgCl_2$. **QSPA:** 3 µl. **LSA:** C. **TE:** RT.

72

IV: 200 V. **DE:** 2 h and 30 min.

Staining. RM-1: 1.0 ml 0.3 M Na_2HPO_4–HCl (pH 7.5), 0.25 ml Venom, 0.2 ml POD, 0.5 ml substrate A (valine-leucine, 7.5 mg/ml), 0.2 ml O-dianisidine (25 mg dissolved in 1.5 ml 1.0% HCl and diluted to 10 ml with H_2O). **IG:** 20–40 min at 37°C.

Result: brown bands against a white background indicate the PEPA activity (Fig. 29).

Peptidase-B (PEPB) (van Someren et al. 1974)

Electrophoresis. BS: 0.02 M Tris 0.02 M maleic acid, pH adjusted to 7.4 with NaOH, to which is added 0.4 ml/l 1 M $MgCl_2$. **QSPA:** 1 μl. **LSA:** C. **TE:** RT. **IV:** 200 V. **DE:** 2 h and 30 min.

Staining. RM: 0.6 ml 0.3 M Na_2HPO_4–HCl (pH 7.5), 0.2 ml Venom, 0.1 ml POD, 0.75 ml substrate B (leucine-glycine, 5 mg/ml), 0.2 ml O-dianisidine (for preparation see PEPA). **IG:** 15–30 min at 37°C.

Result: brown bands against a white background indicate PEPB activity (Fig. 30).

Peptidase-C (PEPC) (van Someren et al. 1974)

Electrophoresis. BS: 0.07 M boric acid-Tris, pH adjusted to 7.4 with saturated Tris solution; 0.4 ml/l 1 M $MgCl_2$ added before use. **QSPA:** 4 μl (applied in a thin streak). **LSA:** C. **TE:** RT. **IV:** 200 V. **DE:** 1 h and 50 min.

Staining. RM-1: 1.0 ml 0.3 M Na_2HPO_4–HCl (pH 7.5), 0.2 ml Venom, 0.1 ml POD, 0.4 ml substrate C (lysine-leucine, 5 mg/ml), 0.2 ml O-dianisidine (for preparation see PEPA). **IG:** 20–45 min at 37°C.

Result: brown bands appear against a white background indicating PEPC activity (Fig. 31).

Peptidase-D (PEPD) (van Someren et al. 1974)

Electrophoresis. BS: 0.07 M boric acid-Tris, pH adjusted to 7.4 with saturated Tris solution; 0.4 ml/l 1 M $MgCl_2$ is added before use. **QSPA:** 4 μl. **LSA:** C. **TE:** 4°C (cold room). **IV:** 200 V. **DE:** 1 h and 30 min.

Staining. RM-1: 1.0 ml 0.3 M Na_2HPO_4–HCl (pH 7.5), 0.5 ml substrate D (leucine-proline, 7.5 mg/ml), 0.2 ml Venom, 0.2 ml POD, 0.2 ml O-dianisidine (for preparation see PEPA). **IG:** 30 min to 1 h at 37°C.

Result: brown bands against a white background indicate PEPD activity (Fig. 32).

Phosphofructokinase (PFK)

Electrophoresis. BS: 0.01 M NaH_2PO_4, pH adjusted to 7.0 with NaOH; 10 ml/l 0.1 M Na_2EDTA and 0.2 ml/l 1 M β-mercaptoethanol added before use. **QSPA:** 4 μl. **LSA:** C. **TE:** RT. **IV:** 200 V. **DE:** 2 h.

Staining. RM-1: 1.0 ml 0.36 M Tris–HCl (pH 8.0), 0.1 ml 1 M $MgCl_2$, 2 mg F6P, 2 mg NADH, 5 mg ATP, 5 μl αGPD, 1 μl TPI, 10 μl ALD. **IG-1:** 10–30 min at 37°C. The appearance of bands is monitored under long wave UV light. Quenched areas in the fluorescent background indicate the PFK activity. **RM-2:**

1 ml 1 M Tris–HCl–Na$_2$EDTA (4 mM) (pH 8.6), 0.5 ml MTT, 0.5 ml PMS. **IG-2:** about 5 min at RT.

Result: white bands against a blue background (Fig. 33).

Phosphogluconate dehydrogenase (PGD) (Meera Khan and Rattazzi 1968)
Electrophoresis. BS: 0.02 M Tris, 0.002 M citric acid, 0.004 M Na$_2$EDTA, pH 7.5. **QSPA:** 2 µl. **LSA:** C. **TE:** RT. **IV:** 200 V. **DE:** 4 h.
Staining. RM-1: 1.0 ml 1 M Tris–HCl–Na$_2$EDTA (4 mM) (pH 8.6), 0.2 ml NADP, 0.2 ml 6PG, 0.2 ml MTT, 0.2 ml PMS, 0.2 ml 0.25 M CoCl$_2$. **IG:** 10–30 min at 37°C.
Result: blue bands against a white background (Fig. 34).

Phosphoglycerate kinase-1 (PGK1) (Meera Khan 1971)
Electrophoresis. BS: 0.01 M Tris, pH adjusted to 7.5 with citric acid. **QSPA:** 2 µl [if the samples are too old treat each of them (5 µl) with (1 µl) 1 M β-mercaptoethanol]. **LSA:** 1/3 C. **TE:** RT. **IV:** 200 V. **DE:** 3 h.
Staining. RM-1: 1.6 ml 0.36 M Tris–HCl (pH 8.0), 0.6 ml 3PGA, 2 mg NADH, 7 mg ATP, 0.2 ml 1 M MgCl$_2$, 0.2 ml 0.054 M Na$_2$EDTA (pH 7.0), 10 µl GAPD. **IG-1:** 5–15 min at 37°C. Appearance of bands monitored under long wave UV light. Quenched areas in the fluorescent background indicate the PGK1 activity. **RM-2:** 1.0 ml 1 M Tris–HCl–Na$_2$EDTA (4 mM) (pH 8.6), 0.5 ml MTT, 0.5 ml PMS. **IG-2:** about 5 min at RT.
Result: white bands against a blue background (Fig. 35).

Phosphoglucomutase-1 (PGM1) (Meera Khan 1971)
Electrophoresis. BS: 0.01 M Na$_2$HPO$_4$, adjust pH to 6.5 with citric acid. **QSPA:** 2 µl. **LSA:** C. **TE:** RT. **IV:** 200 V. **DE:** 4 h.
Staining. RM: 0.8 ml 0.36 M Tris–HCl (pH 8.0), 0.2 ml 1 M MgCl$_2$, 0.2 ml 0.054 M Na$_2$EDTA (pH 7.0), 0.2 ml 0.003 M KCN, 0.2 ml NADP, 0.5 ml G1P$^+$, 5 µl G6PD, 0.2 ml MTT, 0.2 ml PMS. **IG:** 5–20 min at 37°C.
Result: blue bands against a white background (Fig. 36).

Phosphopentomutase (PPM) (Wijnen et al. 1977)
Electrophoresis. BS: 0.008 M Tris, 0.008 M maleic acid, 0.0008 M MgCl$_2$, 0.0008 M Na$_2$EDTA, pH adjusted to 7.4 with NaOH. **QSPA:** 2 µl. **LSA:** C. **TE:** RT. **IV:** 200 V. **DE:** 3 h and 30 min.
Staining. RM-1: 1.5 ml 0.002 M cysteine-HCl (pH 8.0), 0.3 ml 0.2 M Tris–HCl (pH 8.0), 10 µl Gl.6diP, 0.1 ml RlP, 0.1 ml 1 M MgCl$_2$. **IG-1:** 10 min at 37°C.
RM-2: 50 mg ascorbic acid in 1 ml 1.25% ammoniumheptamolybdate in 2 N H$_2$SO$_4$. **IG-2:** 15 min at RT.
Result: white bands against a blue background (Figs. 37 and 38). Note: see ITPA.

Phosphoglucomutase-2 (PGM2)
Electrophoresis. BS: 0.008 M Tris, 0.008 M maleic acid, 0.0008 M MgCl$_2$, 0.0008 M Na$_2$EDTA, pH adjusted to 7.4 with NaOH. **QSPA:** 2 µl. **LSA:** C. **TE:** RT. **IV:** 200 V. **DE:** 2 h and 30 min.

Staining. RM: 0.8 ml 0.36 M Tris–HCl (pH 8.0), 0.2 ml 1 M MgCl$_2$, 0.2 ml 0.054 M Na$_2$EDTA (pH 7.0), 0.2 ml 0.003 M KCN, 0.2 ml NADP, 0.5 ml G1P$^+$, 5 μl G6PD, 0.2 ml MTT, 0.2 ml PMS. **IG:** 30–60 min at 37°C.
Result: blue bands against a white background (Fig. 39). Note: the gel will separate and stain simultaneously the PGM1, PGM2 and PGM3 activities in white cell lysates and tissue extracts.

Phosphoglycolate phosphatase (PGP)
Electrophoresis. BS: 0.02 M Tris, 0.02 M maleic acid, 0.002 M Na$_2$EDTA, 0.002 M MgCl$_2$, pH adjusted to 6.0 with NaOH. **QSPA:** 5 μl. **LSA:** C. **TE:** RT. **IV:** 200 V. **DE:** 4 h.
Staining. RM-1: 1.0 ml 0.2 M Tris–HCl (pH 7.5), 4 mg phosphoglycolic acid, 2 mg MgSO$_4$. **IG-1:** 1 h at 37°C. **RM-2:** 50 mg ascorbic acid in 1 ml 1.25% ammoniumheptamolybdate in 2 N H$_2$SO$_4$. **IG-2:** 15 min at RT.
Result: blue bands against a white background (Fig. 40). Note: see ITPA.

Pyruvate kinase-1 (PK1) (van Someren et al. 1974)
Electrophoresis. BS: 0.02 M Tris–citric acid, pH 7.0. **QSPA:** 6 μl. **LSA:** 1/3 C. **TE:** RT. **IV:** 200 V. **DE:** 4 h.
Staining. RM-1: 1.0 ml 0.36 M Tris–HCl (pH 8.6), 1 mg PEP, 5 mg ADP, 2 mg NADH, 0.2 ml 1 M KCl, 0.3 ml 1 M MgCl$_2$, 5 μl LDH. **IG-1:** 1–2 h at 37°C. Appearance of bands monitored under long wave UV light. Quenched areas in the fluorescent background indicate the PK1 activity. **RM-2:** 1.0 ml 1 M Tris–HCl–Na$_2$EDTA (4 mM) (pH 8.6), 0.5 ml MTT, 0.5 ml PMS. **IG-2:** about 5 min at RT.
Result: white bands against a blue background (Fig. 41).

Pyrophosphatase (PP)
Electrophoresis. BS: 0.01 M Tris, 0.01 M maleic acid, 0.001 M MgCl$_2$, 0.001 M Na$_2$EDTA, pH adjusted to 7.4 with NaOH. **QSPA:** 4 μl. **LSA:** C. **TE:** RT. **IV:** 200 V. **DE:** 2 h.
Staining. RM-1: 1.0 ml 0.2 M Tris–HCl (pH 7.5), 1.0 ml 1 M MgCl$_2$, 2 mg pyrophosphate tetrasodium salt. **IG-1:** 15 min at 37°C. **RM-2:** 50 mg ascorbic acid in 1.0 ml 1.25% ammoniumheptamolybdate in 2 N H$_2$SO$_4$. **IG-2:** 15 min at RT.
Result: blue bands against a white background (Fig. 42). Note: see ITPA.

Superoxide dismutase-1 (SOD1) (Meera Khan 1971)
Electrophoresis. BS: 0.06 M Tris, 0.014 M citric acid, 0.004 M Na$_2$EDTA, pH 7.5. **QSPA:** 2 μl. **LSA:** C. **TE:** RT. **IV:** 200 V. **DE:** 3 h.
Staining. RM: 1.0 ml 1 M Tris–HCl–Na$_2$EDTA (4 mM) (pH 8.6), 0.5 ml MTT, 0.5 ml PMS. **IG:** 30–60 min in a moist chamber at room temperature exposing the gel constantly to daylight, but not to direct sun, and to air now and then.
Result: white bands against a blue background indicate the SOD1 activity (Fig. 43).

Triosephosphate isomerase (TPI) (van Someren et al. 1974)
Electrophoresis. BS: 0.01 M barbituric acid, 0.01 M Tris, pH 8.0; add 0.2 ml/l

1 M β-mercaptoethanol and 0.4 ml/l 1 M MgCl$_2$ before use. **QSPA:** 1 µl of 1 to 100 diluted haemolysate. **LSA:** 1/3 C to M. **TE:** RT. **IV:** 200 V. **DE:** 5 h.
Staining. RM-1: 0.8 ml 0.36 M Tris–HCl (pH 8.0), 2 mg NADH, 0.8 ml G3P, 5 µl GPD (preparation of G3P: dissolve 25 mg of the Ba salt of G3P in 1.0 ml H$_2$O and add 0.4 ml Dowex 50-acid form 200 mesh; shake thoroughly; remove the supernatant by suction and rinse the resin twice with 0.5 ml H$_2$O; heat for 2 min in a boiling water bath). **IG:** 1–5 min at 37°C. Monitor the appearance of bands under long wave UV light. Quenched areas in the fluorescent background indicate the TPI activity. **RM-2:** 1 ml 1 M Tris–HCl–Na$_2$EDTA (4 mM) (pH 8.6), 0.5 ml MTT, 0.5 ml PMS. **IG-2:** about 5 min at RT.
Result: white bands against a blue background (Fig. 44).

Uridine monophosphate kinase (UMPK)
Electrophoresis. BS: 0.02 M Na$_2$HPO$_4$, 0.02 M NaH$_2$PO$_4$, pH 7.0; add 0.4 ml/l 1 M MgCl$_2$ before use. **QSPA:** 4 µl. **LSA:** C. **TE:** 4°C (cold room). **IV:** 200 V. **DE:** 3 h.
Staining. RM-1: 0.8 ml 0.1 M Tris–HCl (pH 7.8), 4 mg UMP, 4 mg ATP, 2 mg PEP, 4 mg NADH, 0.2 ml 1 M MgCl$_2$, 0.2 ml K$_2$SO$_4$ (87 mg/ml), 10 µl PK, 10 µl LDH. The RM-1 is mixed thoroughly and left at room temperature for about 5 min prior to use. **IG-1:** 15 min to 1 h at 37°C. Monitor the appearance of bands under long wave UV light. Quenched areas in the fluorescent background indicate the UMPK activity. **RM-2:** 1.0 ml 1 M Tris–HCl–Na$_2$EDTA (4 mM) (pH 8.6), 0.5 ml MTT, 0.5 ml PMS. **IG-2:** about 5 min at RT.
Result: white bands against a blue background (Fig. 45).

Results

Of the 45 red cell enzyme marker systems (Table 2) 12 (ADA, AK1, CA1, CA2, ENO1, FH, G6PD, GUK3, NP, PEPC, PGM1 and PGM2) exhibited one or more variants among the 16 orang utans screened by the electrophoretic procedures described above. Zymogram patterns of all the individual systems are presented below. The channel numbers in the figures simply represent the serial numbers of the orang utans tested without the prefix 'OU' (see Table 1).

In naming the loci, genes, alleles and phenotypes of the orang utan enzymes presumed to be homologous to those of man, we follow the guidelines laid down by the International System for Human Gene Nomenclature (1979).

ACP1: The orang utan zymogram patterns resemble somewhat the human phenotype B and exhibit no variation (Fig. 1).

ADA: As in the case of human ADA the ADA in the orang utan was found to exhibit storage effect. In the presence of β-mercaptoethanol the ADA in the individuals OU11–16 formed a single major band and a very faint anodal band. The situation reversed when the sample was incubated with GSSG for 2 hours at room temperature: a single sharp intense anodal band coinciding with the minor band of the β-mercaptoethanol treated samples was formed. The OU4, a full Bornean-Sumatran hybrid orang utan was an exception in which the thiol reagent treatment resulted in a two-banded pattern, indicating that it was a heterozygote with a variant probably carrying a substituted amino acid for a cystinyl group imparting a reactive sulphydryl group to the wild form. It is not certain at present whether this variant is peculiar to Sumatran orang utans or whether it is the same variant designated as 103 reported by Bruce and Ayala (1979). Bruce and Ayala found a slow ADA variant in three Bornean orang utans and a fast variant in eight Sumatran animals. In our experiments, however, the ADA of OU16, a pure-bred Sumatran orang utan, was found to be indistinguishable from that of the Borneans (Fig. 2). We tentatively designate the usual form of ADA found in OU11–16 as phenotype 1, that found in OU4 as phenotype 2–1 (genotype $ADA*1/ADA*2$). The ADA of the remaining orang utans could not be studied adequately.

AK1: The OU8 is presumably heterozygous for a variant with an electrophoretic pattern similar to that of human phenotype 2–1 (Harris and Hopkinson 1976). This being the first variant of AK1 known in the orang utan, we designate the corresponding phenotype as AK1 2–1 and the genotype as $AK1*1/AK1*2$ and the phenotype and the allele of the usual form as AK1 1 and $AK1*1$ respectively. Unlike most other enzymes, the AK1 in orang utans like in man migrates cathodal to the origin in the gel under the acidic condition of electrophoresis (Fig. 3).

ALAD (Fig. 4), **ALD** (Fig. 5) and **BLVR** (Fig. 6) did not show variation in the sample of orang utans studied.

CA1: Three electrophoretic forms of CA1 were found, all of them migrating to the cathode under the conditions of electrophoresis employed. In accordance with Tashian (1965, 1969) and Tashian and Carter (1971), but capitalizing the letters, we designate these forms as CA1 A (slowest form), CA1 B (intermediate) and CA1 C (fastest form) (Fig. 7). Four phenotypes were encountered among the 16 animals studied: the homozygotes A (OU1, 3, 4, 8, 9, 12 and 14) and B (OU16), and the heterozygotes AB (OU2 and 6) and AC (OU5, 7, 10, 11, 13 and 15). It is noteworthy that the $CA1*B$ allele in our sample occurs only in Sumatran orang utans, the OU2 (heterozygous) and OU16 (homozygous). OU2 apparently transmitted the B allele to its hybrid offspring OU6. The B allele does not occur in any of the pure-bred Bornean orang utans studied. The C allele, on the other hand, was found only in Bornean orang utans. However, our sample of Sumatran animals is much too small to draw conclusions on the general patterns of distribution of the various alleles in the Sumatran population.

Tashian (1965, 1969) and Tashian and Carter (1971) studied the CA1 of 33 orang utans. They found allelic frequencies of 0.69 for A, 0.21 for B and 0.10 for C. Unfortunately, the previous reports did not indicate the origins of the orang utans studied.

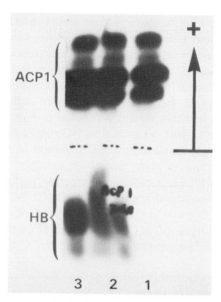

Fig. 1. Zymogram patterns of acid phosphatase-1 (ACP1) of the orang utans OU1–3.

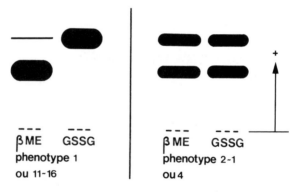

βME GSSG
phenotype 1
ou 11-16

βME GSSG
phenotype 2-1
ou 4

Fig. 2. Zymodiagram showing the patterns of adenosine deaminase (ADA) of phenotypes 1 and 2–1 after β-mercaptoethanol and GSSG treatment (for further explanation see text).

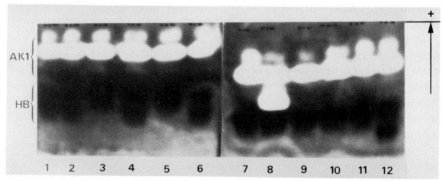

Fig. 3. Zymogram patterns of adenylate kinase-1 (AK1) of the orang utans OU1–12. Note that the OU8 is a variant (phenotype 2–1).

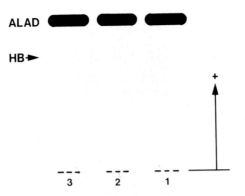

Fig. 4. Zymodiagram showing the patterns of δ-aminoleuvilinate dehydrase (ALAD) of the orang utans OU1–3. HB indicates the position of haemoglobins in the gel.

Fig. 5. Zymodiagram showing the patterns of aldolase (ALD) of the orang utans OU1–3. HB indicates the position of haemoglobins in the gel.

Fig. 6. Zymodiagram showing the patterns of biliverdin reductase (BLVR) of the orang utans OU1–3. HB indicates the position of haemoglobins in the gel.

79

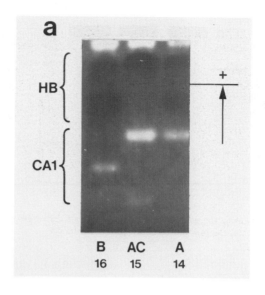

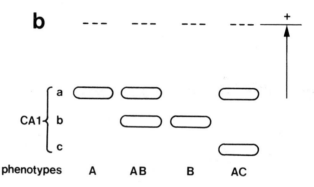

Fig. 7. Carbonic anhydrase-1 (CA1): a, zymogram patterns of the orang utans OU14–16 (phenotypes A, AC and B respectively); b, diagrammatic representation of phenotypes A, AB, B and AC.

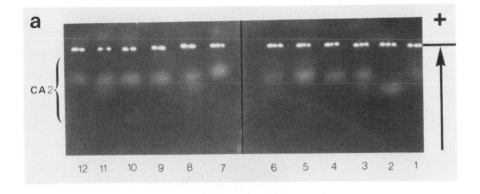

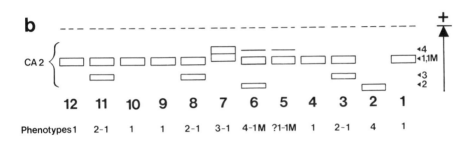

Fig. 8. Carbonic anhydrase-2 (CA2): a, zymogram patterns of the orang utans OU1–12; b, diagrammatic representation of the patterns with phenotype designations.

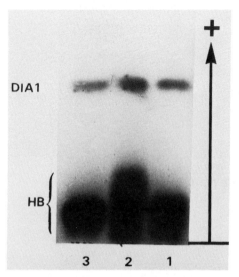

Fig. 9. Zymogram patterns of diaphorase-1 (DIA1) of the orang utans OU1-3.

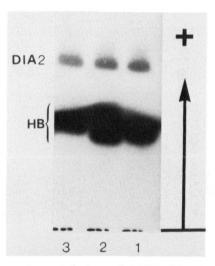

Fig. 10. Zymogram patterns of diaphorase-2 (DIA2) of the orang utans OU1-3.

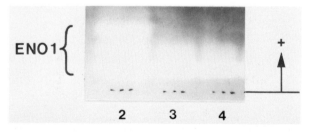

Fig. 11. Zymogram patterns of enolase-1 (ENO1) of the orang utans OU2 (phenotype 2-1) and OU3 and 4 (phenotype 1).

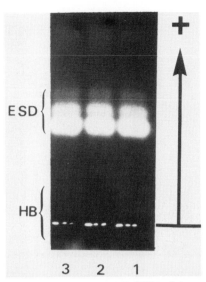

Fig. 12. Zymogram patterns of esterase-D (ESD) of the orang utans OU1-3.

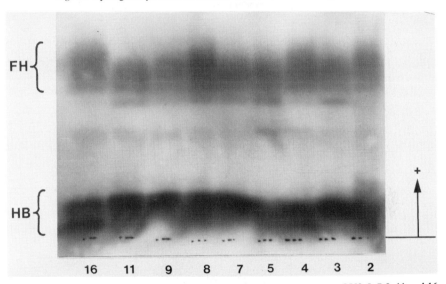

Fig. 13. Zymogram patterns of fumarate hydratase (FH) of the orang utans OU2-5, 7-9, 11 and 16. The patterns of OU2, 4 and 16 are somewhat different from the rest, but no conclusions could be drawn as to their genetic background.

Fig. 14. Zymodiagram showing the patterns of fructokinase (FK) of the orang utans OU1-3. HB indicates the position of haemoglobins in the gel.

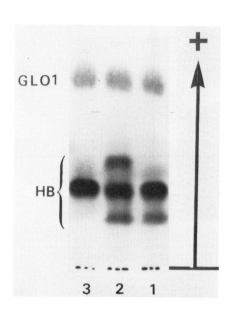

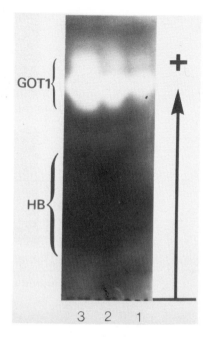

Fig. 15. Zymogram patterns of glyoxalase-1 (GLO1) of the orang utans OU1-3.

Fig. 16. Zymogram patterns of glutamic oxaloacetic transaminase-1 (GOT1) of the orang utans OU1-3.

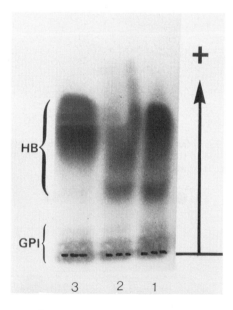

Fig. 17. Zymogram patterns of glucosephosphate isomerase (GPI) of the orang utants OU1-3.

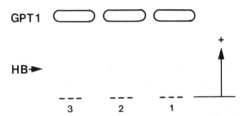

Fig. 18. Zymodiagram showing the patterns of glutamic pyruvic transaminase-1 (GPT1) of the orang utans OU1-3. HB indicates the position of haemoglobins in the gel.

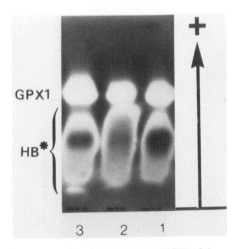

Fig. 19. Zymogram patterns of glutathione peroxidase-1 (GPX1) of the orang utans OU1-3. Note that the haemoglobins exhibit peroxidase activity.

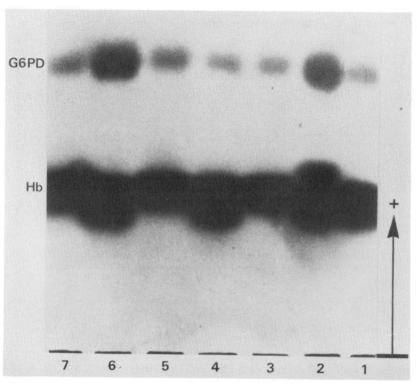

Fig. 20. Zymogram patterns of glucose-6-phosphate dehydrogenase (G6PD) of the orang utans OU1-7. Note high activity of G6PD in OU2 and 6 (for further explanation see text and Wijnen et al., 1982, this volume).

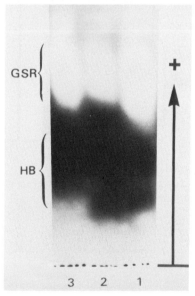

Fig. 21. Zymogram patterns of glutathione reductase (GSR) of the orang utans OU1-3. Note different relative activities.

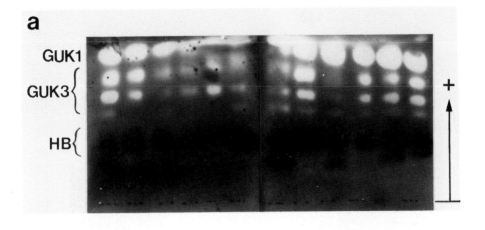

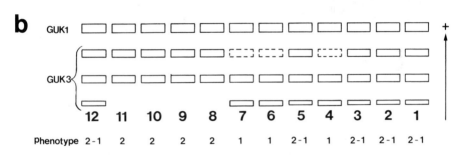

Fig. 22. Guanylate kinase-1 and -3 (GUK1 and GUK3): a, zymogram patterns of the orang utans OU1-12; b, diagrammatic representation of the patterns with GUK3 phenotype designations.

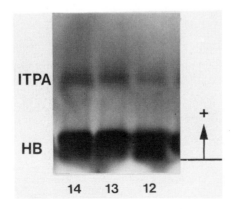

Fig. 23. Zymograms showing the patterns of inosine triphosphatase (ITPA) of the orang utans OU12-14, HB indicates the position of haemoglobins in the gel.

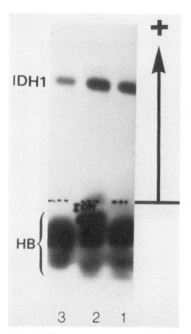

Fig. 24. Zymogram patterns of isocitrate de-hydrogenase-1 (IDH1) of the orang utans OU1-3.

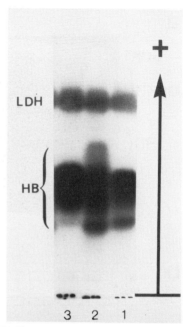

Fig. 25. Zymogram patterns of isozyme 3 of lactate dehydrogenase (LDHA and LDHB) of the orang utans OU1-3.

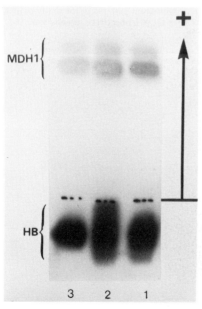

Fig. 26. Zymogram patterns of malate de-hydrogenase-1 (MDH1) of the orang utans OU1-3.

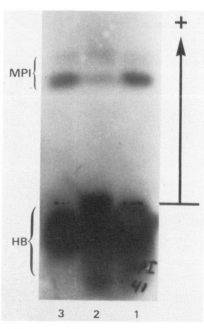

Fig. 27. Zymogram patterns of mannosephos-phate isomerase (MPI) of the orang utans OU1-3. Note difference in relative intensities of the two bands in OU2 compared to OU1 and 3 (see text).

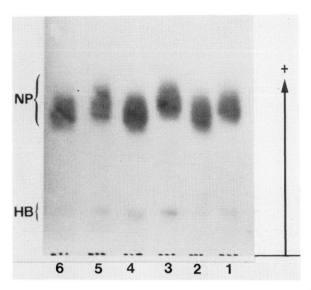

Fig. 28. **Zymogram** patterns of nucleoside phosphorilase (NP) of the orang utans OU1-6. OU2, 4 and 6 are variants of phenotype 2-1. The others are phenotype 1.

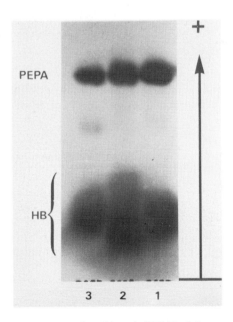

Fig. 29. **Zymogram** patterns of peptidase-A (PEPA) of the orang utans OU1-3.

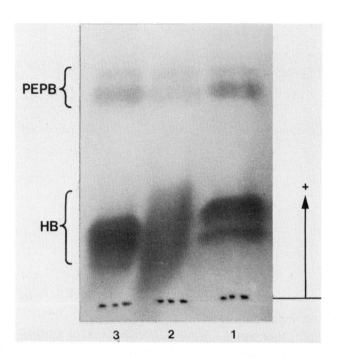

Fig. 30. Zymogram patterns of peptidase-B (PEPB) of the orang utans OU1-3.

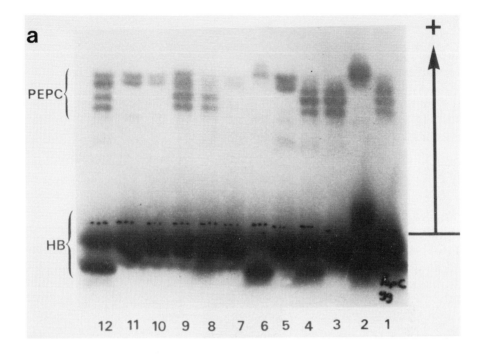

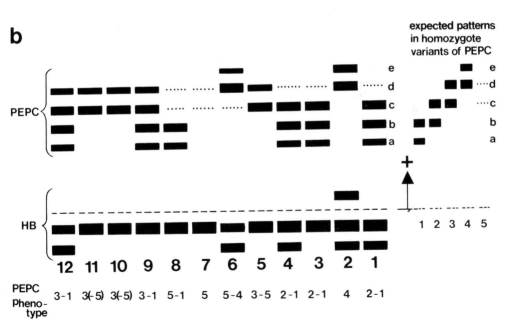

Fig. 31. Peptidase-C (PEPC): a, zymogram patterns of the orang utans OU1-12; b, diagrammatic representation of the patterns with phenotype designations (on the right the expected homozygote patterns of phenotypes 1 through 5 are shown).

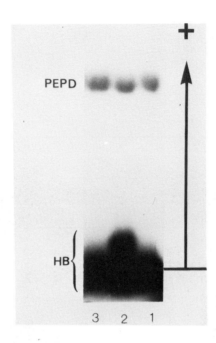

Fig. 32. Zymogram patterns of peptidase-D (PEPD) of the orang utans OU1-3.

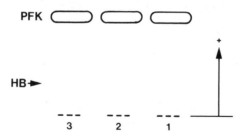

Fig. 33. Zymodiagram showing the patterns of phosphofructokinase (PFK) of the orang utans OU1-3. HB indicates the position of haemoglobins in the gel.

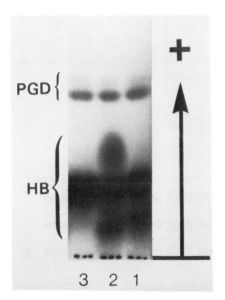

Fig. 34. Zymogram patterns of phosphogluconate dehydrogenase (PGD) of the orang utans OU1-3.

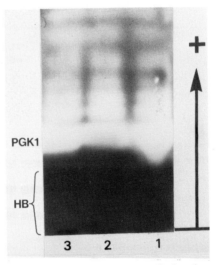

Fig. 35. Zymogram patterns of phosphoglycerate kinase-1 (PGK1) of the orang utans OU1-3.

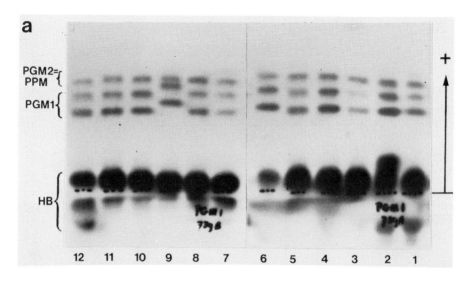

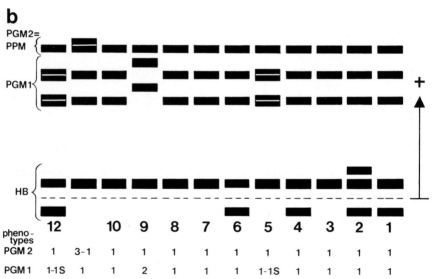

Fig. 36. Phosphoglucomutase-1 (PGM1): a, zymogram patterns of OU1-12; b, diagrammatic representation of the patterns with phenotype designations (the phenotypes of PGM2 are added; see also Fig. 37).

94

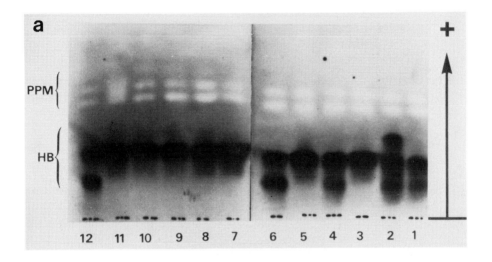

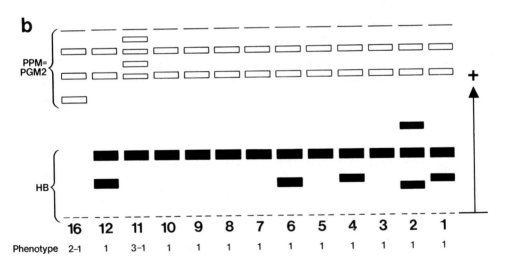

Fig. 37. Phosphopentomutase (PPM = PGM2): a, zymogram patterns of OU1-12 (for explanation of the pattern of OU11 see text and Fig. 39); b, diagrammatic representation of the patterns with phenotype designations (the pattern of OU16, the only animal with phenotype 2–1 in our series, is added; see also Fig. 38).

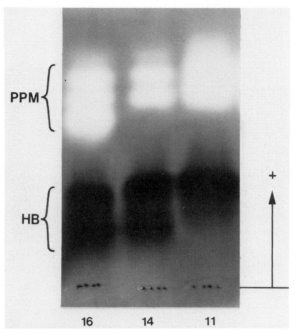

Fig. 38. Zymogram patterns of phosphopentomutase (PPM=PGM2) of the orang utans OU11 (phenotype 3–1), OU14 (phenotype 1) and OU16 (phenotype 2–1).

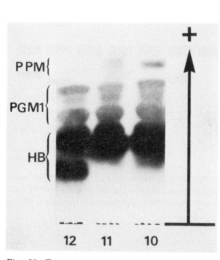

Fig. 39. Zymogram patterns of PGM1 and PGM2 (=PPM) of the orang utans OU10–12. Note the split in the PPM band of OU11.

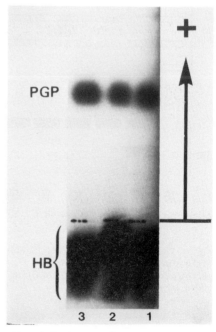

Fig. 40. Zymogram patterns of phosphoglycolate phosphatase (PGP) of the orang utans OU1-3.

Fig. 41. Zymodiagram showing the patterns of pyruvate kinase-1 (PK1) of the orang utans OU1-3. HB indicates the position of haemoglobins in the gel.

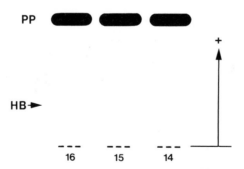

Fig. 42. Zymodiagram showing the patterns of pyrophosphatase (PP) of the orang utans OU14-16. HB indicates the position of haemoglobins in the gel.

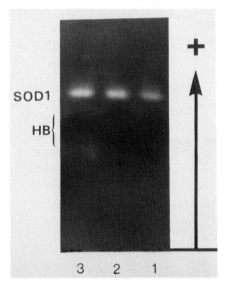

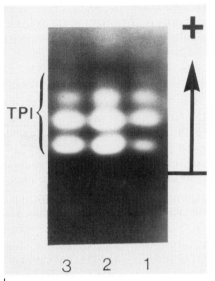

Fig. 43. Zymogram patterns of superoxide dismutase-1 (SOD1) of the orang utans OU1-3.

Fig. 44. Zymogram patterns of triose phosphate isomerase (TPI) of the orang utans OU1-3.

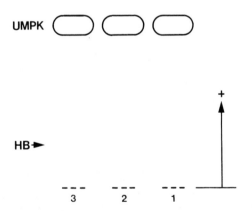

Fig. 45. Zymodiagram showing the patterns of uridine monophosphate kinase (UMPK) of the orang utans OU1-3. HB indicates the position of haemoglobins in the gel.

98

CA2: Fluorescein diacetate was used as a substrate to detect CA2 polymorphism in man (Hopkinson et al. 1974). The procedure of Hopkins et al. revealed an extreme degree of CA2 variation in the orang utan (Fig. 8). Under the conditions employed the orang utan CA2, either variant or usual, migrated cathodal to the origin in the gel. The OU individuals 1, 4, 9, 10 and 11 form a single sharp band and are designated as phenotype CA2 1 (genotype *CA2*1/CA2*1*). The most common variant, presumed to be a heterozygote and called CA2 2–1 (genotype *CA2*1/CA2*2*), was found in OU3, 8 and 11. The OU7 appears to be another variant, most probably a heterozygote between the usual form and a form with somewhat slower cathodal rate of migration than the former. This variant is named CA2 3–1 (genotype *CA2*1/CA2*3*). The OU5, a wild Bornean female, formed a major band indistinguishable from the band of phenotype 1, and a minor band appearing in between the origin and the major band. The OU5 was mated to OU2, a wild Sumatran male with a unique CA2 phenotype, called CA2 4. The OU6, the offspring of this cross, exhibited a typical heterozygote pattern with a band corresponding to that of CA2 4 and the minor band combination of OU5. Based upon this pattern we interpret that OU5 is at least a heterozygote between CA2 1 and a 1-like variant with a tendency of forming a secondary (minor) isozyme. We designate, therefore, the OU5 as CA2 1–1M (M for 'modified') phenotype (genotype *CA2*1/CA2*1M*), and the OU6 as phenotype CA2 4–1M (genotype *CA2*4/CA2*1M*).

In their electrophoretic procedure Tashian (1965, 1969) and Tashian and Carter (1976) did not find variation in CA2 of a sample of 33 orang utans.

DIA1 (Fig. 9) and **DIA2** (Fig. 10) did not show variation in the present sample of orang utans.

ENO1: All orang utans except OU2 exhibited a two-banded pattern (Fig. 11), designated as phenotype ENO1 1. The variant OU2 (one of the two Sumatran animals in the series) formed a three-banded pattern with an intermediate heteropolymeric band, suggesting that the functional molecule is a dimer as in the case of man (Harris and Hopkinson 1976). The phenotype of OU2 is designated as ENO1 2–1 (genotype *ENO1*1/ENO1*2*).

ESD (Fig. 12) did not show variation in our sample of orang utans.

FH: The banding patterns of FH demonstrate that this enzyme occurs in multiple forms. The pattern of the Sumatran orang utans OU2 and OU16 and the hybrid OU4 differ somewhat from the rest (the hybrid OU6 was not tested) (Fig. 13). In the absence of further investigations it is hard to make acceptable interpretations of the observed variant patterns. The phenotypes of the remaining orang utans are designated as FH 1.

Bruce and Ayala (1979) tested eight Sumatran orang utans for FH and no Borneans. They found no variation in this system. Therefore, it is not known whether their Sumatran animals conform the phenotype of our Sumatran orang utans or that of the Borneans.

FK (Fig. 14), **GLO1** (Fig. 15), **GOT1** (Fig. 16), **GPI** (Fig. 17), **GPT1** (Fig. 18) and **GPX1** (Fig. 19) did not show variation in the orang utans studied.

G6PD: The two wild-born Sumatran animals in our sample (OU2 and 16) exhibited red cell G6PD activity levels of 5 to 7 times that of most of the other orang utans tested. The hybrid daughter of OU2 (OU6) showed an activity almost as high as that of her Sumatran father. All Bornean orang utans exhibited low activity levels, except OU14 whose G6PD activity was two times the normal. These findings and their implications are discussed in more detail by Wijnen et al. (1982, this volume). G6PD zymograms are shown in Fig. 20.

GSR: Relative intensities of the enzyme activity on the gel suggest that the red cell GSR in the OU2 is somewhat less compared to the others (Fig. 21). It requires a quantitative study to verify this suggestion.

GUK1: Jamil and Fisher (1977) identified one variant in a sample of 12 orang utans. No variant was found in our series (Fig. 22).

GUK3: All three phenotypes described by Jamil and Fisher (1977) in the orang utan (GUK3 1, 2–1 and 2) have been identified in our series (Fig. 22). Moreover, the OU6, offspring of parents with the GUK 2–1 phenotype (OU2 and 5), has been identified as a homozygous GUK 2 phenotype, supporting the hypothesis that GUK3 variation is genetically determined. Both alleles *GUK3∗1* and *GUK3∗2* apparently occur in the Sumatran as well as in the Bornean population.

ITPA (Fig. 23) and **IDH1** (Fig. 24) showed no electrophoretically distinguishable variation.

LDHA and **LDHB:** All samples were screened for the isozyme 3 (BBAA) of the LDH system, which facilitates the identification of both fast and slow variants of the locus A as well as locus B (see Herbschleb-Voogt and Meera Khan 1981). Neither LDHA nor LDHB showed variation (Fig. 25).

MDH1: In our series of orang utans no variant was encountered (Fig. 26). Bruce and Ayala (1979) found a fast variant in a sample of eight Sumatran orang utans. Giving no other details on the nature of this variant they reported simply its allelic frequency to be 0.14.

MPI: All animals tested formed a sharp dense major band and a faint anodal minor band, which is presumed to represent a secondary isozyme (Fig. 27). This pattern is interpreted as the homozygous phenotype MPI 1. In the pure Sumatran OU2, the only exception, there seemed to be a tendency to form two major bands, the second major band being in the position of the minor band seen in the other animals. Further studies would be required to trace whether or not we are dealing with a true variant in this case.

NP: Three orang utans, one Sumatran and two hybrids, exhibited a variant pattern (Fig. 28). We call the usual phenotype as NP 1, and the variant NP 2–1 with the assumption that the pattern is due to the occurrence of a variant allele in polymorphic frequency. This variant, possibly a Sumatran trait, requires further study. Unfortunately, white blood cells which form unambiguous and clear isozyme patterns and facilitate the verification of the above hypothesis, could not be studied in these orang utans.

PEPA (Fig. 29) and **PEPB** (Fig. 30) did not exhibit variation in our series of orang utans.

PEPC: Seven patterns interpreted as seven distinct phenotypes determined by five alleles at an autosomal locus were identified (Fig. 31). The OU7 formed a faint zone of activity with the mobility of phenotype 3. The phenotype of this animal is designated as PEPC 5. We are not sure at present whether the PEP zone of this phenotype is PEPC-specific or non-specific like others occasionally appearing between the zones of Hb and PEPC in the gel (see Fig. 31). If the latter were true, the *PEPC*5* represents a silent allele. The PEPC phenotype of the offspring of the cross between OU2 and OU5 suggests that, most probably, the father (OU2) is a homozygote for the allele *PEPC*4* while the mother (OU5) is a 3–5 heterozygote, giving rise to a 4–5 heterozygous daughter (OU6). We tentatively interpret the OU8 as a 5–1 phenotype. It is not possible at present to rule out that OU10, 11, 14 and 15 with apparent phenotypic patterns of PEPC 3, are not carriers of the allele *PEPC*5* as well.

It is noteworthy that both pure Sumatran orang utans (OU2 and OU16) in the present series probably are homozygous for the allele *PEPC*4*. The full hybrid OU6 apparently inherited this allele from its father. The partial hybrid OU4 (3/4 Sumatran, 1/4 Bornean), however, exhibits the phenotype PEPC 2–1. The chance that it does not carry a Sumatran PEPC allele shows that *PEPC*4* probably is not the only Sumatran allele.

PEPD (Fig. 32), **PFK** (Fig. 33), **PGD** (Fig. 34) and **PGK1** (Fig. 35) did not show variation.

PGM1: The majority of the orang utans in our series are of phenotype 1, forming two sharp bands cathodal to the main zone of PGM2. A clear exception is OU9, a wild Bornean female, in which both the bands maintaining the same interband distance between themselves are displaced anodally, close to the main band of PGM2 (Fig. 36). This pattern is typical of a homozygote; we designate it as a PGM1 2 phenotype. In the absence of any evidence for the occurrence of a null allele at the PGM1 locus in the orang utan, this isolated instance of an obvious homozygote (and no heterozygote 2–1) in a total of 40 individuals (10 animals studied by Schmitt et al. 1971, 3 by Barnicot and Cohen 1971, 11 by Bruce and Ayala 1979, and 16 in the present series) may indicate the existence of further heterogeneity in the orang utan. Possibly, the animal concerned (OU9) originates from an area in Borneo from which so far no other specimens have been studied.

The OU5, OU12 and OU14 appear to carry a slow moving *PGM1*1*-like allele, designated as *PGM1*1S*, in addition to 1. This allele is probably similar to one of those seen in the human PGM1 subtypes of 1 and 2 (Bissbort et al. 1978).

PGM2: The individual OU11 exhibited a pattern indicating a split in the most cathodal band of PGM2, but not in the others (Fig. 37). A more basic buffer system and longer run of electrophoresis exhibited the formation of an additional anodal band in the PGM2 zone of this specimen (Fig. 39). The labile phosphate detection technique for phosphopentomutase activity described by Wijnen et al. (1977) confirmed that the OU11 is a PGM2 variant, a heterozygote between the usual allele *PGM2*1* and a variant allele coding for a PGM2 migrating faster than 1. OU11 is a male born to wild Bornean orang utans in captivity (Rhenen Zoo). To our knowledge no other Bornean was found to be a variant of PGM2. Bruce and Ayala (1979) reported the occurrence of a slow PGM2 variant in a series of eight Sumatran orang utans, with an allelic frequency of 0.25. If the designation of *PGM2*2* is reserved to describe this slow allele, we call the fast variant found in OU11 allele *PGM2*3*.

One of the two pure Sumatran orang utans in our series (OU16) was found to be heterozygous for a slow PGM2 variant, probably the same as that described by Bruce and Ayala (1979) due to allele *PGM2*2* (Figs. 37 and 38). This allele therefore is probably unique to the Sumatran population of orang utans.

The remaining systems, **PGP** (Fig. 40), **PK1** (Fig. 41), **PP** (Fig. 42). **SOD1** (Fig. 43), **TPI** (Fig. 44) and **UMPK** (Fig. 45) exhibited no variation in the animals studied.

Discussion

Barnicot and Cohen (1970) screened the eryhrocytes of one Bornean and two Sumatran orang utans (called Sya, and Bukit and Ini respectively) from the Yerkes Regional Primate Research Center for the electrophoretic patterns of ACP1, G6PD, AK1, PGD and PGM1. Around the same period Schmitt and his colleagues investigated several enzymes of single individuals comparing orang utans with other primates. In 1970 Schmitt et al. published a study on ACP1, ADA, AK1, PGM1 and PGM2 in an unidentified group of 10 orang utans. Another 33 animals, also unidentified, were studied for red cell CA1 and CA2 variation by Tashian (see Tashian and Carter 1971). More recently, Bruce and Ayala (1979) screened eight Sumatran and three Bornean orang utans for electrophoretic variation of GOT1, IDH1, LDHA, LDHB, MDH1, DIA1, GPI, CAT and FH in addition to ADA, AK1, G6PD, PGM1, PGM2 and PGD. Jamil and Fisher (1977) investigated 12 orang utans, including three pure-bred Sumatrans and seven pure-bred Borneans, for variation in the erythrocyte GUK1 and GUK3. The above studies, which included 20 markers, used starch gel as a supporting medium and revealed one or more variants in each of the following systems: ADA, CA1, GUK1, GUK3, MDH1, G6PD and PGM2.

Table 4. Phenotype distribution of 12 informative enzyme marker systems, two major haemoglobin chains and the chromosomes 2, 9 and 14 in the orang utans OU1–OU16 (Table 1).

Animal number	Enzyme systems												Haemoglobins[b]		Chromosomes[c]		
	ADA	AK1	CA1	CA2	ENO1	FH	G6PD[a]	GUK3	NP	PEPC	PGM1	PGM2	HBA	HBB	2	9	14
OU1	nt	1	A	1	1	1	+	2–1	1	2–1	1	1	AB	A	BB	NV	NN
OU2	nt	1	AB	4	2–1	variant	+++++	2–1	2–1	4	1	1	AB	AC	SS	NN	NV
OU3	nt	1	A	2–1	1	1	+	2–1	1	2–1	1	1	A	A	BB	NV	NN
OU4	2–1	1	A	1	1	variant	+	1	2–1	2–1	1–1S	1	AB	A	BS	NV	NN
OU5	nt	1	AC	?1–1M	1	1	+	2–1	1	3–5	1	1	A	A	BB	NV	NN
OU6[d]	nt	1	AB	4–1M	1	nt	++++	1	2–1	5–4	1	1	AB	A	BS	NV	NN
OU7	nt	1	AC	3–1	1	1	+	2	1	5	1	1	A	A	BB	NN	NN
OU8	nt	2–1	A	2–1	1	1	+	2	1	5–1	2	1	A	A	BB	NV	NN
OU9	nt	1	A	1	1	1	+	2	1	3–1	1	1	A	A	BB	NV	NN
OU10	nt	1	AC	1	1	1	+	2	1	3(–5)	1	3–1	A	A	BB	NV	NN
OU11[e]	1	1	AC	2–1	1	1	+	2–1	1	3(–5)	1–1S	1	AB	A	BB	NV	NN
OU12[f]	1	1	A	1	1	1	+	2–1	1	3–1	1–1S	1	AB	A	BB	NN	NN
OU13	1	1	AC	nt	1	1	++	2	1	3(–5)	1	1	AB	A	BB	NN	NN
OU14	1	1	A	nt	1	1	+	2	1	3(–5)	1–1S	1	A	A	BB	NN	NN
OU15	1	1	AC	nt	1	1	+	2	1	3(–5)	1	1	A	A	BB	NN	NN
OU16	1	1	B	nt	1	variant	+++++++	1	1	4	1	2–1	B	AC	SS	NN	NN

[a] The data on G6PD are discussed by Wijnen et al. 1982 (this volume). +'s indicate relative activities.

[b] The data on haemoglobins are reported by De Boer and Meera Khan 1982 (this volume). HBA = α-chain locus (A and B indicate the presence of α^A and/or α^B chains); HBB = β-chain locus (A and C indicate the presence of β^A and/or β^C chains).

[c] The data on the chromosomes are given in De Boer and Seuánez 1982 (this volume). Under chromosome 2 B indicates the Bornean type of this chromosome, S the Sumatran type. Under chromosomes 9 and 14 N indicates the normal chromosome, V the variant chromosome.

[d,e,f] See Table 1.

103

Table 5. Red cell enzyme variation in the orang utan.

Enzyme system	Number of alleles identified				Phenotypes with total numbers of animals in brackets
	Schmitt et al. 1970 (n = 12)	Barnicot and Cohen 1971 (n = 3)	Bruce and Ayala 1979 (n = 11)	This report[a] (n = 7–16)	
ACP1	1	1	nt	1 (12)	1(25)
ADA	1	nt	2	2 (7)	see the text
AK1	1	1	1	2 (16)	1(39), 2–1(1)
ALAD	nt	nt	nt	1 (16)	1(16)
ALD	nt	nt	nt	1 (16)	1(16)
BLVR	nt	nt	nt	1 (16)	1(16)
CA1[b]	nt	nt	nt	3 (16)	A(7), AB(2), AC(6), B(1)
CA2[b]	nt	nt	nt	5 (12)	1(5), 2–1(3), 3–1(1), 4(1), 4–1M(1), ?1–1M(1)
CAT	nt	nt	1	nt	1(11)
DIA1	nt	nt	1	1 (12)	1(23)
DIA2	nt	nt	nt	1 (12)	1(12)
ENO1	nt	nt	nt	2 (16)	1(12)
ESD	nt	nt	nt	1 (12)	1(12)
FH	nt	nt	1 (n = 8)	2?(16)	see the text
FK	nt	nt	nt	1 (16)	1(15)
GLO1	nt	nt	nt	1 (12)	1(12)
GOT1	nt	nt	1	1 (12)	1(23)
GPI	nt	nt	1	1 (12)	1(23)
GPT1	nt	nt	nt	1 (16)	1(16)
GPX1	nt	nt	nt	1 (12)	1(12)
G6PD	nt	2	1	2 (16)	see Wijnen et al. 1982
GSR	nt	nt	nt	1 (12)	1(12)
GUK1[c]	nt	nt	nt	1 (12)	1(27), 2–1(1)[c]
GUK3[c]	nt	nt	nt	2 (16)	1(10), 2–1(9), 2(9)[c]
ITPA	nt	nt	nt	1 (12)	1(12)
IDH1	nt	nt	nt	1 (12)	1(12)
LDHA	nt	nt	1	1 (12)	1(23)
LDHB	nt	nt	1	1 (12)	1(23)
MDH1	nt	nt	2	1 (12)	1(21), 2–1(2)[d]
MPI	nt	nt	nt	1 (16)	1(16)
NP	nt	nt	nt	2 (16)	1(13), 2–1(3)
PEPA	nt	nt	nt	1 (12)	1(12)
PEPB	nt	nt	nt	1 (12)	1(12)
PEPC	nt	nt	nt	5 (16)	2–1(3), 3–1(3), 3–5(3), 4(2), 5(1), 5–1(1), 5–4(1), 3(–5) (2)
PEPD	nt	nt	nt	1 (12)	1(12)
PFK	nt	nt	nt	1 (16)	1(16)
PGD	nt	1	1	1 (12)	1(26)
PGK1	nt	nt	nt	1 (12)	1(12)
PGM1	1	1	1	3 (16)	1(36), 1–1S(3), 2(1)
PGM2	1	nt	2	3 (16)	1(31), 2–1(5)[d], 3–1(1)
PGP	nt	nt	nt	1 (12)	1(12)
PK1	nt	nt	nt	1 (12)	1(12)
PP	nt	nt	nt	1 (12)	1(12)
SOD1	nt	nt	nt	1 (12)	1(12)
TPI	nt	nt	nt	1 (12)	1(12)
UMPK	nt	nt	nt	1 (12)	1(12)

nt = not tested. A fuller account of the alleles and phenotypes of individual systems appears in the text.

[a] In the present report enzyme systems were tested in 7 to 16 orang utans. The number of animals tested for each system is given in brackets after the number of alleles found (see Table 2 for the OU numbers of the individuals tested for each system).

The present work, on the other hand, used Cellogel for screening 45 enzyme marker systems in 16 orang utans of known geographic origin and revealed one or more variants of ADA, AK1, CA1, CA2, ENO1, FH, G6PD, GUK3, NP, PEPC, PGM1 and PGM2 (Table 4). A summary of the results of the present study and the previous ones mentioned above is given in Table 5.

It must be stressed that the results of our study reported above have certain obvious limitations. The number of individuals screened is extremely small, especially that of Sumatran orang utans. For practical reasons, repeat samples could not be collected within the study period. Family studies on the observed variants were not possible. The enzyme systems could be investigated only in the red blood cells, while some of the systems require studies on other tissues to obtain supportive evidence for the red cell pattern based interpretations (e.g. NP). For establishing the nature of the observed electrophoretic variation, studies involving partial purification and/or further characterization of individual enzymes could not be performed, because the quantities of blood collected from the individual orang utans were too small. Therefore we consider the findings and their interpretations reported in this paper to be very tentative and subject to confirmation. In fact, with the exception of carbonic anhydrase-1, almost all the studies reported so far on the enzyme variation in the orang utan suffer from one or more of the limitations mentioned above. It may be noted, however, that the variants of CA1, GUK1 and PGM2 observed during the present study conform to those reported earlier.

Regarding the possible differences between the Bornean and Sumatran populations of the orang utan, three different levels have to be distinguished:
1. Enzyme marker systems with different alleles fixed in the two populations (comparable to the situation of the structure of chromosome no. 2, which distinguishes all Bornean from all Sumatran orang utans, cf. De Boer and Seuánez 1982, this volume). There are only two indications of the possible existence of such differences in the enzyme systems studied. Firstly, the FH of the two unrelated Sumatran animals and the one 3/4 Sumatran hybrid was found to be different from that of the 12 Borneans screened. Secondly, the G6PD activity level might distinguish between the two subspecies of orang utan (Wijnen et al. 1982, this volume). Much more detailed investigations are required to test these possibilities. The existence of a consistent difference between the ADA of Bornean and Sumatran orang utans as reported by Bruce and Ayala (1979) could not be confirmed by us.
2. Enzyme marker systems with polymorphic variants unique to one of the two populations (variant alleles occurring only in one of both populations

---

[b] CA1 and CA2 were typed by Tashian in 33 orang utans. The results of these studies are summarized in Tashian and Carter (1971). They found 3 CA1 alleles with frequencies 0.69 (allele A), 0.21 (B) and 0.10 (C). Since these authors do not give numbers of individual phenotypes, their data could not be included in the last column. In CA2 they did not find variation probably because they did not use the specific substrate proposed by Hopkinson et al. (1974).

[c] GUK1 and GUK3 were also studied by Jamil and Fisher in a sample of 12 orang utans. In GUK1 they found one variant phenotype (2–1). In GUK3 they found six animals of phenotype 1, 3 of 2–1 and 3 of 2. These figures are included in the numbers of the last column.

[d] Since Bruce and Ayala (1979) only gave allelic frequencies, the exact numbers of animals with variant phenotypes (homozygous or heterozygous) are not known.

but not in the other; compare the possible uniqueness of the β^C allele of the haemoglobin system to Sumatran orang utans, cf. De Boer and Meera Khan 1982, this volume). In this respect our sample only allows to point to possible unique Sumatran alleles, since the sample of Sumatran orang utans is too small to exclude the possible occurrence of alleles so far only found in Borneans. The following alleles were found in Sumatran animals but in none of the Borneans: *CA1∗B, CA2∗4, PEPC∗4, PGM2∗2, ADA∗2* and *ENO1∗2*. The study of a larger sample is needed to confirm such possibilities.

3.   Polymorphic enzyme marker systems with different allelic frequencies in Borneo and Sumatra (comparable to the reversed frequencies of the haemoglobin α^A and α^B alleles, cf. De Boer and Meera Khan 1982, this volume). It goes without saying that nothing can be said on such differences on the basis of the very small sample screened.

When the data on haemoglobin α and β locus are added to the above data on 45 enzyme marker systems (see Table 4), the series of orang utans studied shows polymorphic variation in 14 of 47 loci (30%). It must be stated, however, that this figure is based on the Bornean and Sumatran populations together. Taken separately the figures may be somewhat lower, but, on the other hand, investigation of larger samples may reveal additional polymorphisms. When the orang utans OU1-12, which were screened for all the systems, are considered, the average heterozygozity is 0.078 (excluding the G6PD system, the genetics of which are not yet clear, cf. Wijnen et al. 1982, this volume). This figure, though preliminary, appears to be within the normal range, and does not make the orang utans appear relict populations, in which lower indices of heterozygozity are usually expected.

In general, the present data indicate that the orang utan is an excellent subject among the higher primates to study the molecular and population genetics aspects of speciation and microevolution. Further investigations should supply enough data to estimate the genetic distance between the two forms of orang utan. Nevertheless, the available data clearly indicate that the results of any kind of studies on orang utans should be treated separately and lumping of the data from Sumatrans and Borneans should be avoided as much as possible. Finally, the existence of a fairly high degree of polymorphism indicates that interesting information on geographical intrasubspecific variability may be obtained from the study of red cell enzymes in orang utans. Apart from the scientific aspects, such information may be useful in planning the conservation strategies for feral orang utans.

Acknowledgements

The authors are grateful to the zoological gardens of Arnhem, Rotterdam, Antwerp, Hannover, Rhenen and Wuppertal for generous cooperation in providing the blood samples of the orang utans.

References

Barnicot, N.A. and P. Cohen (1970). Red cell enzymes of primates (Anthropoidea). Biochem. Genet., 4: 41–57.

Boer, L.E.M. de (1982). Genetics and conservation of the orang utan. In: L.E.M. de Boer (ed.), The orang utan. Its biology and conservation. Junk, The Hague.

Boer, L.E.M. de and P. Meera Khan (1982). Haemoglobin polymorphisms in Bornean and Sumatran orang utans. In: L.E.M. de Boer (ed.), The orang utan. Its biology and conservation. Junk, The Hague.

Boer, L.E.M. de and H.N. Seuánez (1982). The chromosomes of the orang utan and their relevance to the conservation of the species. In: L.E.M. de Boer (ed.), The orang utan. Its biology and conservation. Junk, The Hague.

Bissbort, S., H. Ritter and J. Kömpf (1978). PGM1 subtyping by means of acid starch gel electrophoresis. Hum. Genet., 45: 175–177.

Bruce, E.J. and F.J. Ayala (1979). Phylogenetic relationships between man and the apes: electrophoretic evidence. Evolution, 33: 1040–1056.

Cavalli-Sforza, L.L. and W. Bodmer (1971). The genetics of human populations. Freeman and Co., San Francisco.

Ebeli-Struijk, A.C., E.M. Wurzer-Figurelli, F. Ajmar and P. Meera Khan (1976). The distribution of esterase-D variants in different ethnic groups. Hum. Genet., 34: 299–306.

Estop, A., J.J. Garver, P.L. Pearson, T.M. Dijksman, L.M.M. Wijnen and P. Meera Khan (1978). Gene assignments to the presumptive homologs of human chromosomes 3, 4, 5, 7, 8, 18, 19, 20, and 21 in the Pongidae and Cercopithecidae. Cytogenet. Cell Genet., 22: 558–563.

Garver, J.J., P.L. Pearson, A. Estop, T.M. Dijksman, L.M.M. Wijnen, A. Westerveld and P. Meera Khan (1978). Gene assignments to the presumptive homologs of human chromosomes 1, 6, 11, 12 and X in the Pongidae and Cercopithecidae. Cytogenet. Cell Genet., 22: 564–569.

Harris, H. and D.A. Hopkinson (1976). Handbook of enzyme electrophoresis in human genetics. North Holland Publishing Company, Amsterdam.

Herbschleb-Voogt, E. and P. Meera Khan (1981). Defining the locus of origin of a genetically determined electrophoretic variant of a multilocus enzyme system; the Calcutta-1 of human LDH system is a B-locus variant. Hum. Genet., 57: 290–295.

Herbschleb-Voogt, E., P.L. Pearson, J.M. Vossen and P. Meera Khan (1981). Basic defect in the expression of adenosine deaminase in ADA⁻SCID disease investigated through the cells of an obligate heterozygote. Hum. Genet., 56: 379–386.

Hopkinson, D.A., J.S. Coppock, M.E. Mühlenmann and J.H. Edwards (1974). The detection and differentiation of the products of the human carbonic anhydrase loci CA1 and CA2 using fluorogenic substrates. Ann. Hum. Genet., 38: 155.

International System for Human Gene Nomenclature (1979). Shows et al. Cytogenet. Cell Genet., 25: 96–116.

Jamil, P. and R.A. Fisher (1977). An investigation of the homology of guanylate kinase isozymes in mammals and further evidence for multiple GUK gene loci. Biochem Genet., 15: 847–858.

Los, W.R.T. and P. Meera Khan (1977). NADH-Diaphorase polymorphism in *Macaca mulatta*. J. Hum. Evol., 6: 227–230.

Meera Khan, P. (1971). Enzyme electrophoresis on cellulose acetate gel: zymogram patterns in man-mouse and man-Chinese hamster somatic cell hybrids. Arch. Biochem. Biophys., 145: 483–490.

Meera Khan, P. and H. Balner (1972). Polymorphic enzymes in rhesus monkeys and chimpanzees. Medical Primatol. part I: 363–371 (Karger, Basel).

Meera Khan, P. and B.A. Doppert (1976). Rapid detection of glyoxalase I (GLO) on cellulose acetate gel and the distribution of GLO variants in a Dutch population. Hum. Genet., 34: 53–56.

Meera Khan, P. and M.C. Rattazzi (1968). Rapid detection of 6-phosphogluconate dehydrogenase variants by electrophoresis on cellulose acetate gel. Biochem. Genet., 2: 231–235.

Meera Khan, P., A. Westerveld, K-H. Grzeschik, B.F. Deys, O.M. Garson and M. Siniscalco (1971). X-linkage of human phosphoglycerate kinase confirmed in man-mouse and man-Chinese hamster somatic cell hybrids. Am. J. Hum. Genet., 23: 614–623.

Meera Khan, P., H. van Someren, W.W. de Jong and M. Vervloet (1972). Red cell enzyme polymorphisms in rhesus monkeys and chimpanzees. Transpl. Proc., 4: 137–140.

Meera Khan, P., L.M.M. Wijnen, J.Th. Wijnen and K-H. Grzeschnik (1982). Electrophoretic characterization and genetics of human biliverdin reductase (BLVR E.C. number 1.3.1.24); assignment of BLVR to P14→Cen region of human chromosome 7 in the mouse-human somatic celll hybrids. Biochem. Genet., in press.

Meera Khan, P., P.L. Pearson, L.M.M. Wijnen, B.A. Doppert, A. Westerveld and D. Bootsma (1976). Assignment of inosine triphosphatase gene to gorilla chromosome 13 and to human chromosome 20 in primate-rodent somatic cell hybrids. In: Proc. 3rd Internl. Conf. on Human Gene Mapping, Baltimore 1975. Birth Defects: Original Article Series, National Foundation, New York, vol. XII (7): 420–421.

Meera Khan, P., W.R.T. Los, P.L. Pearson, A. Westerveld and D. Bootsma (1974). Genetical studies on multiple forms of human guanylate kinase in man-Chinese hamster somatic cell hybrids. Hum. Hered., 24: 415–423.

Pearson, P.L., J.J. Garver, A. Estop, T.M. Dijksman, L.M.M. Wijnen and P. Meera Khan (1978). Gene assignments to the presumptive homologs of human chromosomes 2, 9, 13, 14 and 15 in the Pongidae and Cercopithecidae. Cytogenet. Cell Genet., 22: 588–593.

Pearson, P.L., T.H. Roderick, M.T. Davisson, J.J. Garver, D. Warburton, P.A. Lalley and S.J. O'Brien (1979). Report of the committee on comparative mapping. Cytogenet. Cell Genet., 25: 82–95.

Rattazzi, M.C., L.F. Bernini, G. Fiorelli and P.M. Mannucci (1967). Electrophoresis of glucose-6-phosphate dehydrogenase: a new technique. Nature (London), 213: 79–80.

Schmitt, J., K.H. Lichte and W. Fuhrmann (1970). Red cell enzymes of the Pongidae. Humangenetik, 10: 138–144.

Selenka, E. (1898). Menschenaffen: Rassen, Schädel und Bezahnung des Orang-utan. Kreidel, Wiesbaden.

Someren, H. van, H.B. van Henegouwen, W. Los, E. Wurzer-Figurelli, B. Doppert, M. Vervloet and P. Meera Khan (1974). Enzyme electrophoresis on cellulose acetate gel, II. Zymogram patterns in man-Chinese hamster somatic cell hybrids. Humangenetik, 25: 189–201.

Tashian, R.E. (1965). Genetic variation and evolution of the carboxylic esterases and carbonic anhydrases of primate erythrocytes. Am. J. Hum. Genet., 17: 257–272.

Tashian, R.E. (1969). The esterases and carbonic anhydrases of human erythrocytes. In: J.J. Yunis (ed.), Biochemical methods in red cell genetics. Academic Press, New York.

Tashian, R.E. and N.D. Carter (1971). Biochemical genetics of carbonic anhydrase. In: H. Harris and K. Hirschhorn (eds.), Advances in human genetics, vol. 2. Plenum Press, New York.

Wijnen, L.M.M., K-H. Grzeschik, P.L. Pearson and P. Meera Khan (1977). The human PGM-2 and its chromosomal localization in man-mouse hybrids. Hum. Genet., 37: 271–278.

Wijnen, L.M.M., M. Monteba-van Heuvel, P.L. Pearson and P. Meera Khan (1978). Assignment of a gene for glutathione peroxidase (GPX1) to human chromosome 3. Cytogenet. Cell Genet., 22: 223–235.

Wijnen, J.Th., H. Rijken, L.E.M. de Boer and P. Meera Khan (1982). Glucose-6-phosphate dehydrogenase (G6PD) variation in the orang utan. In: L.E.M. de Boer (ed.), The orang utan. Its biology and conservation. Junk, The Hague.

Authors' addresses:
P. Meera Khan, H. Rijken, J.Th. Wijnen and L.M.M. Wijnen
Dept. of Human Genetics, Sylvius Laboratories
State University of Leiden
Wassenaarseweg 72
2333 AL Leiden
The Netherlands

L.E.M. de Boer
Biological Research Dept.
Royal Rotterdam Zoological and Botanical Gardens
Rotterdam
The Netherlands

5. Glucose-6-phosphate dehydrogenase (G6PD) variation in the orang utan

Juul Th. Wijnen, Herbert Rijken, Leobert E.M. de Boer and P. Meera Khan

Introduction

Glucose-6-phosphate dehydrogenase (G6PD; D-glucose-6-phosphate; NADP oxidoreductase; E.C. 1.1.1.49) is the first member enzyme in the pentose phosphate shunt pathway of carbohydrate metabolism. It catalyses the conversion of glucose-6-phosphate to 6-phosphogluconolactone in the presence of the coenzyme nucleotide adenine dinucleotide phosphate (NADP). Severe deficiency of G6PD in man is known to cause haemolysis either spontaneously or on exposure to certain therapeutic agents and food stuffs (Beutler 1978).

Biochemical and genetic evidence indicates that only one gene is responsible for the production of G6PD in man (Yoshida 1968) and that the gene is X-linked. Somatic cell genetic studies confirmed these observations (Meera Khan et al. 1971) and demonstrated that the G6PD gene is X-linked also in the chimpanzee, gorilla, orang utan, rhesus monkey and African green monkey (Garver et al. 1978; L.M.M. Wijnen and P. Meera Khan, unpublished observations).

More than one hundred and fifty genetically determined human variants of G6PD have been listed by Yoshida and Beutler (McKusick 1978). Several of the native African and most of the Mediterranean populations exhibit some of these variants in polymorphic frequencies (Mourant et al. 1976). The G6PD variants are largely identified by means of semiquantitative screening procedures (Motulsky and Campbell-Kraut 1961; WHO TRS–366 1967) and electrophoresis followed by specific histochemical staining (Harris and Hopkinson 1972; Rattazzi et al. 1966).

An occasional electrophoretic variant of G6PD had been identified in other primates in the past by a number of investigators (Meera Khan and Balner 1972; Beutler and West 1978; Lucotte 1980). Quantitative and/or qualitative variation of red cell G6PD between different species of primates were reported by others (Huser 1970; Kömpf et al. 1971; Barnicot and Cohen 1970). Huser (1970) reported the data of McCurdy on the physicochemical characteristics of G6PD of two orang utans and those of man and a number of other primates. Barnicot and Cohen (1970) determined the specific activities and electrophoretic mobilities of G6PD in the crude haemolysates of 269 individuals of about 20 primate species belonging to three superfamilies of anthropoidea and noticed that the specific activities were quite variable among species. Three orang utans from Yerkes Regional Primate Research Center were included, in whom they observed the highest activity of G6PD which was on an average eight times the human value. Two of them (Bukit and Ini) were Sumatran and the third (Sya) a Bornean (Jones 1980). Under the conditions of their assay procedure Barnicot and Cohen found that the three orang utans exhibited 10, 43 and 43 units of red cell G6PD activity respectively. Unfortunately the authors did not identify these distinct values with the individual orang utans. However, the electrophoretic mobilities of their G6PD's resembled the normal human G6PD B+ in the starch gel in tris-EDTA-borate buffer pH 8.6.

In this chapter we describe the results of our search for the occurrence of intra- and inter-subspecific variations of G6PD among orang utans. Incidentally, we compare the physicochemical characteristics of G6PD in crude haemolysates of Sumatran orang utans with those of Borneans as an attempt to gain an insight into the basis for the striking differences in the red cell G6PD activities between the two subspecies observed during the present study.

Materials and methods

Details on the possible origins and on the identification of the 16 orang utans included in the present study (indicated as OU1–OU16) are listed in Meera Khan et al. (1982, this volume). The procedures of preparation and preservation of haemolysates used in the present investigations have been described by De Boer and Meera Khan (1982, this volume). For want of sufficient amounts of blood samples even for a partial purification, all the characterization studies described in this paper were performed on crude haemolysates dialysed at 4°C against 1 to 2.5 diluted TEC buffer (Rattazzi et al. 1966) containing 1 mM β-mercaptoethanol and 2×10^{-5} M NADP, overnight with two changes of the buffer.

G6PD activities were measured as per the recommendations of the WHO Scientific Group (WHO TRS-366 1967) and expressed as international units (IU) per gram of haemoglobin. An IU is defined as the amount of G6PD which reduces 1 µ mole of NADP per minute at 37°C (Beutler 1971).

Several electrophoretic procedures were used to characterize the G6PD:
(1) *Cellogel:* (a) tris-EDTA-citrate (TEC) buffer pH 7.5 (Rattazzi et al. 1966) (3 h); (b) TEC buffer pH 6.5 (1 h) and (c) TEMM (0.01 M) pH 7.4 (Wijnen et al. 1977) (1 h). The electrophoresed Cellogels were stained for G6PD as described by Meera Khan (1971);
(2) *Starch gel:* Electrophoresis buffer (0.0614 M) tris-(0.004 M) EDTA-(0.0136 M) citric acid, pH 7.5. This buffer is diluted to 10 for preparing a 13% (Connaught) starch gel. Sixty µl of the sample was applied in the slots made by templates (8 × 1 mm). The samples were prepared by mixing 10 µl of haemolysate with 50 µl of Sepharose G-200 suspension in saccharose (100:40:5, v/w/w of water, saccharose and Sephadex G-200 respectively). The electrophoresis was performed for 16–17 h at 7 V/cm at 4–6°C. The gel was stained for G6PD using the 1% agar overlay procedure (Harris and Hopkinson 1977).

110

Procedures for the determination of Michelis constants, pH optimum, thermostability, and utilization of G6P and NADP analogues have been described elsewhere (Lenzerini et al. 1969; Beutler et al. 1968; WHO TRS-366 1967).

Results

Twelve of the 16 orang utans included in the present study exhibited a red cell G6PD activity ranging from 12.47 to 20.78 IU per gram of haemoglobin when assayed at 37°C (Beutler 1971). Their G6PD formed a sharp band in the starch gel and Cellogel in TEC buffer and a rather broad band in Cellogel in TEMM buffer on electrophoresis and was generally indistinguishable from the most common form of human G6PD known as G6PD B (Table 1; Figs. 1–4). We designate this form of G6PD in the orang utan as G6PD 1. Eleven of these orang utans with phenotype 1 have been identified as Borneans (Jones 1980; De Boer and Seuánez 1982, this volume), the twelfth (OU4) being a hybrid son of a Sumatran female (Pedigree 2, Fig. 5). If the information on the origin of OU4

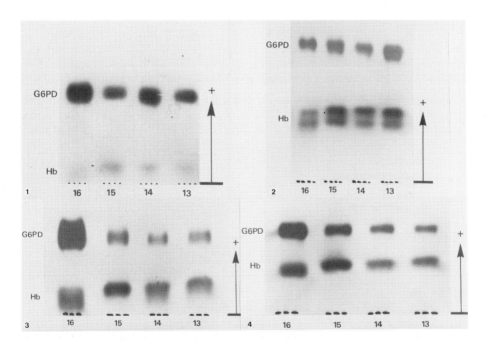

Fig. 1. Zymograms of G6PD in starch gel (TEC buffer pH 7.5). Hb = haemoglobin. The channel numbers correspond to the orang utan (OU) numbers in Table 1.

Fig. 2. Electrophoretic patterns of G6PD in Cellogel (TEC buffer pH 7.5). Channel numbers correspond to the OU numbers.

Fig. 3. G6PD patterns in Cellogel (buffer system TEMM, pH 7.4). Channels are indicated by OU numbers. Note that unlike OU16, the sample of OU14 formed a relatively weak band of activity with a migration intermediate to OU16 and 13 or 15.

Fig. 4. Electropherograms of G6PD in Cellogel (TEC buffer, pH 6.5). Samples in the channels are like in Figures 1 to 3.

Table 1. Activity distribution and electrophoretic behaviour of G6PD in a set of 16 orang utans.[a]

Orang utan No.	Name	Sex[b] and origin	Specific activity[c] IU/g Hb		Patterns of electrophoresis of G6PD[d]			
			G6PD	6PGD	Starch gel TEC, pH 7.5 (Fig. 1)	Cellogel TEC, pH 7.5 (Fig. 2)	Cellogel TEMM, pH 7.4 (Fig. 3)	Cellogel TEC, pH 6.5 (Fig. 4)
OU1	Geert	MB	16.35	7.40	100 (sh)	100 (sh)	100 (br)	100 (sh)
OU2	Joop[e]	MS	70.58	6.46	97 (sh)	99 (br)	94 (br)	100 (sh)
OU3	Piku	MB	14.96	6.78	100 (sh)	100 (sh)	100 (br)	100 (sh)
OU4	Timo[e]	MH	16.52	5.32	100 (sh)	100 (sh)	100 (br)	100 (sh)
OU5	Petra[e]	FB	16.23	6.19	100 (sh)	100 (sh)	100 (br)	100 (sh)
OU6	Connie[e]	FH	57.76	5.12	(100) 97 (sh)	(100) 99 (br)	(100) 94 (br)	100 (sh)
OU7	Fred[e]	MB	13.90	4.50	100 (sh)	100 (sh)	100 (br)	100 (sh)
OU8	Joke	FB	20.78	7.15	100 (sh)	100 (sh)	100 (br)	100 (sh)
OU9	Sarita	FB	16.79	7.17	100 (sh)	100 (sh)	100 (br)	100 (sh)
OU10	Siljo	FB	12.47	5.60	100 (sh)	100 (sh)	100 (br)	100 (sh)
OU11	Anak[e]	FB	13.42	6.55	100 (sh)	100 (sh)	100 (br)	100 (sh)
OU12	Bernadine[e]	FB	16.20	7.22	100 (sh)	100 (sh)	100 (br)	100 (sh)
OU13	Bongo[e]	MB	15.63	3.84	100 (sh)	100 (sh)	100 (br)	100 (sh)
OU14	Sjaan[e]	FB	30.67	9.31	(100) 97 (sh)	(100) 99 (br)	(100) 98 (br)	100 (sh)
OU15	Ori[e]	MB	12.81	7.42	100 (sh)	100 (sh)	100 (br)	100 (sh)
OU16	Djambi	FS	112.03	10.75	97 (sh)	99 (br)	94 (br)	100 (sh)

[a] 6-phosphogluconate dehydrogenase (6PGD), the second enzyme reducing NADP in the pentose phosphate shunt pathway, was assayed following Beutler (1971) in all the samples, as a control. The activities of G6PD and 6PGD presented in this table are average values of two consecutive measurements.

[b] M = male; F = female; B = Bornean; S = Sumatran; H = hybrid.

[c] Average ± S.D.: G6PD (N = 16); 28.57 ± 27.92
6PGD (N = 16): 6.67 ± 1.71

[d] Relative mobility with respect to that of human G6PD B+ as 100; sh = sharp, br = broad (banding patterns); (100) indicates the presence of a presumed second allelic product of G6PD in the heterozygote OU6 and possibly in OU14.

[e] Appears in the pedigrees of Fig. 5.

(Jones 1980) is unequivocal, the structural gene for G6PD expressed in OU4 should be of Sumatran origin.

Of the remaining four, the OU2 and OU16, the only two pure bred Sumatran orang utans in the present series, have shown 4.5 to 7.5 times the average activity found in the above 12. The OU6, a hybrid daughter of OU2 and OU5, also exhibited an unusually high G6PD activity (Pedigree 1, Fig. 5). The enzymes of OU2, OU16 and OU6 appear to migrate somewhat slower and form a broader band than G6PD 1 on electrophoresis in certain media (Table 1; Figs. 1–4). We call this variant as G6PD 2. The fourth orang utan with high G6PD activity, the OU14, a pure-bred Bornean female (Pedigree 4, Fig. 5; Jones 1980) with twice the average activity usually found in the Borneans, exhibited a G6PD 2-like pattern of electrophoretic migration not only in the media reported in this paper (Material and Methods; Figs. 1–4) but also in starch gels in EBT buffer pH 8.0 and phosphate buffer pH 6.5 described by Dern et al. (1969) (personal observations). However, it is slightly faster than G6PD 2 in TEMM buffer (only one experiment; Fig. 3). It is not possible at present to establish whether OU14 is an additional variant or simply a carrier of one or both of the above variants.

The results summarised in Table 2 show that G6PD 1 and G6PD 2 are largely similar in their substrate and substrate analogue utilizations and exhibit identical pH optima which are slightly higher and narrower than that of human G6PD B. Nevertheless, they show a clear-cut difference in their temperature

Table 2. Comparison of orang utan G6PDs 1 and 2 and human G6PD B in the dialysed samples of crude hemolysates.[a]

Physico-chemical characteristics	G6PD 1	G6PD 2	G6PD B[b]
Specific activity (IU/g Hb)	13.90 (OU7)	112.03 (OU16)	9.12 (1)
(Table 1)			10.17 (2)
Electrophoretic mobility (Table 1)			
Starch gel (TEC, pH 7.5) (Fig. 1)	100	97	100
Cellogel (TEC, pH 7.5) (Fig. 2)	100	99	100
(TEC, pH 6.5) (Fig. 4)	100	100	100
(TEMM, pH 7.4) (Fig. 3)	100	94	100
pH Optimum (Fig. 6)	8.5	8.5	7.0–8.0
Thermostability (Fig. 7)	normal	labile	normal
Km: G6P (μM)	83.11 (OU7)	65.08 (OU16)	77.61 (1)
	93.34 (OU5)		70.84 (2)
Km: NADP (μM)	4.31 (OU7)	14.50 (OU16)	12.14 (1)
	13.86 (OU5)		14.37 (2)
Substrate analogue utilization			
2 deoxy-G6P (% of G6P rate)	3.3 (OU7)	3.6 (OU16)	5.4 (1)
	4.6 (OU5)		5.9 (2)
Gal-6-P (% of G6P rate)	5.0 (OU7)	4.9 (OU16)	10.8 (1)
	7.2 (OU5)		8.6 (2)
deamino-NADP (% of NADP rate)	58 (OU7)	62.5 (OU16)	71.0 (1)
	81 (OU5)		70.5 (2)
NAD (% of NADP rate)	0 (OU7)	0 (OU16)	0 (1)
	0 (OU5)		0 (2)

[a] G6PD 1 and G6PD 2 are G6PDs of the orang utan; G6PD B = normal human G6PD phenotype originally designated as B+ (WHO TRS-366 1967).
[b] Two adult male Caucasian volunteers [(1) and (2)] known to be normal for G6PD.

sensitivities (Fig. 7), in addition to the specific activity and electrophoretic pattern differences described above.

Unlike G6PD, the gene coding for 6-phosphogluconate dehydrogenase (PGD) is syntenic with the linkage group of human chromosome 1 and its homologues in other primates including that of orang utan (Garver et al. 1978). The PGD activity did not show any gross intersubspecific or intraspecific variation (Table 1). The electrophoretic patterns of PGD are identical in all the samples (Meera Khan et al. 1982, this volume).

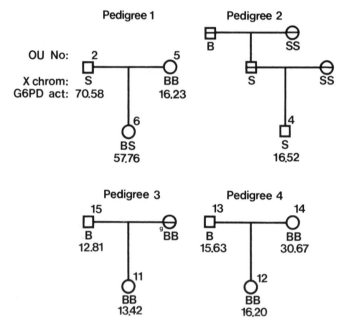

Fig. 5. G6PD being X-linked the males will be hemizygous and the females either heterozygous or homozygous for the variants. In the pedigrees depicted above the origin of the X-chromosome(s) in the respective individuals (OU2 etc.) is indicated by letters B and S for Bornean and Sumatran respectively and the specific activity of red cell G6PD (IU/g of Hb) in each of the individuals studied is presented immediately below the letters. Barred symbol of an individual indicates that it was not tested for G6PD.

Discussion

It is interesting to note that the high activity form is more thermolabile than the usual form of G6PD in the orang utan. The $4\frac{1}{2}$ to $7\frac{1}{2}$ times difference in the specific activities between the two forms raises the question whether it is due to a regulatory phenomenon independent of the structural gene or it is determined by the inherent nature of the variant of the structural gene itself. From the results of the present study, it is not possible to decide whether the increase of activity in the G6PD 2 is due to an increase in the rate of formation of the active molecules or due to an increase in the absolute specific activity of the G6PD molecules. Moreover, we do not know at present whether the increase found in

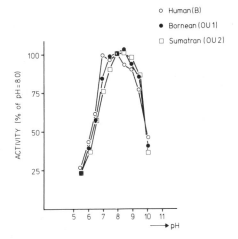

Fig. 6. Comparison of pH optima of G6PDs in the two phenotypes of orang utan and a human male with B type enzyme. The OU1 and OU2 are G6PD 1 and G6PD 2 respectively.

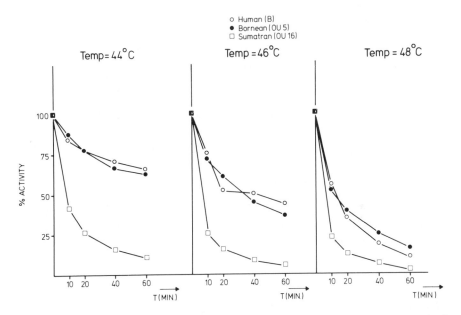

Fig. 7. Heat stability of G6PD in the two phenotypes of orang utan and a normal human male of phenotype G6PD B. The OU5 and OU16 are phenotypically G6PD 1 and G6PD 2 respectively.

the erythrocytes is present also in other cells and tissues of the G6PD 2 individuals.

In order to explain the quantitative increases of enzyme activities in higher organisms several mechanisms have been proposed by different authors in the past. Valentine et al. (1977) reported an unusual variant of human adenosine deaminase (ADA) associated with 45 to 70 times the normal activity in the erythrocytes but not in other tissues. The trait was found to segregate as an autosomal dominant condition in a large pedigree of three generations. Their studies indicated that the greatly increased ADA activity represents simply an overproduction of normal enzyme. The authors have considered, therefore, a defect in feedback regulation concerned with induction and suppression of ADA synthesis in the nucleated red cell precursor as a possibility. On the other hand, Coleman (1971) has discussed the possibility of the involvement of one and the same gene in the control of the rate of synthesis as well as the structure of δ-aminolevulinate dehydratase in mice. Yoshida (1970) demonstrating the occurrence of substitution of histidine in the human G6PD B by tyrosine as the basis for the Hektoen variant of G6PD (Dern 1966; Dern et al. 1969) associated with four times the normal activity, proposed that 'a single step base change in a structural gene resulting in an aminoacid substitution may also increase production of the variant protein'.

The differences in the specific activity, electrophoretic mobility and thermo-stability of G6PD studied in the extensively dialysed crude extracts of the erythrocyte samples from different individuals can be explained in terms of one or more aminoacid differences due to base pair substitutions in the nucleotide sequence of the structural gene. We therefore assume that the two distinct forms observed in the present series of orang utans are determined by alleles at the structural locus for G6PD and designate the alleles as G6PD*1 and G6PD*2. Thus, for example, the genotypes of the hemizygotes OU1 and OU2 become G6PD*1 and G6PD*2 respectively while that of the heterozygote OU6 G6PD*1/G6PD*2 (see International System for Human Gene Nomenclature 1979).

If we leave out the case of OU14 which is uncertain at present, the results presented and discussed above indicate that G6PD*1 and G6PD*2 are alleles peculiar to Bornean and Sumatran orang utans respectively. The red cell G6PD activities reported by Barnicot and Cohen (1970) in two Sumatran (Bukit and Ini) and one Bornean (Sya) orang utan from Yerkes Primate Center will be in agreement with this conclusion if the two high and one low activity variant reported by them are the two Sumatrans and the single Bornean respectively. If the studbook (Jones 1980) information stands correct the OU4 (Pedigree 2, Fig. 5) of our series forms an exception.

Admittedly, the above discussion is largely based upon a number of simple assumptions whose confirmation requires further biochemical and genetical studies. Nevertheless, the present study clearly shows that the G6PD system offers an additional genetic marker in the orang utan and an excellent animal model to investigate the mechanism of regulation of G6PD activity in higher organisms evolutionarily very close to man. Moreover, further studies may unravel the biological and physiological significance of the unusually high activities of G6PD found in the Sumatran subspecies of orang utans in their natural habitats.

Acknowledgements

We are grateful to the colleagues at the zoos of Arnhem, Rotterdam, Antwerp, Hannover, Rhenen and Wuppertal, who generously cooperated in obtaining the orang utan samples in an excellent condition. We thank Mrs L.M.M. Wijnen and Mr C. Vreeken for laboratory assistance, Mr B. Simons for the drawings and Mrs H.M. Emanuelson-Oostvriesland for preparing the manuscript.

References

Barnicot, N.A. and P. Cohen (1970). Red cell enzymes of primates (Anthropoidea). Biochem. Genet., 4: 41–57.

Beutler, E., C.K. Mathai and J.E. Smith (1968). Biochemical variants of glucose-6-phosphate dehydrogenase giving rise to congenital non-spherocytic hemolytic disease. Blood, 31: 131–150.

Beutler, E. (1971). Red cell metabolism. A manual of biochemical methods. 2nd Edition (pp. 66–69). Grune and Stratton, New York.

Beutler, E. (1978). Glucose-6-phosphate dehydrogenase deficiency. In: J.B. Stanbury et al. (eds.), The metabolic basis of inherited disease. 4th Edition (pp. 1430–1451). McGraw-Hill, New York.

Beutler, E. and C. West (1978). Glucose-6-phosphate dehydrogenase variants in the chimpanzee. Biochem. Med., 20: 364–370.

Boer, L.E.M. de and P. Meera Khan (1982). Haemoglobin polymorphisms in Bornean and Sumatran orang utans. In: L.E.M. de Boer (ed.), The orang utan. Its biology and conservation. Junk, The Hague.

Boer, L.E.M. de and H.N. Seuánez (1982). The chromosomes of the orang utan and their relevance to the conservation of the species. In: L.E.M. de Boer (ed.), The orang utan. Its biology and conservation. Junk, The Hague.

Coleman, D.L. (1971). Linkage of genes controlling the rate of synthesis and structure of aminolevulinate dehydratase. Science, 173: 1245–1246.

Dern, R.J. (1966). A new hereditary quantitative variant of glucose-6-phosphate dehydrogenase characterized by a marked increase in enzyme activity. J. Lab. Clin. Med., 68: 560–565.

Dern, R.J., P.R. McCurdy and A. Yoshida (1969). A new structural variant of glucose-6-phosphate dehydrogenase with a high production rate (G6PD Hektoen). J. Lab. Clin. Med., 73: 283–290.

Garver, J.J., P.L. Pearson, A. Estop, T.M. Dijksman, L.M.M. Wijnen, A. Westerveld and P. Meera Khan (1978). Gene assignments to the presumptive homologs of human chromosomes 1,6,11,12 and X in the Pongidae and Cercopithecoidea. Cytogenet. Cell Genet. 22: 564–569.

Harris, H. and D.A. Hopkinson (1976). Handbook of enzyme electrophoresis in human genetics. Chapter 4. The enzymes, 1.1.1.49, glucose-6-phosphate dehydrogenase. North Holland Publishing Company, Amsterdam.

Huser, H.-J. (1970). Atlas of comparative primate hematology (pp. 60–65). Academic Press, New York and London.

International System for Human Gene Nomenclature (1979). Shows et al. Cytogenet. Cell Genet., 25: 96–116.

Jones, M.L. (1980). Studbook of the orang utan. Zoological Society of San Diego, San Diego.

Kömpf, J., H. Ritter and J. Schmitt (1971). Zur transspezifischen Variabilität der Glucose-6-phosphatdehydrogenase (E.C. 1.1.1.49) der Primaten. Humangenetik, 11: 342–344.

Lenzerini, L., P. Meera Khan, G. Filippi, M.C. Rattazzi, A.K. Ray and M. Siniscalco (1969). Characterization of glucose-6-phosphate dehydrogenase variants. I. Occurrence of G6PD 'Seattle-like' in Sardinia and its interaction with the Mediterranean variant. Amer. J. Hum. Genet., 21: 142–153.

Lucotte, G. (1980). Polymorphisme électrophorétique des proteines et enzymes sériques et érythrocytaires chez le Chimpanzée. Hum. Genet., 54: 97–102.

McKusick, V.A. (1978). Mendelian inheritance in man. 5th Edition. The Johns Hopkins University Press, Baltimore and London. (pp. 728–758: 30590. glucose-6-phosphate dehydrogenase).

Meera Khan, P. (1971). Enzyme electrophoresis on cellulose acetate gel. Arch. Biochem. Biophys., 145: 470–483.

Meera Khan, P., A. Westerveld, K.-H. Grzeschik, B.F. Deys, O.M. Garson and M. Siniscalco (1971). X-linkage of human phosphoglycerate kinase confirmed in man-mouse and man-chinese hamster somatic cell hybrids. Amer. J. Hum. Genet., 23: 614–623.

Meera Khan, P. and H. Balner (1972). Polymorphic enzymes in Rhesus monkeys and chimpanzees. Med. Primatol., part I (pp. 363–371), Karger, Basel.

Meera Khan, P., H. Rijken, J.Th. Wijnen, Lucie M.M. Wijnen and L.E.M. de Boer (1982). Red cell enzyme variation in the orang utan; electrophoretic characterization of 45 enzyme systems in Cellogel. In: L.E.M. de Boer (ed.), The orang utan. Its biology and conservation. Junk, The Hague.

Motulsky, A.G. and J.M. Campbell-Kraut (1961). Population genetics of glucose-6-phosphate dehydrogenase deficiency of the red cell. In: B.S. Blumberg (ed.), Proceedings of the conference on genetic polymorphism and geographical variation in disease (pp. 159–191). Grune and Stratton, New York.

Mourant, A.E., A.C. Copec, K. Domaniewska-Sobczak (1976). The distribution of the human blood groups and other polymorphisms. Oxford Monographs on Medical Genetics (pp. 725–752; see also pp. 36–38), Oxford University Press, London.

Rattazzi, M.C., L.F. Bernini, G. Fiorelli and P.M. Mannucci (1967). Electrophoresis of glucose-6-phosphate dehydrogenase: a new technique. Nature (Lond.), 213: 79–80.

Valentine, W.N., D.E. Paglia, A.P. Tartaglia and F. Gilsanz (1977). Hereditary hemolytic anemia with increased red cell adenosine deaminase (45- to 70-fold) and decreased adenosine triphosphate. Science, 195: 783–785.

WHO TRS-366 (1967). Standardization of procedures for the study of glucose-6-phosphate dehydrogenase. World Health Organization Technical Report Series No. 366, Geneva.

Wijnen, L.M.M., K.-H. Grzeschik, P.L. Pearson and P. Meera Khan (1977). The human PGM-2 and its chromosomal localization in man-mouse hybrids. Hum. Genet., 37: 271–278.

Yoshida, A. (1968). Subunit structure of human glucose-6-phosphate dehydrogenase and its genetic implications. Biochem. Genet., 2: 237–243.

Yoshida, A. (1970). Aminoacid substitution (histidine to tyrosine) in a glucose-6-phosphate dehydrogenase variant (G6PD Hektoen) associated with over-production. J. Mol. Biol., 52: 483–490.

Authors' addresses:
J.Th. Wijnen, H. Rijken and P. Meera Khan
Dept. of Human Genetics, Sylvius Laboratories
State of University of Leiden
Wassenaarseweg 72
2333 AL Leiden
The Netherlands

L.E.M. de Boer
Biological Research Dept.
Royal Rotterdam Zoological and Botanical Gardens
Rotterdam
The Netherlands

118

6. Orang utan haemoglobins: a short review

Leobert E.M. de Boer and P. Meera Khan

Haemoglobins are among the best known and most extensively investigated proteins of the vertebrates. They are easily obtained from peripheral blood erythrocytes and their variation can be analysed by relatively simple physico-chemical procedures. Both common and rare variants of haemoglobins, some of which lead to disease, have been described in man and other species. The study of primary structures of haemoglobin chains has shed much light on the evolutionary mechanisms operating at the molecular level. Several studies were performed on primate haemoglobins during the past decades. For reviews see Sullivan (1971), Nute (1974) and several contributions in Goodman and Tashian (1976).

The major haemoglobin found in the adult primates including man is a tetramer consisting of two α- and two β-globin (polypeptide) chains. (In man these chains consist of 141 and 146 amino acid residues, respectively.) The human α- and β-chains are known to be coded for by genes at two separate autosomal loci; the β-locus is probably evolved from a duplicate of the (original) α-locus (Dixon 1966). There are indications that the α-chain of human as well as that of some other primate species is produced by two closely linked genes originated by complete duplication during evolution (Nute 1974). There is no evidence that the human or primate β-chains are determined by genes at more than one locus.

Apart from the major haemoglobin in the adults of man and other higher primates a minor haemoglobin component is found which has been designated as HbA_2. This protein is composed of two α-globin chains and two δ-globin chains. The δ-chain differs only by 10 amino acid residues from the β-chain, and the δ-locus is believed to have originated from the β-locus by gene duplication. The amount of HbA_2 in the normal adult is approximately 1/40 of the total haemoglobin.

The human foetal red cells contain a different major haemoglobin, called HbF, consisting of two α- and two γ-globin chains. In very early human embryos two other haemoglobins, designated HbGower1 (consisting of two γ- and two ε-globin chains) and HbGower2 (consisting of two α- and two ε-globin chains), were found. (The HbF of the orang utan was studied by Schroeder et al. 1978.) The present paper, however, deals only with the adult forms of haemoglobin.

In man only a single major haemoglobin (HbA) occurs. Apart from a few local exceptions, such as the sickle cell polymorphisms (due to a β-chain variant) in certain parts of Africa and Asia, generally the haemoglobins do not occur as polymorphisms in man. However, extensive population studies and clinical investigations revealed the occurrence of many rare variants (with gene frequencies below 0.01), resulting usually from amino acid substitutions in α- or β-globin chains. A few of these substitutions may affect the functional properties of the resultant haemoglobin molecules, or the stability or the rate of their formation, and lead to pathologic conditions. Individuals heterozygous for one variant chain (v) show two instead of one major haemoglobin components: $\alpha_2\beta_2^v$ plus $\alpha_2\beta_2$ when heterozygous for a β-chain variant, or $\alpha_2^v\beta_2$ plus $\alpha_2\beta_2$ when heterozygous for an α-chain variant. Double heterozygotes would form four major components, normal $\alpha_2\beta_2$, two abnormals with one variant chain each, $\alpha_2^v\beta_2$ and $\alpha_2\beta_2^v$, and one with two variant chains, $\alpha_2^v\beta_2^v$. The α-chain variants affect also the minor components (e.g. $\alpha_2\delta_2$ and $\alpha_2^v\delta_2$), while the β-chain variants do not.

In contrast to man, some of the non-human primates exhibit extensive haemoglobin variation. Such variation has been extensively studied in some Old World primates of the genera *Macaca* and *Papio* (Hewett-Emmet 1976; Sullivan et al. 1976). The anthropoids on the other hand generally conform to the human situation; the chimpanzee (*Pan troglodytes*) the gorilla (*Gorilla gorilla*) and the gibbons (*Hylobates lar*, *H. agilis* and *H. concolor*) do not exhibit such haemoglobin polymorphisms. They all form a single major component (with the exception of *H. concolor*, which consistently shows two major components, probably due to the occurrence of duplication of one of the Hb genes and its fixation). Variant haemoglobins were found very rarely (Sullivan 1971). Among the anthropoids the orang utan (*Pongo pygmaeus*), however, forms an exception.

The first study of haemoglobins in the orang utan was reported by Buettner-Janusch and Buettner-Janusch (1963). They observed two major components in each of the two individuals (one Bornean and one Sumatran) investigated by them. The existence of haemoglobin polymorphism in the species, however, was first noted by Barnicot and Jolly (1966). They studied 10 specimens, four of which showed a single electrophoretic band of haemoglobin (designated HbA)*, two showed a band migrating slower than HbA (designated HbB), three others formed two major haemoglobins (HbA and HbB), and the last one three. The slowest band of the last animal corresponded to HbB, the middle zone, twice as intense as the other two, had the same electrophoretic mobility as HbA, and the fastest one was designated HbC. Hybridisation experiments with canine and human haemoglobins indicated that the HbB represented a haemoglobin carrying an α-chain variant ($\alpha_2^B\beta_2^A$) and that HbC is a β-chain variant ($\alpha_2^A\beta_2^C$). Because of its frequent occurrence, Barnicot and Jolly regarded the HbA ($\alpha_2^A\beta_2^A$) as the normal haemoglobin of the orang utan. The presence of α^B chains results in a variant with slower electrophoretic mobility and that of β^C chains in a faster variant. Thus, a haemoglobin molecule

---

*The nomenclature for orang utan haemoglobins used in this paper is that of Barnicot and Jolly (1966). This nomenclature was followed also by Sullivan and Nute (1968), Buettner-Janusch et al. (1969) and Nute (1974). See Fig. 1.

composed of two α^B and two β^C chains would have the same electrophoretic mobility as the HbA. Since the band in the region of HbA in the specimen with three bands was twice as intense as the other two (HbB and HbC) it was interpreted that this specimen had four major haemoglobins in equal proportions, HbA, HbB, HbC and HbD (the last being $\alpha_2^B\beta_2^C$, electrophoretically inseparable from the first with the structure $\alpha_2^A\beta_2^A$), and thus that it was a double heterozygote possessing two alleles at the α-chain locus and two at the β-chain locus (i.e. α^A, α^B, β^A and β^C).

All animals studied by Barnicot and Jolly (1966) showed a single minor component (probably $\alpha_2^A\delta_2$) in the HbA individuals and an electrophoretically slower component (probably due to $\alpha_2^B\delta_2$) in the HbB carriers. In the heterozygous animals (HbAB and HbABCD) both these minor components are expected to be present, but the $\alpha_2^A\delta_2$ haemoglobin band would be masked by HbB which has the same mobility. (The existence of a globulin chain in the orang utan analogous to the human δ was later confirmed by Buettner-Janusch et al. (1969) using urea gel electrophoresis.)

The results of Barnicot and Jolly (1966) were confirmed by Sullivan and Nute (1968) who studied 31 orang utans (most of which were from the Yerkes Regional Primate Research Center and which included all the specimens previously reported by Barnicot and Jolly) and encountered the same four phenotypes, HbA, HbB, HbAB and HbABCD. Dissociation experiments, however, revealed the existence of a fifth phenotype, HbBD. This phenotype, which is indistinguishable from the HbAB in its native form, on electrophoresis forms two major components (HbB = $\alpha_2^B\beta_2^A$ and HbC = $\alpha_2^A\beta_2^C$) on the gel. The dissociation experiments also confirmed the earlier findings that HbB contains a more positively charged α-chain than that of HbA, and that HbC contains a more negatively charged β-chain (see Fig. 1). Thus, two alleles occur at each of the two loci, the α and the β. Theoretically this should result in four possible homozygous phenotypes (HbA, HbB, HbC and HbD), four heterozygous phenotypes (HbAC, HbAB, HbCD and HbBD) and one double heterozygous phenotype (HbABCD). Neither the homozygotes HbC and HbD, nor the heterozygotes HbAC and HbCD were encountered in the material of Sullivan and Nute. Since the α^B and β^C chains exhibited the same electrophoretic mobilities as the α- and β-chains of human HbA, they supposed that the α^B and β^C were not the variant chains in the orang utan, but rather the α^A and β^A chains. The α^B and β^C (and thus HbD) would be the original forms in the species. They regarded the situation in the orang utan as a case of gradual gene replacement, in which genes α^B and β^C would be in the process of being replaced by α^A and β^A (a first mutation would have led to the α^A allele and selection favouring the animals heterozygous and homozygous for this allele; a second mutation would then have given rise to the β^A allele and selection in favour of animals homozygous for both mutant genes). However, they did not exclude the possible existence of a balanced polymorphism, favouring heterozygous animals (HbAB). The non-existence of specimens with phenotypes HbAC, HbCD and HbC was explained by possible selection against animals with the HbC ($\alpha_2^A\beta_2^B$) molecule. This haemoglobin was found to occur only in the double heterozygotes (HbABCD), in which its effect would be diluted by the presence

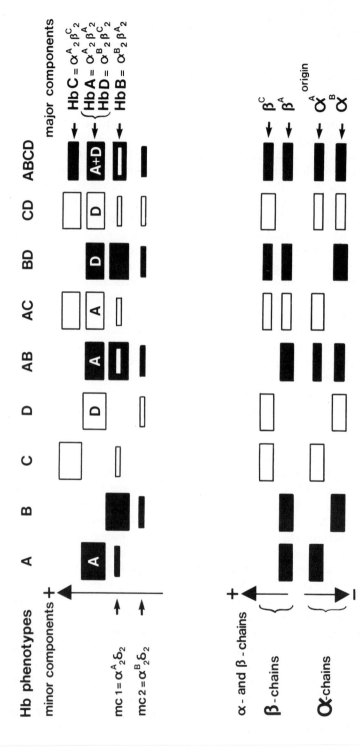

Fig. 1. Schematic representation of electrophoretic patterns of haemoglobins of orang utans. Above: the electrophoretic patterns of the major and minor components of native haemoglobins. Below: the electrophoretic patterns of the subunits resulting from dissociation of the major components. Theoretical patterns are shown in white, observed ones being depicted in black. Data according to Sullivan and Nute (1968), but adding the information on minor components (the expected minor components masked by major components due to identical mobility are shown as white bars) and indicating the approximate relative band intensities. The nomenclature used is that of Barnicot and Jolly (1966), which was also followed by Sullivan and Nute (1968), Buettner-Janusch et al. (1969) and Nute (1974).

122

of three other major components. The orang utan's original haemoglobin (HbD) seems to occur only in the heterozygotes (HbBD) and double heterozygotes (HbABCD) (Sullivan and Nute 1968). Studies on the functional properties of the various orang utan haemoglobins by Sullivan and Nute did not reveal significant differences, on the basis of which selective forces could act in favour of certain chains or haemoglobin molecules.

Based on the presence of equal amounts of HbA, HbB, HbC and HbD in double heterozygotes and on the occurrence of several animals with only one type of α-chain, Nute (1974) concluded that multiple α-chain loci are not characteristic in the orang utan.

While Sullivan and Nute (1968) did not discuss or mention the geographic origin of the animals they studied, Bruce and Ayala (1979) treated Bornean and Sumatran orang utans separately, but they did not make a distinction between α- and β-chains and treated all the variants as products of genes at a single locus. In both Bornean (three individuals studied) and Sumatran orang utans (eight individuals studied) they found two types of haemoglobin, an electrophoretically slow one (referred to as due to allele 98) and a faster one (referred to as due to allele 100). In spite of the small sample size Bruce and Ayala calculated the allelic frequencies and found 29% for allele 100 and 71% for allele 98 in Bornean orang utans, and 67% for allele 100 and 33% for allele 98 in Sumatran orang utans. Most probably the haemoglobin of 'allele 100' is HbA, that of 'allele 98' — HbB. In other words, the frequency of allele 100 would represent that of the α^A allele, that of allele 98 the frequency of the α^B allele. Thus, there would be significant differences in the frequencies of the two alleles between Borneo and Sumatra. Bruce and Ayala's sample, however, is much too small to draw any reasonable conclusion in this respect.

Two studies involved the primary structures of orang utan haemoglobin chains. Firstly, Buettner-Janusch et al. (1969) determined the amino acid composition and terminal amino end groups of the α^A and β^C chains (from an animal with phenotype HbABCD) and the α^B and β^A chains (from an individual with phenotype HbB; both animals were included in the study of Sullivan and Nute 1968). They concluded that an asparagine residue in the α^A chain was substituted by a glycine residue to form the α^B chain with a net gain of one unit of positive charge. The β^C chain would differ from the β^A chain in having one more glutamyl and one less alanyl residue, with a net gain of one negative charge in the former. These data conform to the electrophoretic mobilities of the four possible haemoglobin molecules (HbA and HbB having the same mobility; HbB moving slower and HbC faster in the gels).

Secondly, Maita et al. (1978) studied the amino acid sequences of the haemoglobins of an orang utan with two major haemoglobins, containing two types of α-chains (designated α-I and α-II) and a single type of β-chain. They did not provide particulars of the animal studied, nor of the electrophoretic behaviour of the haemoglobin components. The α-II chain was found to differ from the α-I chain by the replacement of one glycyl residue of the former by an aspartyl residue at position 57. The α-II chain differs from human α-chain by two substitutions (at positions 12 and 23), while α-I differs in addition by the aspartyl residue at position 57 (glycyl in the human α as in α-II). A comparison

123

of these data with those of Buettner-Janusch et al. (1969) indicates that α-I and α-II probably represent α^A and α^B, respectively. In that case the earlier suggestion of Sullivan and Nute (1968), that α^B ($= \alpha$-II) is the original α-chain in the orang utan, would be confirmed since the glycine residue at position 57 is found in all primate α-chains sequenced so far (Matsuda 1976). The α^A chain would then be a mutant chain in the orang utan.

The β-chain sequenced by Maita et al. (1978) differs in only two amino acid residues from human β-chain (at positions 87 and 125). It is not clear which type of β-chain was studied (it seems more likely that the β^A chain was concerned since HbAB orang utans are rather common, while if the β^C chain was concerned the haemolysate used would have originated from an HbCD animal, a phenotype which so far has never been found). It would be of interest to collect information on the sequences of both the orang utan β-chains to see which of the two is original to the species and which originated from mutation.

References

Barnicot, N.A. and C.J. Jolly (1966). Haemoglobin polymorphism in the orang utan and an animal with four major haemoglobins. Nature (London), 210: 640–642.

Bruce, E.J. and F.J. Ayala (1979). Phylogenetic relationships between man and the apes: electrophoretic evidence. Evolution, 33: 1040–1056.

Buettner-Janusch, J. and V. Buettner-Janusch (1963). Hemoglobins of primates. In: J. Buettner-Janusch (ed.), Evolutionary and genetic biology of primates, vol. II. Academic Press, New York.

Buettner-Janusch, J., V. Buettner-Janusch and G.A. Mason (1969). Amino acid composition and amino-terminal end groups of α and β chains from polymorphic hemoglobins of *Pongo pygmaeus*. Arch. Biochem. Biophys., 133: 164–170.

Dixon, G.H. (1966). Mechanisms of protein evolution. In: P.N. Campbell and G.D. Greville (eds.), Essays in biochemistry, Vol. 2. Academic Press, New York.

Goodman, M. and R.E. Tashian (eds.) (1976). Molecular anthropology. Plenum Press, New York.

Hewett-Emmet, D., C.N. Cook and N.A. Barnicot (1976). Old World monkey hemoglobins: deciphering phylogeny from complex patterns of molecular evolution. In: M. Goodman and R.E. Tashian (eds.), Molecular anthropology. Plenum Press, New York.

Maita, T., A. Araya, M. Goodman and G. Matsuda (1978). The amino acid sequences of the two main components of adult hemoglobin from orangutan (*Pongo pygmaeus*). Hoppe-Seyler's Z. Physiol. Chem., 359: 129–132.

Nute, P.E. (1974). Multiple hemoglobin α-chain loci in monkeys, apes and man. Ann. New York Acad. Sci., 241: 39–60.

Schroeder, W.A. (1978). The γ-chain of fetal haemoglobin of the orangutan. Biochem. Genet., 16.

Sullivan, B. (1971). Comparison of the hemoglobins in non-human primates and their importance in the study of human hemoglobins. In: A.B. Chiarelli (eds.), Comparative genetics of monkeys, apes and man. Academic Press, London.

Sullivan, B. and P.E. Nute (1968). Structural and functional properties of polymorphic hemoglobins from orang utans. Genetics, 58: 114–124.

Authors' addresses:
L.E.M. de Boer
Biological Research Department
Royal Rotterdam Zoological and Botanical Gardens
Rotterdam
The Netherlands

P. Meera Khan
Dept. of Human Genetics, Sylvius Laboratories
State University of Leiden
Wassenaarseweg 72
2333 AL Leiden
The Netherlands

7. Haemoglobin polymorphisms in Bornean and Sumatran orang utans

Leobert E.M. de Boer and P. Meera Khan

Introduction

Although it has been known for some time that the orang utan is the only anthropoid which exhibits a considerable degree of haemoglobin polymorphism (for a review see De Boer and Meera Khan 1982, this volume), so far no serious attempts have been made to assess the possible differences in haemoglobin patterns between the Bornean and Sumatran populations of the species. In fact, such differences are rather likely to exist, since here we are dealing with two subspecies that have been completely isolated geographically for a large number of generations. Admittedly, Bruce and Ayala (1979) who assessed genetic distances between a number of higher primate species using electrophoretic data, including haemoglobins, distinguished between Bornean and Sumatran orang utans, but their sample (three Borneans and eight Sumatrans) was too small and too poorly documented to allow sound conclusions. The present communication presents new data on the haemoglobins of another sample of 16 orang utans of known geographic origin and combines these with the data on 29 animals previously reported by Sullivan and Nute (1968). The subspecific identity of the latter animals could be traced retrospectively with the aid of the Studbook of the Orang Utan (Jones 1980).

Material and methods

The haemoglobins of 16 orang utans (OU1–OU16) were studied. Names and studbook numbers of the animals are given in Table 1 of Meera Khan et al. (1982, this volume), who also list the other protein markers studied in the same individuals. All animals were karyotyped by De Boer and Seuánez (1982; this volume), who also give the pedigrees of the animals. Based on the chromosome analyses the sample consists of 12 Bornean orang utans (OU1, 3, 5, 7–15), two Sumatran orang utans (OU2, 16), one full hybrid (OU6) and one second generation hybrid (OU4, $\frac{3}{4}$ Sumatran, $\frac{1}{4}$ Bornean). The sample includes three sets of related individuals: OU2 and OU5 are the father and mother of OU6, OU13 and OU14 are the father and mother of OU12, OU15 is the father of OU11.

The orang utan. Its biology and conservation, edited by L.E.M. de Boer
© *1982, Dr W. Junk Publishers, The Hague, ISBN 90 6193 702 7*

Peripheral blood was collected in lithium-heparin tubes (approx. 200 units/ml of whole blood) from the antecubital vein. Anaesthesia was not required in animals of three years or less (OU4, 6, 11, 12 and 16), while the others were given a dose of 5 mg of hetamin-HCl per kg body weight prior to the drawing of blood. All blood samples were transported (room temperature) to the laboratory within 8 h of collection, except that of OU4 from Hannover which took about 20 h (cooled) to reach the laboratory.

The erythrocytes were washed twice in normal saline and the buffy coat was removed. An equal quantity of distilled water was added to the pellet of packed red cells, mixed on a Vortex for 5 min and left at 4°C for about 15 min. About 1 volume of carbon tetrachloride to 4 volumes of the mixture was added and shaken for 1 more minute and centrifugated in a Sorvall centrifuge at 4°C at 15,000 rpm for 30 min. The clear supernatant haemolysate was collected in small plastic (Beckmann) tubes and stored under liquid nitrogen until used.

Electrophoresis was performed in Cellogel, following the general procedure of Cellogel electrophoresis described elsewhere (Meera Khan 1971). The electrophoresis buffer (pH 8.0) consisted of 0.03 M barbituric acid, 0.03 M Tris, 2 mM β-mercaptoethanol and 4 mM $MgCl_2$. About 2 µl of haemolysate (concentration approx. 10 g haemoglobin/100 ml) was applied in a narrow streak about 8 mm long, near the cathodic shoulder, and electrophoresed for 2 h 30 min to get a satisfactory separation of the various haemoglobins. All the forms of haemoglobins identified during the present study migrated anodal to the origin, the variant of A_2 (mc2 $= \alpha_2^A \delta_2$) being the most cathodal (see Fig. 1).

Results and discussion

Barnicot and Jolly (1966) and Sullivan and Nute (1968) reported the existence of four types of major haemoglobins in orang utans: HbA (subunit structure $\alpha_2^A \beta_2^A$), HbB ($= \alpha_2^B \beta_2^A$), HbC ($= \alpha_2^A \beta_2^C$) and HbD ($= \alpha_2^B \beta_2^C$). The presence of α^B-chains results in an electrophoretically slower haemoglobin molecule (HbB) as compared to HbA containing the α^A-chains; the presence of β^C-chains results in an electrophoretically faster haemoglobin molecule (HbC) as compared to that of β^A-chains. The $\alpha_2^A \beta_2^A$ molecule (HbA) has the same electrophoretic mobility as the $\alpha_2^B \beta_2^C$ molecule (HbD). Thus, HbD and HbA can be distinguished only when dissociated haemoglobin chains are studied. Using this procedure, Sullivan and Nute (1968) found five haemoglobin phenotypes among a sample of 31 orang utans: HbA (animals homozygous α^A/α^A and β^A/β^A), HbB (homogygous α^B/α^B and β^A/β^A), HbAB (heterozygous α^A/α^B, homogygous β^A/β^A). HbBD (homozygous α^B/α^B, heterozygous β^A/β^C) and HbABCD (heterozygous α^A/α^B, heterozygous β^A/β^C). Four phenotypes, theoretically possible, were not represented in the sample: HbC (homozygous α^A/α^A and β^C/β^C), HbD (homozygous α^B/α^B and β^C/β^C), HbCD (heterozygous α^A/α^B, homozygous β^C/β^C) and HbAC (homozygous α^A/α^A, heterozygous β^A/β^C). Because of the occurrence of two allelic forms of α-chains in the orang utan, two different minor components of haemoglobin are expected: the $\alpha_2^A \delta_2$ (designated mcl in the present paper), which has the same electrophoretic mobility as HbB and which is present in all animals with at least one α^A-gene; and the $\alpha_2^B \delta_2$ (designated mc2), which is the most cathodal of all haemoglobin components and which is present in all animals with at least one α^B-gene. Heterozygous animals (α^A/α^B) would show both minor components, but in the HbAB and HbABCD phenotypes the mc1 is masked by the HbB band (see Fig. 1).

In the present study only native, undissociated haemoglobins were electro-phoresed and thus, in view of the reasons discussed above, the interpretation of the results presented in this paper has to be regarded as purely tentative. Five different patterns were found among the haemoglobins of the 16 orang utans studied. Their individual patterns are shown in Fig. 1, and represented schematically in Fig. 2. Three of these are identical to the patterns already described by Barnicot and Jolly (1966) and Sullivan and Nute (1968) and they represent the HbA, HbAB and HbABCD phenotypes, respectively.

The HbA pattern was found in eight orang utans (OU3, 5, 7–11 and 15). It forms one major haemoglobin band, the HbA ($=\alpha_2^A \beta_2^A$) and one minor component band (mc1$=\alpha_2^A \delta_2$).

The HbAB pattern which was found in two animals (OU5 and 12) shows two major bands of equal intensity representing HbA ($=\alpha_2^A \beta_2^A$) and the more cathodal HbB ($=\alpha_2^B \beta_2^A$). A faint minor component band (mc2$=\alpha_2^A \delta_2$) is seen most cathodally, while the mc1 ($=\alpha_2^A \delta_2$) is masked by HbB. As discussed

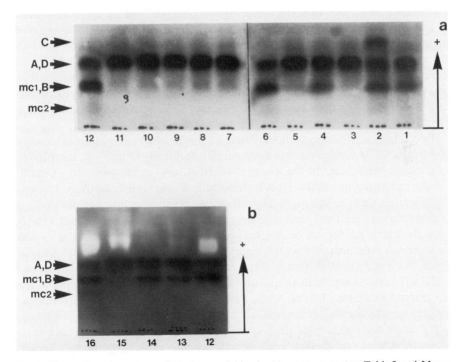

Fig. 1. Electrophoretic patterns of the haemoglobins in 16 orang utans (see Table 2 and Meera Khan et al. 1982, this volume). Electrophoresis in series **a** (OU1–OU12) was performed according to the techniques described in this paper; electrophoresis in series **b** (OU12–OU16) was performed according to the techniques described for GSR in Meera Khan et al. (1982, this volume). White bands in **b** represent GSR, dark bands haemoglobins.

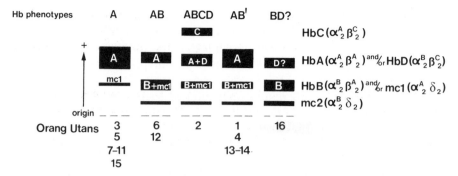

Fig. 2. Schematic representation of the haemoglobin patterns of Fig. 1, with interpretations on the possible subunit compositions of the haemoglobins (see also Table 2).

earlier (De Boer and Meera Khan 1982, this volume), the HbAB phenotype cannot be distinguished from the HbBD phenotype. Nevertheless, the pedigree data indicate that in our case we are dealing with the HbAB phenotypes. OU6 is the offspring of a mother (OU5) with an unambiguous HbA phenotype and a father (OU2) with the HbABCD phenotype, a cross which could not result in an HbBD offspring. Both parents of OU12 (OU13 and OU14) possess a phenotype which probably represents a variant of the HbAB phenotype (as will be explained below), and thus the offspring's phenotype is likely to be HbAB.

The HbABCD pattern was found in a single specimen (OU2). It shows three major haemoglobin bands, from cathode to anode: HbB ($=\alpha_2^B\beta_2^A$), HbA ($=\alpha_2^A\beta_2^A$) comigrated with HbD ($=\alpha_2^B\beta_2^C$), and HbC ($=\alpha_2^A\beta_2^C$). The central band is twice as intense as the other two, which agrees with the supposition that this phenotype possesses HbA, B, C and D in equal proportions (Barnicot and Jolly 1966). A faint minor component band (mc2 $=\alpha_2^B\delta_2$) is seen cathodal to the HbB band, while the latter masks the mc1 ($=\alpha_2^A\delta_2$) band.

Two additional patterns encountered in the present study are not readily identifiable. The first, present in four animals (OU1, 4, 13 and 14), consists of two major bands, one in the position of HbA, one in the position of HbB. Unlike the situation in the HbAB pattern, however, the band A is approximately three times as intense as band B. When the illustrations of Sullivan and Nute (1968) are watched carefully, the relative band intensities of their HbAB animals which proved to be heterozygous α^A/α^B and homozygous β^A/β^A in dissociation experiments, vary from almost equal amounts of HbA and HbB to much more intense HbA bands. Therefore, we tentatively assume that in our OU1, 4, 13 and 14 we are dealing with a variant of the HbAB phenotype, which we designate HbAB'. In that case, there are two possible alternative explanations for the marked difference in the amounts of HbA and HbB produced:

(1) there are two loci for the α-chain, and the animals concerned carry two α^A-alleles, one α^B-allele and one α-thallasemia allele (genotype $\alpha^A\alpha^A/\alpha^B\alpha^{th}$ — β^A/β^A or $\alpha^A\alpha^B/\alpha^A\alpha^{th}$ — β^A/β^A). The inactiveness of the α-thallasemia allele would result in the production of roughly 2/3 $\alpha_2^A\beta_2^A$ (HbA) and 1/3 $\alpha_2^B\beta_2^A$ (HbB);

(2) there are two loci for the α-chain, and the animals concerned are

128

homozygous α^A/α^A at one of them and heterozygous α^A/α^B at the other (genotype $\alpha^A\alpha^A/\alpha^A\alpha^B — \beta^A/\beta^A$). Consequently, they would carry three α^A-alleles and one α^B-allele, resulting in the production of 3/4 $\alpha_2^A\beta_2^B$ (HbA) and 1/4 $\alpha_2^B\beta_2^A$ (HbB).

It is notable that OU13 and OU14, both with the HbAB′ phenotype, are the parents of OU12, which exhibits the normal HbAB pattern with equal amounts of HbA and HbB. This is in agreement with both the above explanations when the offspring inherited the $\alpha^A\alpha^B$-allelic combination from both its parents. Its genotype would then be $\alpha^A\alpha^B/\alpha^A\alpha^B — \beta^A/\beta^A$, resulting in 50/50 production of both types of α-chains. At present it is not possible to verify these the explanations. Nevertheless, we believe that the HbAB′ phenotype and its mode of inheritance will provide evidence for the existence of multiple α-loci in the orang utan.

The last electrophoretic haemoglobin pattern encountered in our sample of orang utans occurred in a single animal (OU16). This pattern resembles the HbAB pattern, but differs from it in that the faster major haemoglobin band is slightly more cathodal than HbA (the other band of major haemoglobin is exactly in the position of HbB), and in that it shows a relatively intense minor haemoglobin band in the position of mc2 ($\alpha_2^B\delta_2$). It is possible that our electrophoretic system, unlike that of Sullivan and Nute (1968), distinguishes HbA from HbD (the latter being slightly slower in our system), and that in the case of OU16 we are dealing with the HbBD phenotype. Theoretically the HbBD phenotype would have twice as much mc2 relative to the HbAB and HbABCD phenotypes, since the first is homozygous for the (slow) α^B-chain, while the latter two are heterozygous α^A/α^B. This may be the cause of the relatively intense mc2 band in OU16. In the absence of dissociation experiments, however, the above explanation has to be regarded as highly hypothetical. It needs confirmation by detailed structural studies.

Comparison between the haemoglobins of Bornean and Sumatran orang utans

In their study of the haemoglobins of 31 orang utans Sullivan and Nute (1968) recorded the 'house names' of the animals studied along with their haemoglobin phenotypes. With the aid of the Studbook of the Orang Utan (Jones 1980) it was possible to recover the subspecific status of 29 of these orang utans. In addition, many of these specimens in the meantime have been karyotyped by Seuánez et al. (1979) in order to confirm their geographic origin. The results of the karyotype analyses are also presented by De Boer and Seuánez (1982, this volume), together with the pedigrees of the individuals studied. All appropriate data on Sullivan and Nute's sample are listed in Table 1.

When the haemoglobin phenotypes of Bornean and Sumatran orang utans (11 and 18 animals, respectively in Sullivan and Nute's sample) are considered separately, the following tentative conclusions may be drawn:

1. The α^A-frequency is high in Borneo (0.68) and low in Sumatra (0.31), and consequently the α^B-frequency is low in Borneo (0.32) and high in Sumatra (0.69).

Table 1. Haemoglobin phenotypes of 29 orang utans studied by Sullivan and Nute (1968) whose subspecific status could be traced using the Studbook of the Orang Utan (Jones 1980). The numbers given in De Boer and Seuánez (1982, this volume) to those animals that have been karyotyped, or whose relatives have been karyotyped, are shown in a separate column. Their pedigrees are shown in Fig. 7 of De Boer and Seuánez. Names and numbers of animals whose subspecific status was confirmed by chromosome studies (of the animals themselves or of their relatives), are printed in italics. The origin of two of the orang utans studied by Sullivan and Nute could not be identified, either by means of the studbook, or by chromosome data. These are 'Kitchee' (studbook number 00536– F, Hb phenotype A) and 'Delta orang' (studbook number unknown, Hb phenotype AB).

Phenotype	Probable genotype	Name	Studbook number	Number in De Boer and Seuánez	Totals
Sumatran orang utans (18 animals)					
A	$\alpha^A/\alpha^A — \beta^A/\beta^A$	*Sungei*	*00549S F*	*SI10*	1
B	$\alpha^B/\alpha^B — \beta^A/\beta^A$	*Ini*	*00613S F*	*SI9*	5
		Sibu	*00550S F*	*SI12*	
		Sampir[a]	*00545S M*	*SI1*	
		Jowata[a]	*00615S F*	*SI12*	
		Elsy	00446B F[b]	—	
AB	$\alpha^A/\alpha^B — \beta^A/\beta^A$	*Lada*	*00611S F*	*SI4*	6
		Bagan	*00547S M*	*SI3*	
		Dyak	*00607S M*	*SI6*	
		Din Ding	*00605S M*	*SI15*	
		Tupa	*00612S F*	*SI5*	
		Kampong	*00603S M*	*SI27*	
BD	$\alpha^B/\alpha^B — \beta^A/\beta^C$	*Bukit*	*00605S M*	—	3
		Lipis	*00534S M*	*SI11*	
		Antu	00616S F	—	
ABCD	$\alpha^A/\alpha^B — \beta^A/\beta^C$	Djambi	00551S F	—	3
		Tuan	*00604S M*	*SI8*	
		Bali	*00618S F*	*SI7*	

Bornean orang utans (11 animals)

A	$\alpha^A/\alpha^A - \beta^A/\beta^A$	Sya	00537B F	—	5
		Padang	00664B M	Bl5	
		Kuching	00601B M	—	
		Datu[c]	00617B M	Bl34	
		Jala	00784B F	—	
B	$\alpha^B/\alpha^B - \beta^A/\beta^A$	Kyan	00608B M	Bl35	1
AB	$\alpha^A/\alpha^B - \beta^A\beta^A$	Lembak	00548B M	—	5
		Padali	00614B F	Bl6	
		Rajah	00602B M	—	
		Durian	00663B M	—	
		Niko[d]	00153B M	—	

[a] Sampit is the father of Jowata.
[b] This animal is listed as Bornean in the studbook. On karyotyping it proved to be Sumatran (see Addendum, De Boer and Seuánez 1982, this volume).
[c] Datu was listed as Sumatran in Seuánez et al. 1976, but proved to be Bornean on karyotyping (see De Boer and Seuánez 1982, this volume).
[d] For chromosome information see Addendum, De Boer and Seuánez 1982, this volume.

2. The β^C-allele has not been found in Bornean orang utans and all the Bornean animals are homozygous for the β^A-allele. In Sumatra the β^C-allele occurs with a frequency of 0.17. It is therefore possible that β^C is unique to the Sumatran population.

3. If the second conclusion proves to be true, the theoretically expected phenotypes HbC ($\alpha_2^A\beta_2^C$), HbD ($\alpha_2^B\beta_2^C$) and HbCD ($\alpha_2^A\beta_2^C$, $\alpha_2^B\beta_2^C$) would occur only in Sumatra, but their frequencies would be rather low because all three are homozygous for the relatively infrequent β^C-allele. The HbAC phenotype, also to be expected in Sumatra must be rare as well since it is homozygous for the less frequently occurring allele at the α-locus (α^A) and heterozygous for β^C. Thus, the absence of these phenotypes in the small sample of orang utans studied so far (Sullivan and Nute 1968) is not surprising.

Sullivan and Nute (1968) pointed out that the α^B and β^C-chains might be the original haemoglobin chains in the orang utan, a suggestion which in part is supported by the amino acid sequences of α^A and α^B-chains (Maita et al. 1978). If this is true, the frequency distributions of the α and β-alleles would indicate that the Sumatran population represents an earlier stage of the process of replacement of α^B and β^C by the mutant alleles α^A and β^A. In Sumatra the α^B is still the more frequent allele at the α-locus, while at the β-locus the β^C continues to occur with a considerable frequency. In Borneo on the other hand, the frequency of α^B would have dropped considerably, while β^C disappeared completely from the population.

The results of our haemoglobin studies on the 16 orang utans seem to support the above conclusions in several ways. Firstly, eight of the 12 Bornean animals are of the HbA phenotype (Table 2), confirming the high incidence of the α^A-allele in this population. Secondly, there is no indication of the occurrence of the β^C-allele among the Bornean animals, which is an additional indication that this allele is probably absent in Borneo. Thirdly, of the two Sumatran orang utans studied at least one (OU2 with the double heterozygous HbABCD pattern) if not both possess the β^C-allele, supporting the idea that this may be a Sumatran specific character.

At present nothing can be said on possible differences in the number of α-chain loci in Bornean and Sumatran orang utans. It is known that in certain human populations chromosomes with two α-loci as well as those with one α-locus do exist, and that there are populations in which the duplicate is completely absent. Because of the indication for the existence of two α-loci in orang utans (found so far in three Bornean and one hybrid animal in our series), it would be profitable to study this phenomenon in detail in this species in order to assess the intraspecific differences and to detect the possible occurrence of thalassemia genes.

Acknowledgements

The authors are greatly obliged to the following zoological gardens which placed orang utan blood samples at their disposal: Antwerp, Amsterdam, Arnhem, Hannover, Rotterdam, Rhenen and Wuppertal.

Table 2. Haemoglobin phenotypes of the 16 orang utans of the present study. The data listed conform to those in Table 1. Further details on the HbAB' phenotype are given in the text. Subspecific status of all animals was confirmed by karyotyping (De Boer and Seuánez 1982, this volume).

Phenotype	Probable genotype	Number and name	Studbook number	Number in De Boer and Seuánez	Totals
Sumatran orang utans (2 animals)					
ABCD	$\alpha^A/\alpha^B — \beta^A/\beta^C$	OU2 Joop[a]	01172S M	SI16	1
?BD[e]	$\beta^B/\beta^B — \beta^A/\beta^C$	OU16 Djambi	01684S F	SI38	1
Bornean orang utans (12 animals)					
A	$\alpha^A/\alpha^A — \beta^A/\beta^A$	OU3 Piku	00379B M	BI38	8
		OU5 Petra[a]	01093B F	BI41	
		OU7 Fred	00738B M	BI20	
		OU8 Joke	00828B F	BI17	
		OU9 Sarita	01461B F	BI19	
		OU10 Siljo	01460B M	BI18	
		OU11 Anak[b]	01503B F	BI16	
		OU15 Ori[b]	00650B M	BI12	
AB	$\alpha^A\alpha^B/\alpha^A\alpha^B — \beta^A/\beta^A$	OU12 Bernadine	01521B F[c]	BI17	1
AB[e]	$\alpha^A\alpha^A/\alpha^A\alpha_2^B — \beta^A/\beta^A$ or $\alpha^A\alpha^B/\alpha^A\alpha^{,th} — \beta^A/\beta^A$	OU1 Geert	01294B M	BI6	3
		OU13 Bongo[c]	01255B M	BI4	
		OU14 Sjaan[c]	01256B F	BI5	
Hybrid orang utans (2 animals)					
AB	$\alpha^A/\alpha^B — \beta^A/\beta^A$	OU6 Connie[a]	01660X F	HI21	1
AB[e]	see above	OU4 Timo[d]	01664X M	HI2	1

[a] Joop (OU2) and Petra (OU5) are the father and the mother, respectively of Connie (OU6).
[b] Ori (OU15) is the father of Anak (OU11).
[c] Bongo (OU13) and Sjaan (OU14) are the father and the mother, respectively of Bernadine (OU12).
[d] Timo (OU4) has one captive-born, pure-bred Sumatran mother, and one captive-born hybrid father.
[e] This phenotype is discussed in the text.

133

References

Barnicot, N.A. and C.J. Jolly (1966). Haemoglobin polymorphism in the orang utan and an animal with four major haemoglobins. Nature (London), 210: 640–642.

Boer, L.E.M. de and P. Meera Khan (1982). Orang utan haemoglobins: a short review. In: L.E.M. de Boer (ed.), The orang utan. Its biology and conservation. Junk, The Hague.

Boer, L.E.M. de and H.N. Seuánez (1982). The chromosomes of the orang utan and their relevance to the conservation of the species. In: L.E.M. de Boer (ed.), The orang utan. Its biology and conservation. Junk, The Hague.

Bruce, E.J. and F.J. Ayala (1979). Phylogenetic relationships between man and the apes: electrophoretic evidence. Evolution, 33: 1040–1056.

Jones, M.D. (1980). Studbook of the orang utan, Pongo pygmaeus. Zoological Society of San Diego, San Diego.

Maita, T., A. Araya, M. Goodman and G. Matsuda (1978). The amino acid sequences of the two main components of adult hemoglobin from orangutan (Pongo pygmaeus). Hoppe-Seyler's Z. Physiol. Chem., 359: 129–132.

Meera Khan, P., H. Rijken, J. Wijnen, L.M.M. Wijnen and L.E.M. de Boer (1982). Red cell enzyme variation in the orang utan; electrophoretic characterization of 45 enzyme systems in cellogel. In: L.E.M. de Boer (ed.), The orang utan. Its biology and conservation. Junk, The Hague.

Seuánez, H.N., J. Fletcher, H.J. Evans and D.E. Martin (1976). A polymorphic structural rearrangement in the chromosomes of two populations of orangutans. Cytogenet. Cell Genet., 17: 327–337.

Seuánez, H., H.J. Evans, D.E. Martin and J. Fletcher (1979). An inversion in chromosome 2 that distinguishes between Bornean and Sumatran orangutans. Cytogenet. Cell Genet., 23: 137–140.

Sullivan, B. and P.E. Nute (1968). Structural and functional properties of polymorphic hemoglobins from orang utans. Genetics, 58: 114–124.

Authors' addresses:
L.E.M. de Boer
Biological Research Department
Royal Rotterdam Zoological and Botanical Gardens
Rotterdam
The Netherlands

P. Meera Khan
Dept. of Human Genetics, Sylvius Laboratories
State University of Leiden
Wassenaarseweg 72
2333 AL Leiden
The Netherlands

8. The chromosomes of the orang utan and their relevance to the conservation of the species

Leobert E.M. de Boer and Hector N. Seuánez

Introduction

Since the first decades of this century the chromosomes of mammalian species, including primates, have been studied using tissues with spontaneously high rates of cell division, such as bone marrow or testicular material. However, there are two developments which have greatly accelerated chromosome research during the past 20 years. Firstly, the development during the late 1950s of good tissue culturing techniques, particularly those for culturing lymphocytes from peripheral blood (Moorhead et al. 1960), has made it possible to investigate chromosomal morphology without necessitating the sacrifice of the animals under study. Secondly, the development in the late 1960s and early 1970s of the so-called chromosome banding techniques has been a tremendous stimulus to cytogenetics due to the fact that these techniques have opened the way to studying chromosomes in much greater detail. Conventional methods of staining chromosomes usually determined only two characteristics of each individual chromosome: its length and the position of the centromere. Chromosome banding techniques, on the other hand, produce sequences of alternating dark and light bands characteristic of each individual chromosome pair. Therefore, these techniques enabled recognition not only of individual chromosomes, but also of tiny parts of them.

The composition and structure of the chromosome set, the karyotype, of a species is simply a morphological trait. It is determined by the number of chromosomes (diploid number) and the length, centromeric position and banding pattern(s) of the members of the individual chromosome pairs. Since morphological traits may be subject to change during the process of speciation and evolution, the comparative study of karyotypes may contribute to a better understanding of phylogenetic relationships. However, studying chromosomes, in a way 'one is looking directly at the genetic material of an animal (i.e. a species)'. It may have been this view in the early 1960s, after methodology had made many primate species accessible to investigation, which led to the overestimation of the value of chromosome studies as a method of resolving phylogenetic problems. Today, many of these early studies appear rather superficial, because their conclusions did not take into consideration much of the information coming from other fields of biology and evolution. It was not

The orang utan. Its biology and conservation, edited by L.E.M. de Boer
© *1982, Dr W. Junk Publishers, The Hague, ISBN 90 6193 702 7*

sufficiently understood that with the light microscope we can only see the external, gross morphological appearance of the genetic material of the chromosome, not the genetic material itself.

However, karyotype structure constitutes a rather unusual morphological trait. Large karyotypic differences sometimes may exist between closely related species, while in other cases only very slight or apparently no differences are seen between very distantly related forms. In addition, indicating the peculiar character of karyotypic structure as a morphological trait, small changes in chromosome structure may have large effects in terms of the reproductive performance of individuals or in producing reproductive isolation between populations (see White 1977). Thus, while phylogenetic conclusions can be based on comparative karyotype studies, they should be drawn extremely carefully.

The much greater degree of resolution of chromosome morphology, facilitated by the introduction of banding techniques, opened unexpected possibilities not only for phylogenetic studies, but also more particularly for intraspecific investigations assessing chromosomal polymorphisms and geographical differences between populations of the same species. It will be clear that the latter aspect of chromosome research is relatively young, and apart from man, so far comparatively few mammalian species have been studied with this purpose (for some examples among primates see Ma et al. 1974; De Boer 1982b; De Boer and Reumer 1979; Reumer and De Boer 1980).

With respect to both phylogeny and intraspecific differentiation the orang utan is the most interesting of the great apes. It is the only extant species of the many great apes that once inhabited Asia. Though formerly distributed throughout large parts of South East Asia, it is currently confined to the islands of Borneo and Sumatra (see Von Koenigswald 1982). These two islands have been geographically isolated from each other, from the continent and from other Indonesian islands for the last 8,000 years, a reason why two separate populations of orang utan exist today. The Bornean orang utan, *Pongo pygmaeus pygmaeus*, has been described as the type form, while the Sumatran orang utan, *P. pygmaeus abelii*, has been described as a paratype. Since both forms differ somewhat in physical and behavioural characteristics (e.g. Chasen 1940; Van Bemmel 1968; Jones 1969; MacKinnon 1974) it has been generally accepted that they represent two different subspecies or races. In practice, however, subspecific identification sometimes causes considerable problems, and generally much experience is necessary for the proper identification of living specimens using external characteristics. This is not surprising since both Bornean and Sumatran orang utans seem to exhibit considerable variability, a fact which led Selenka (1896, 1898) to describe two Sumatran and eight Bornean forms of this ape. Other 19th century taxonomists also distinguished various numbers of orang utan species or subspecies, and though currently their number has been reduced to two, the variation still is unmistakable (see Rijksen 1978).

The original description of the orang utan's chromosomes was reported by Chiarelli (1961), and its diploid number was found to be 48. This is the same as that found in the chimpanzee, *Pan troglodytes* (Yeager et al. 1940; Young et al.

1960), the pygmy chimpanzee, *Pan paniscus* (Hamerton et al. 1963; Chiarelli 1962b) and the gorilla, *Gorilla gorilla* (Hamerton et al. 1961). The early comparisons of human and ape chromosomes showed clear similarities between these species (Chiarelli 1962a; Hamerton et al. 1963; Klinger et al. 1963, and others). The difference in chromosome number between man (46) and the great apes (48) was explained as the result of a fusion, occurring in the human lineage, of two pairs of subtelocentric chromosomes to form the large metacentric chromosomes of the second pair in the human karyotype. The chromosome complements of the gibbons (Hylobatidae) were found to differ markedly from those of man and the great apes, having diploid numbers of 44 in *Hylobates*, 50 in *Symphalangus* and 52 in *Nomascus* (Chu and Bender 1961; Chiarelli 1962b, 1963; Bender and Chu 1963; Hamerton et al. 1963; Klinger et al. 1963; Wurster and Benirschke 1969; Hösli and Lang 1970; Markvong 1973; Warburton et al. 1975; De Boer and Van Oostrum-Van der Horst 1975; Tantravahi et al. 1975). Thus, as was expected, the orang utan karyotypically constitutes a close group with man and the African apes, while being more remotely related to other primate groups.

Phylogenetic relationships of the orang utan karyotype

When the karyotypes of man and the great apes were compared using modern chromosome banding techniques, similarities appeared so striking that many chromosomes were recognised as 'interspecific homologues'. This term is used for chromosomes that show similar morphology and banding pattern in spite of belonging to different species or genera. Initially, different criteria of chromosome nomenclature and of presumed homologies were proposed for each of the great ape species (see Pearson et al. 1971; Turleau et al. 1972; De Grouchy et al. 1973; Lucas et al. 1973; Dutrillaux et al. 1975; Turelau et al. 1975, 1976 for orang utan chromosomes), until standard criteria were finally recommended at the Paris Conference on the Standardisation of Human Cytogenetics (Paris Conference 1971; Supplement 1975). A standard chromosome nomenclature was proposed for the karyotypes of all the great apes with the exception of the pygmy chimpanzee, and presumed interspecific homologies between these species and man were recorded. These criteria were later modified in the report of the Stockholm Conference (1977). Two chromosomes of the orang utan karyotype (nos. 13 and 19) and two pairs of the gorilla karyotype (nos. 13 and 18) formerly not matched were then recognised as interspecific homologues to corresponding chromosome pairs in man and the other great apes, including the pygmy chimpanzee for which a standardisation was also provided (see Table 1). These two reports have established that complete interspecific homology exists between human, chimpanzee, pygmy chimpanzee, gorilla and orang utan chromosomes, in view of the findings of similar morphology and banding patterns of chromosomes or assignment of similar structural genes.

A thorough comparison of human and great ape chromosomes has shown that hominoid chromosomes have been conserved evolutionarily during the period of phyletic divergence of man and the great apes from a common ancestral stock. An example of such conservation is illustrated in Fig. 1, in

Table 1. Presumptive chromosome homologies in the Hominoidea.*

Homo sapiens	Pongo pygmaeus	Pan paniscus	Pan troglodytes	Gorilla gorilla
1	1	**1**	1	1
2p, 2q	12, 11	**12, 13**	12, 13	12, 11
3	2	**2**	2	2
4	3	**3**	3	3
5	4	**4**	4	4
6	5	**5**	5	5
7	10	**6**	6	6
8	6	**7**	7	7
9	(13)	**11**	11	(13)
10	7	**8**	8	8
11	8	**9**	9	9
12	9	**10**	10	10
13	14	**14**	14	14
14	15	**15**	15	(18)
15	16	**16**	16	15
16	18	**18**	18	17
17	(19?)	**19**	19	19
18	17	**17**	17	16
19	20	**20**	20	20
20	21	**21**	21	21
21	22	**22**	22	22
22	23	**23**	23	23
X	X	**X**	X	X
Y	Y	**Y**	Y	Y

* Data according to the Paris Conference 1971 (Supplement 1975) in normal figures; data according to the Stockholm Conference 1977, in bold figures. Question mark in *Pongo pygmaeus* chromosome 19 indicates that banding homology between this chromosome and the others has not been completely resolved. Homologies do not indicate complete structural identity of the chromosomes in the various species. Part of the interspecific homologues are believed to have originated from each other by chromosome rearrangements (see text for further explanation).

which human and orang utan chromosomes are matched side by side. It can be seen that many chromosomes are similar or even identical, whereas others can be matched by assuming that a rearrangement (changing band sequence and morphology, but not the content of the chromosome) has occurred in one of them to form the other. Comparisons of this kind have led to the suggestion that chromosome rearrangements have occurred in primate radiation, thus modifying the original complement of the ancestral forms and giving rise to different karyotypes, each of which is characteristic of the primate species currently living. Within the great ape/human radiation, the most common chromosome rearrangement seems to have been pericentric inversion (inversion of a chromosome segment including the centromere) (Turleau et al. 1972; Bobrow and Madan 1973; Seuánez 1979). Other types of chromosome rearrangement such as telomeric fusion, paracentric inversion (inversion of a chromosome segment not including the centromere) or translocation (relocation of a chromosome segment from one chromosome to another) seem to have occurred significantly less frequently.

Tracing phylogenetic relationships on the basis of chromosome morphology is confused by the fact that the rate of chromosomal change during evolution is not constant as is often supposed to be the case with molecular changes (e.g. Sarich and Cronin 1976). Examples of clearly different rates of chromosomal

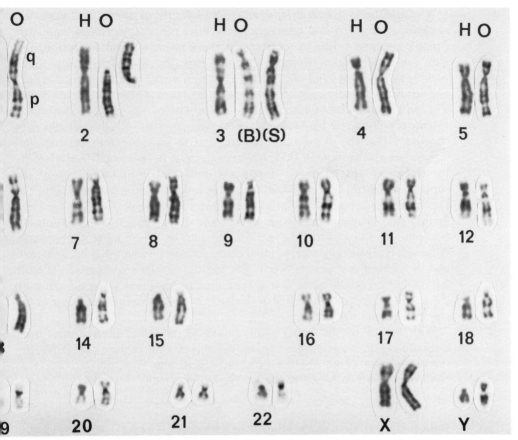

Fig. 1. Presumed homologies between the G-banded human and orang utan karyotype. In each pair or triplet of homologous elements the human chromosome (H) is placed on the left, the corresponding orang utan element (O) on the right. Only human chromosomes are numbered and grouped according to the Paris Conference (1971) rules. The numbers of the corresponding orang utan chromosomes are given in Table 1. In the first pair, p and q indicate short and long chromosome arm, respectively. In the third pair, B and S indicate Bornean and Sumatran orang utan chromosome number 2.

evolution in various mammalian taxa are legion. In general terms, with the primate group it is the New World primates which show rapid change of karyotype structure (De Boer 1974, 1982b), while, for example, the Cynopithecinae within the Old World primates seem to have undergone relatively little change during the latter part of their radiation. An extreme example of rapid evolution is to be found in the New World owl monkey (*Aotus trivirgatus*), which apparently comprises at least seven different populations with vastly differing karyotypes and chromosome numbers ranging from 46 to 56 (De Boer and Reumer 1979; Reumer and De Boer 1980) despite the fact that all these animals are generally considered as belonging to one species. Here, within a close cluster of populations, subspecies or semispecies, the karyological differences are as great as or even greater than those within the hominoids whose radiation started at a much earlier date in history.

A comparison of human and great ape karyotypes shows that man and chimpanzee differ by four rearrangements (three pericentric inversions and one fusion), man and gorilla by six (five pericentric inversions and one fusion) and man and orang utan (Bornean) by five (one paracentric inversion, three pericentric inversions and one fusion). However, in view of the above, this does not straightforwardly reflect phylogenetic distances within the Hominoidea. The problem of chromosomal relationships within this group recently was extensively discussed by Seuánez (1979), who in view of all available data concluded that the orang utan was the first to separate from the common lineage (see also Goodman 1975; Hoyer et al. 1972; Beneviste and Todaro 1976, for molecular genetic data). It retained most of the characteristics of the supposed ancestral hominoid karyotype (the Bornean orang utan retained 19 of the chromosome pairs of the primitive hominoid karyotype), while not sharing apomorphic characteristics with any of the other species. The relationship between the chimpanzee, the gorilla and man is much more problematic. Although it is generally accepted that there is a closer relationship between man and the chimpanzee than between man and the gorilla, a comparison of their karyotypes, especially of the shared apomorph characteristics, seems to indicate a more recent common ancestor of the latter two, or a common ancestor of the chimpanzee and the gorilla.

With regard to the orang utan, however, the conclusions drawn from comparative chromosome morphology are unequivocal: of the extant great apes the orang utan is the one most distantly related to man, an inference that confirms the generally accepted view arising from many other fields of study.

Intraspecific variation of the orang utan karyotype

Neither the Paris Conference (1971, Supplement 1975), nor the report of the Stockholm Conference (1977) pointed to any chromosomal differences between Bornean and Sumatran orang utans, but in fact proposed a single karyotype for the species, *Pongo pygmaeus*, without specifying which subspecies. This lack of specification is a problem in cytogenetic studies of most primate species. Even though the advanced banding techniques are most useful in detecting intra-specific differences, the search for the ancestry of man has overshadowed interest in the structure of the individual species and their populations. As a consequence, most studies (including chromosome studies) have not even attempted to give more detailed information (such as subspecific status or geographic origin) of the specimens analysed. In addition, most reports on primate chromosomes are based on very few animals, sometimes even members of the same family, a circumstance that does not easily allow the detection of intraspecific variation. Finally, the fact that the orang utan is, and always has been regarded as the great ape most distantly related to man led to the situation that (until recently) it was the least studied anthropoid because of the anthropocentric approach in primatology.

The study of intraspecific variation is, however, of great interest in several respects. Firstly, it enables the possible detection of differences between subspecies or populations. Secondly, when such differences appear to exist, valuable information is obtained on chromosomal evolution at the species level and on possible mechanisms of speciation. Thirdly, studies of chromosomal polymorphisms, variations within randomly breeding populations, render information as to population structure and inheritance of rearranged chromosomes in the heterozygous condition.

In fact, Turleau et al. (1975, 1976) were the first to discover intraspecific variation in orang utan karyotypes. They found a small difference in the structure of the second chromosome pair (their no. 3) between two Sumatran and one Bornean animal, and heterozygosity for the two types of chromosome 2 in two offspring from a Bornean × Sumatran cross. Since one of both types was only found in a single Bornean specimen (the father of both hybrids), they stated that it was impossible to conclude whether the difference in the second chromosome pair was subspecies-bound, or would represent a polymorphism or whether they were dealing with an individual variation. In addition, they found two types of chromosome 9 (their no. 12) in the same family of orang utans.

In view of these findings and the above, the present authors analysed the karyotypes of a larger number (72) of orang utans of known subspecific status (35 Bornean, 25 Sumatran and 12 Bornean × Sumatran hybrids). Information on 30 of these specimens has been published previously (Seuánez 1977; Seuánez et al. 1976a, 1976b, 1978, 1979; Seuánez and Fletcher 1978), while that on the remaining 42, studied by De Boer and Belterman (1982), appears here for the first time. The chromosome preparations were obtained by routine lymphocyte culturing techniques for mammalian chromosomes, while most of the presently available banding techniques were used for staining the chromosome material. Examples of karyotypes stained with the various techniques are shown in Figs. 2–6. The pedigrees of all animals studied, including their parents and offspring are shown in Fig. 7. All studbook numbers and names are listed in Table 2.

These studies revealed information on intraspecific differences involving four chromosome pairs, numbers 2, 9, 14 and 22, of the orang utan, and clearly demonstrated the value of such larger scale investigations. Below, special attention is payed to these four chromosome pairs exhibiting structural variation; the structure of the other pairs is clearly shown in the illustrations and has been extensively reported on in previous publications.

Chromosome number 2 structure distinguishing between Bornean and Sumatran orang utans

The two different types of chromosome 2 described by Turleau et al. (1975, 1976) were found among the 72 orang utans studied (see Figs. 2–6 and 8). The first type, which was carried in the homozygous condition by all Bornean specimens, with a centromeric index (length of the short arm expressed as percentage of the total chromosome length) of 10.2, shows one positive G- or Q-band in the short arm and two in the subcentromeric region. As against this, in all pure Sumatran animals the homologue to this chromosome is a more submetacentric element with a clearly longer short arm and consequently a higher centromeric index (18.9), but with the same total length as that of the Bornean chromosome 2. This chromosome shows two positive G- or Q-bands in the short arm and only one in the subcentromeric region. The patterns observed with Q- and G-banding were confirmed with R-, C- and Acridine Orange banding (after 6-hour Budr pulse). Each of these two types of chromosome 2 can be derived from the other by a simple pericentric inversion involving two breaks, one close to the terminal end of the short arm and one just below the subcentromeric Q- or G-band(s), followed by inversion of the intermediate centromere containing segment (Fig. 8).

Each of the 12 cross-bred orang utans studied showed one Bornean type chromosome 2 and one Sumatran type chromosome 2 (Figs. 3, 4, 5 and 6) and in some of them it was possible to trace from which parent (Bornean or Sumatran) each of them had been transmitted (Fig. 7). Heterozygous individuals were only found among captive-bred animals, not among wild-caught ones.

In view of these findings it may be concluded that either of the two types of chromosome 2 is characteristic of one of the orang utan populations, the Bornean or the Sumatran. This means that the proposition of Turleau et al. (1975, 1976), that we might be dealing with a structural chromosome 2 polymorphism, can be excluded. If this were the case, both homozygous types as well as heterozygotes would have occurred within the same population. Since this does not happen, we must indeed be dealing with a subspecific difference, a reason why it has been suggested (Seuánez et al. 1978; Seuánez 1979) that the types of chromosome 2 should be designated as 'Bornean' and 'Sumatran' chromosome 2 of *Pongo pygmaeus*, rather than considering one of them as 'normal' and the other as 'variant' chromosome in the species. In the future, the standardization of hominoid chromosome nomenclature (see Paris Conference 1971; Supplement 1975) should be adapted according to these findings. This chromosomal difference allows unequivocal identification between Bornean and Sumatran orang utans. It also allows distinction of hybrids from pure bred Bornean and Sumatran specimens. The value of this characteristic for captive breeding and conservation of the orang utan will be discussed later on.

Table 3 gives a summary of the numbers of orang utans tested for the chromosome number 2 difference.

Structural polymorphism of chromosome number 9

A variant chromosome 9 in the orang utan was first shown in the report of Lucas et al. (1973). Turleau et al. (1975, 1976) have described two different types of chromosome number 9 (their no. 12) in a family of orang utans. The difference between these types can only be demonstrated with certainty in the banding sequences, since both are submetacentric, with only slightly different centromeric indexes (40.9 and 34.0). The difference was believed to be the result of a replacement of the centromere from its original position to a new position in the long arm (a rearrangement theoretically involving three chromosome

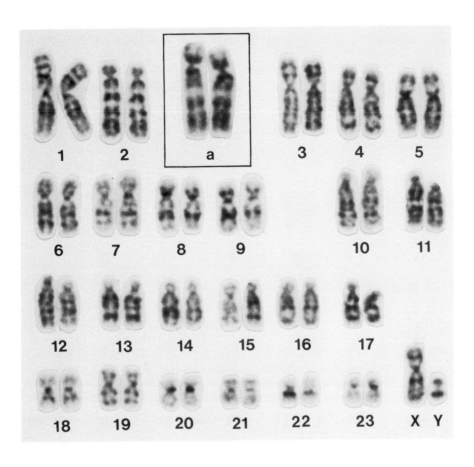

Fig. 2. G-banded karyogram of a male Bornean orang utan, *Pongo pygmaeus pygmaeus*. The inset (a) shows the second chromosome pair of a Sumatran orang utan, *Pongo pygmaeus abelii*. Note the difference in the length of the short arms of both chromosome pairs 2.

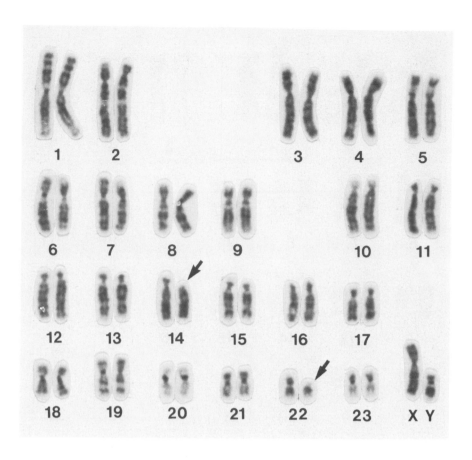

Fig. 3. G-banded karyogram of a male cross-bred Bornean × Sumatran orang utan. In pair 2, one Sumatran (left) and one Bornean chromosome number 2 are seen. Apart from this phenomenon, characteristic of all hybrid orang utans, this particular animal (specimen HI4 in the pedigrees of Fig. 7) shows structural heterozygosity in two more chromosome pairs, numbers 14 and 22. In both cases the short arm of one of the two members of these pairs is deleted. The resultant 'variant 14' and 'variant 22' elements are indicated by arrows (see also Fig. 10).

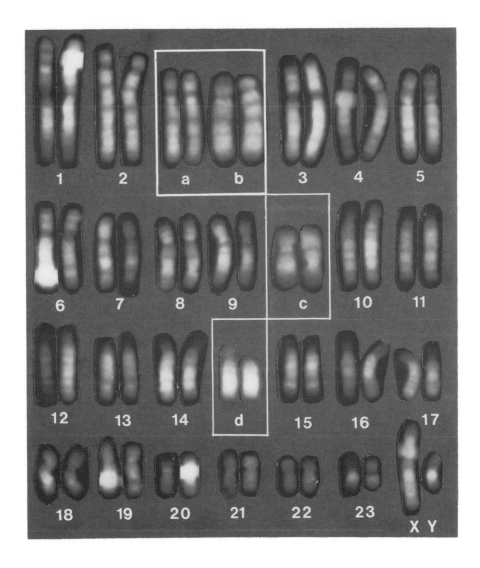

Fig. 4. Q-banded karyogram of a male Bornean orang utan. The insets show chromosome pair 2 of a Sumatran animal (a), chromosome pair 2 of a cross-bred animal (b), chromosome pair 9 of an animal heterozygous for the variant 9 chromosome (c) and chromosome pair 14 of specimen SII 28 in the pedigrees of Fig. 7, which is heterozygous for the variant chromosome 14 (d).

145

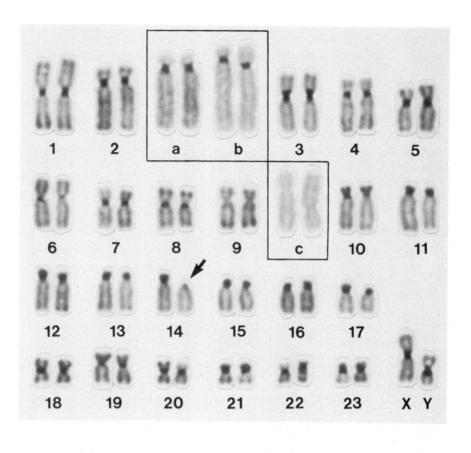

Fig. 5. C-banded karyogram of a male Sumatran orang utan. The arrow points to the variant chromosome 14 which was found heterozygously in this animal (SI16). The insets show chromosome pair 2 of a Bornean animal (a), chromosome pair 2 of a hybrid (b), and chromosome pair 9 of an animal heterozygous for the variant chromosome 9 (c).

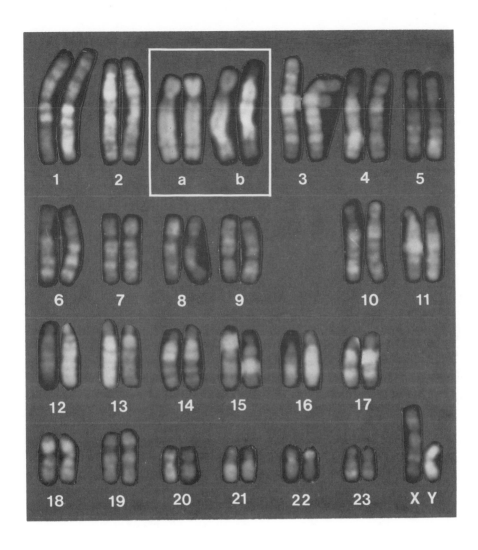

Fig. 6. Acridine-orange stained karyogram (after a 6-hour pulse of the lymphocyte culture with Budr) of a male Bornean orang utan. The inset shows chromosome pair 2 of a Sumatran animal (a) and that of a hybrid (b).

147

Fig. 7. Pedigrees of the orang utans karyotyped in this study. The pedigrees include only parents, sibs and offspring of the specimens studied. B_I and S_I indicate wild-caught Bornean and Sumatran animals, respectively; B_{II} and S_{II} indicate first generation, captive-born, pure Bornean and Sumatran animals, respectively; $B_{II/III}$ indicates half-second generation, captive-born, pure Bornean animals; H_I and H_{II} indicate first and second generation, captive-born hybrid animals. For house names and studbook numbers of all animals included in the pedigrees, see Table 2.

For additional information on the animals B_I42, $S_{II}34$, $H_{II}1$, recently born offspring of B_I42, B_I41, B_I38, B_I16 and B_I17, and five additional orang utans unrelated to the animals of the above pedigrees, we refer to the *Addendum*.

148

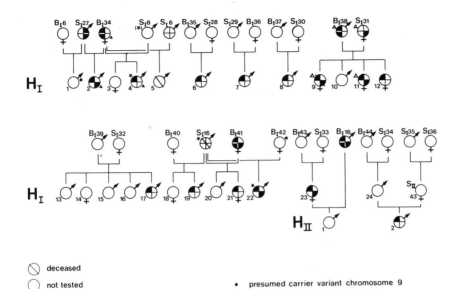

⊘	deceased
◯	not tested
◖	homozygous Bornean chromosome 2
◗	homozygous Sumatran chromosome 2
◖	heterozygous Bornean/Sumatran chromosome 2
◗	homozygous normal chromosome 9
◗	heterozygous normal/variant chromosome 9
◗	homozygous variant chromosome 9
◗	chromosome 9 not tested

•	presumed carrier variant chromosome 9
★	heterozygous normal/variant chromosome 14
(∗)	presumed carrier variant chromosome 14
▲	heterozygous normal/variant chromosome 22
▫	studied by Ryder (pers. comm.)
∘	studied by Anderle et al. (1979) (trisomic 22)
△	studied by Turleau et al. (1976) (B_I38 & S_I31 also by us)

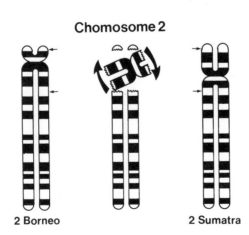

Chomosome 2

2 Borneo   2 Sumatra

Fig. 8. Schematic representation of the G-band pattern of the two types of chromosome 2, distinguishing between Bornean and Sumatran orang utans, and of the pericentric inversion by which these types can be derived from each other.

Table 2. Names and studbook numbers of the orang utans included in the pedigrees of Fig. 7. 'B', 'S' and 'X' in the studbook numbers indicate identification of the animals as Bornean, Sumatran or hybrid, respectively (in a few cases this identification is not confirmed by the karyotype analysis as indicated in the notes at the end of this table). 'F' and 'M' in the studbook numbers indicate female and male. When no studbook numbers are available, only subspecific identification and sex are given. Names and studbook numbers of the orang utans which have been karyotyped are printed in *italics.* Blood samples of these animals were obtained from the Yerkes Regional Primate Research Center (Atlanta) and from the Zoological Gardens of Bristol, London, Jersey, Walton, Rotterdam, Amsterdam, Wassenaar, Rhenen, Arnhem, Antwerp, Wuppertal, Singapore and West Berlin. (Studbook data according to Jones 1980.) Additional data are given in the *Addendum.*

B_I (wild-born Bornean orang utans)

1 Jack	00267B M	16 *Geert*	*01294B M*	31 *Zabu*	*01169B M*
2 *Jill*	*00268B F*	17 *Joke*	*00828B F*	32 Pinky	01180S F[b]
3 *Abang*	*00869B M*	18 *Siljo*	*01460B M*	33 *Girlie*	*01181B F*
4 *Dyang*	*00870B F*	19 *Sarita*	*01461B F*	34 *Datu*	*00617B M[c]*
5 Padang	00664B M	20 *Fred*	*00738B M*	35 Kyan	00608B M
6 Paddi	00614B F	21 Max	00793B M	36 Bunty	00901B F
7 Boy	00896B M	22 Susi	00794B F	37 Giles	00958B M
8 Charli	00269B M	23 *Kajan*	*01610B M*	38 *Pi-ku*	*00379B M[d]*
9 Toli	00176B F	24 *Letchu*	*01489S M[a]*	39 Sam	00130B M
10 Saleh	00899B M	25 *Friday*	*01179B M*	40 Barbara	00494B F
11 Kate	00903B F	26 *Jojo*	*01187B M*	41 *Petra*	*01093B F*
12 *Ori*	*00650B M*	27 *Raman*	*01490B M*	42 Sylvia	00833B F
13 Truus	01045B F	28 *Bobo*	*01178B M*	43 Pongo	00563B M
14 *Bongo*	*01255B M*	29 *Leala*	*01533B F*	44 Karl	00839B M
15 *Sjaan*	*01256B F*	30 Zeena	01170B F		

B_II (first generation captive-born pure Bornean orang utans)

1 *James*	*01244B M*	8 *Bosja*	*01705B F*	15 Benjamin	01601B M
2 Popi	01141B F	9 unnamed	------ B F	16 Duisburg 15	01668B M
3 *Teriang*	*01260B M*	10 Sandra	------ B F	17 Duisburg ---	------ B M
4 Bulu	00436B M	11 Lio	01197B F	18 *Jinak*	*01425B M*
5 *Anak*	*01280B M*	12 Marinda	01366B F	19 *Inoki*	*01495X M[e]*
6 *Anak*	*01503B F*	13 Duisburg 8	01431B M	20 *Tawan*	*01631B M*
7 *Bernadine*	*01521B F*	14 Dua	01534B F		

B_II/III (half-second generation captive-born pure Bornean orang utans)

1 *Suka*	*01155B F*	2 *Jantan*	*01380B M*

S_I (wild-born Sumatran orang utans)

1 Sampit	00546S M	13 *Henry*	*00702S M*	25 *Tarzan*	*00127B M[g]*
2 *Jowata*	*00615S F*	14 Ann	00703S F	26 Suma	00095S F
3 Bagan	00547S M	15 *Dinding*	*00606S M*	27 Kampong	00603S M
4 Lada	00611S F	16 *Joop*	*01172S M*	28 Potts	00218S F
5 Tupa	00612S F	17 Truus	01173S F	29 Gambar	00831S M
6 *Dyak*	*00607S M*	18 *Likoe*	*01600S M*	30 Gina	00959S F
7 Bali	00618S F	19 Charly	00497S M	31 *Pin-chin*	*00380S F[d]*
8 Tuan	00604S M	20 Kiki	00499S F	32 Coco	00256S F
9 Ini	00613S F	21 Brigitte	00723S F	33 Liesje	00394S F
10 Sungei	00549S F	22 ????	------ S M[f]	34 Djambi	00808S F
11 Lipis	00534S M	23 *Ah Meng*	*01188S F*	35 Tarzan	00654S M
12 Sibu	00550S F	24 *Rodney*	*01221S M*	36 Sima	00760S F

S_II (first generation captive-born pure Sumatran orang utans)

1 Kanting	00970S M	16 Gelar	01595S M	31 Hawa	00998S F
2 *Santing*	*01285S M*	17 Patpat	01237S M	32 Mano	01075S M
3 *Jinjing*	*01510S M*	18 Petala	01365S F	33 *Danny*	*01147S F[i]*
4 Kesa	00906S F	19 Mehra	01513S M	34 Sexta	01251S F
5 Bunga	01168S F	20 Hati	01639S F	35 Smuni	01373S F
6 Tukan	00883S M	21 Pensi	00909S F	36 Murid	01457S F
7 *Sah*	*01283S M*	22 *Lunak*	*01194S M*	37 Kwesida	01555S F
8 *Anak*	*01472S M*	23 *Oscar*	*01140S M*	38 Djambi	01684S F

Table 2. (Continued)

S$_{II}$ (Continued)

9 Panjii	01028S F	24 Henrietta	01249S F	39 Wilhelmina	01708S F
10 Biji	01106S F	25 Julitta	01438S F	*40 Hsing Hsing*	*01437X M*[j]
11 Chebek	*01511S M*	26 Anja	01284S F	41 Copenhagen 1	00313X F[k]
12 Yakut	*01070S M*[h]	27 Paulien	01409S F	42 Copenhagen 2	00334X M[k]
13 Ayer	*01323S M*	*28 Mano*	*01573S M*	43 Uta	01165S F
14 Jiran	*01442S M*	29 Samirah	01717S F		
15 Cheli	01209S F	30 Lea	00887S F		

H$_I$ (first generation captive-born hybrid orang utans)

1 Kanak	01055X M	*9 Ursula*	*00864X F*[l]	*17 Djumah*	*01570X M*
2 Chantek	*01617X M*	10 Xeno	01074X M	18 Jacowina	01445X F
3 Yerkes 15	01128X F	*11 Agnes*	*01322X F*[l]	*19 Tom*	*01670X M*
4 Loklok	*01008X M*	*12 Eddy*	*01615X F*	20 Cheemo	01485X M
5 Yerkes 23	01252X M	13 Papan	00635X M	*21 Connie*	*01660X F*
6 Bebas	*01005X M*	14 Njonja	00824X F	*22 Bulu*	*01582X M*
7 Sayang	*01183X M*	15 Barin	01219X M	*23 Lea*	*00895X F*
8 Tunku	*01240X M*	16 Barut	01401X M	24 Schorsch	01134X M

H$_{II}$ (captive-born orang utans with at least one captive-born hybrid parent)

1 Klaasje	------- X M	*2 Timo*	*01664X M*

[a] Registered in the studbook as Sumatran; proved to be Bornean after karyotyping.
[b] Registered in the studbook as Sumatran; appears to have been Bornean according to the karyotype of the offspring (see e).
[c] Listed as Sumatran in a previous publication (Seuánez et al. 1976b), but proved to be Bornean after karyotyping.
[d] Studied by Turleau et al. (1975, 1976), as well as by the present authors.
[e] Registered in the studbook as hybrid; proved to be Bornean after karyotyping. The father (B$_I$31) of this animal proved to be Bornean as well, thus, the deceased mother (Pinky, 001180S F; B$_I$32) must have been misidentified (see b).
[f] This animal is the father of Hsing Hsing, 01437X M (S$_{II}$40), which is registered in the studbook as a hybrid, but which proved to be Sumatran after karyotyping. The mother of this young, Ah Meng, 01188S F (S$_I$23), was karyotyped as well and proved to be Sumatran. The studbook gives as the father of Hsing Hsing a Bornean animal, Jojo, 01187B M (B$_I$26), whose subspecific identification was confirmed by karyotyping. Thus, it has to be concluded that not Jojo, but another (Sumatran) male must have been the father of Hsing Hsing.
[g] Registered in the studbook as Bornean; proved to be Sumatran after karyotyping.
[h] Studied by Ryder (Zoological Society of San Diego) (pers. comm.).
[i] Studied by Anderle et al. (1979).
[j] Registered in the studbook as hybrid; proved to be Sumatran after karyotyping (see f).
[k] Registered in the studbook as hybrids; appear to have been Sumatran after karyotyping of the father (see g).
[l] Studied by Turleau et al. (1975, 1976).

Table 3. Chromosome pair 2 structure in 72 Bornean, Sumatran and cross-bred orang utans.

	homozygous Bornean chromosome 2	homozygous Sumatran chromosome 2	heterozygous Bornean/Sumatran chromosome 2
Bornean, wild-caught	23	—	—
Bornean, captive-born	12	—	—
Sumatran, wild-caught	—	11	—
Sumatran captive-born	—	14	—
Hybrids, captive-born	—	—	12
Totals	35	25	12

151

breaks). Dutrillaux et al. (1975), on the other hand, explained the difference as resulting from the translation of part of the long arm to the short arm. Turleau et al. (1975, 1976) found both types of chromosome 9 in the heterozygous condition in a wild-caught Bornean and a wild-caught Sumatran male orang utan. A Sumatran female was homozygous for one of the types, while two of her hybrid offspring, sired by the heterozygous Bornean male, were also heterozygous (see pedigree with animals SI31, BI38, HI9 and HI11 in our Fig. 7).

Among the 72 orang utans analysed, we found the same type of variation of chromosome 9. However, detailed examination of the G-, R- and Q-banding patterns led to the conclusion that both chromosome types are related through a rather complex rearrangement involving three breaks and a paracentric inversion within a pericentric inversion (Seuánez et al. 1976b) (see Fig. 9). The two types of chromosome 9 were found in Bornean and Sumatran animals, and in both populations there were heterozygous carriers as well as homozygotes of each type of chromosome 9. These findings provide good evidence that we are dealing with similar chromosome polymorphisms in the two populations.

In both orang utan populations one of the chromosome 9 types occurs more frequently (79.4% in Bornean animals, 70.0% in Sumatran animals) and this type therefore is considered to be the 'normal' type. The less frequently occurring type is designated as a 'variant' chromosome 9 (20.6% in Borneo, 30.0% in Sumatra). A summary of the numbers of animals found to be homozygous for either type of chromosome 9 or heterozygous, is given in Table 4 and in the pedigrees of Fig. 7. Those unstudied animals which, on the basis of the karyotypes of their offspring probably carry a variant chromosome 9, are also indicated in Fig. 7. Only four animals have so far been found to be homozygous for the variant (animals BII1, BI2, BI23 and SI25), while as many as 23 were heterozygous carriers. Among the Sumatran × Bornean hybrids, all of which were heterozygous for the chromosome 2 types, three were also heterozygous for the chromosome 9 types.

Chromosome 9

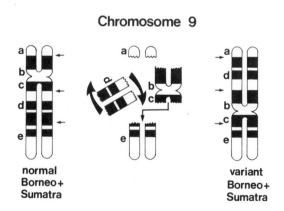

Fig. 9. Schematic representation of the G-band pattern of the normal and the variant chromosome 9, found in Bornean as well as in Sumatran orang utans, and of the double inversion by which both types can be derived from each other.

Table 4. Chromosome pair 9 structure in 71 Bornean, Sumatran and cross-bred orang utans.*

	homozygous normal chromosome 9	heterozygous normal/variant chromosome 9	homozygous variant chromosome 9	frequency 'normal'	frequency 'variant'
(23) **Bornean, wild-caught**	16	5	2	79.4%	20.6%
(11) **Bornean, captive-born**	7	3	1		
(11) **Sumatran, wild-caught**	4	6	1	70.0%	30.0%
(14) **Sumatran, captive-born**	7	7	—		
(12) **Hybrids, captive-born**	9	3	—		
Totals	43	24	4		

* One of the orang utans listed in Table 3 (tested for chromosome number 2) was not tested for chromosome number 9.

Generally, in both Sumatran and Bornean orang utans chromosome number 14 is an acrocentric element with a complex short arm, consisting of a proximal region above the centromere, a secondary constriction and a distal satellite region. In C-banded preparations the entire short arm region is hetero-chromatic. In four of the 72 orang utans studied a variant of this chromosome was found in the heterozygous condition. In this variant almost the entire heterochromatic short arm region is missing, while the remainder of the chromosome is identical to the normal chromosome 14 (see Figs. 3, 4 and 10).

The four animals carrying this variant chromosome belong to two different pedigrees. In one of these the variant is present in a hybrid offspring (HI4, Fig. 7). The Bornean mother of this young (BI34) did not carry the variant, its Sumatran father (SI8) was not studied. Two pure-bred Sumatran offspring of the same male (SII13 and SII14) did not carry the variant. Thus, it can be concluded that either the Sumatran father (SI8) was a heterozygous carrier or that we are dealing with a *de novo* mutation. The latter possibility is rather less likely because of the occurrence of the same variant in another unrelated family: in a Sumatran male (SI16) and in two if its offspring, a hybrid (HI22) and a pure Sumatran young (SII28). It is notable that this hybrid animal is heterozygous for three chromosome pairs, numbers 2, 9 and 14.

Thus, the variant is present in one wild-caught Sumatran orang utan, and probably in a second one, as well as in some of their offspring. Since it was not found in any of the Bornean animals studied, nor in any other hybrid except for HI22, we may perhaps be dealing with a rare variant chromosome which is typical for the Sumatran orang utan population. Its frequency would then be 5.9% (estimated from the pure Sumatran animals only). However, the possibility that this variant chromosome originated twice by independent *de novo* mutations in both pedigrees cannot be excluded.

A possible variant chromosome number 22

A variant chromosome 22 was found in three related orang utans, a Bornean female (BI34) and two of her offspring (HI2 and HI4, both hybrids and half brothers, see Fig. 7) (Seuánez 1979). Chromosome number 22, one of the smallest chromosomes in the complement, generally is acrocentric with a satellited, entirely heterochromatic (C-positive) short arm, similar to that of chromosome 14. In the variant almost the entire short arm is deleted (Fig. 10). The Sumatran father (SI8) of one of these hybrids (HI4) carrying the variant 22 was not studied, while the father (SI27) of the second hybrid (HI2) had two normal chromosomes 22. Although this has not yet been proven, it seems likely that the Bornean mother (BI34), being a heterozygous carrier herself, transmit-ted the variants to both her hybrid offspring. The variant has been found in only one wild-caught orang utan, a reason why it cannot be concluded whether we are dealing with a rare variant typical for the Bornean population, or with a *de novo* mutation in the ancestry of the Bornean mother. It is notable that one of

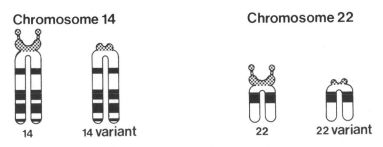

Chromosome 14   Chromosome 22

14   14 variant   22   22 variant

Fig. 10. Schematic representation of the G-band and C-band (dotted areas) patterns of the normal and the deleted variant chromosome number 14, and of the normal and deleted variant chromosome number 22.

her hybrid offspring is heterozygous for three chromosome pairs, numbers 2, 14 and 22.

Recently, Anderle et al. (1979) described the karyotype of an orang utan with a trisomy of the 22nd chromosome pair (2n = 49), showing multiple malformations resembling somewhat those of Down's syndrome in man (caused by trisomy of chromosome 21, which supposedly in homologous to chromosome 22 in the great apes). A similar case had earlier been found in a chimpanzee carrying three chromosomes 22 (McClure et al. 1969, 1970, 1971). The typical mongoloid malformations in this animal, however, were more severe and it died at young age. A male gorilla studied by Turleau et al. (1972) also showed an extra small autosome, probably number 22 with multiple satellites attached to the short arm. This specimen, however, was not reported to exhibit abnormal clinical features.

The short arm regions of the three chromosomes 22 in the trisomic orang utan, born from a pair of Sumatran animals (SI19 and SI20 in the pedigree of Fig. 7) appear to be normal (Anderle et al. 1979; Schweizer et al. 1979), as do those of a sib of the propositus studied by us (animal SII38). Thus, the variant chromosome 22 mentioned above was not found in this family of orang utans.

Evolutionary considerations with regard to the intraspecific chromosome variation in the orang utan

The fact that Bornean and Sumatran orang utans are found to differ from each other by a pericentric inversion in chromosome number 2 is of great interest in several respects. In the first place, it strengthens the thought that both populations indeed must be considered as different subspecies, rather than as races or varieties. It is tempting to accept that these two subspecies indeed are in the process of speciation, and the fact that they proved to be chromosomally distinct is good evidence to believe that they have already diverged from each other more widely than has previously been thought. The initial chromosome 2 rearrangement must have occurred in one of the populations after, or possibly just prior to the geographic isolation of Borneo and Sumatra which took place some 8,000 years or more ago. After the initial occurrence of the chromosome 2 rearrangement in that population a period of chromosomal polymorphism must have existed, during which both types of chromosome 2, the original and the derived one, must have been present homozygously as well as hetero-zygously. Natural selection in favour of the derived chromosome 2 or genetic drift must have led to an increase in its frequency during this period and finally to its fixation, the total disappearance of the original chromosome from the population.

The discovery of fixation of a chromosome rearrangement by pericentric inversion is extremely valuable, since pericentric inversion seems to be the most common type of rearrangement through which interspecific homologue chromosomes can be derived from each other, when comparing the karyotypes of man and the great apes. Thus, pericentric inversion has probably been the main mechanism of chromosome change in the hominoids (e.g. Turleau et al. 1972; Bobrow and Madan 1973; Seuánez 1979). If, however, these re-arrangements were indeed responsible for the reshuffling of the original common ancestral karyotype of man and the great apes to give rise to the present karyotypic divergence, some conditions must have been met. Firstly, it must have been possible for the derived chromosomes to spread in the population after their initial appearance. This must have happened during periods of chromosome polymorphism, which would require full, or only slightly reduced fertility of animals heterozygous for the given rearrangement. Secondly, the derived chromosomes must have had the possibility of reaching fixation within the emerging species, as a result of either natural selection or genetic drift. Otherwise this would not have resulted in a karyotype different from that of the ancestral species and from those of other emerging species. Thus, the process of karyotypic evolution in man and the great apes could be comparable to the mechanism by which the two orang utan subspecies diverged chromosomally.

In the light of these aspects of chromosomal evolution and their bearing on the evolution of primates, it is probable that both orang utan subspecies have been evolving for some considerable period and are still diverging in many aspects as yet unknown. It may well be that they have to be considered in the future as 'cryptic species'. There are already some molecular data indicating a greater genetic distance between the two forms than was previously believed (Bruce and Ayala 1979; Meera Khan et al. 1982; De Boer and Meera Khan 1982; Wijnen et al. 1982). For this reason it would be of great interest to undertake as many other genetic studies as possible on orang utans in order to estimate the actual extent of intersubspecific difference (see also De Boer 1982a).

Comparison of the Bornean and Sumatran type of chromosome 2 does not indicate which of the two might have been the original one, because pericentric inversion is a rearrangement that may be two-directional. Comparison with the human, chimpanzee and gorilla homologue to this chromosome, however, seems to give a clear indication. The human chromosome 3 and the gorilla and chimpanzee chromosome 2, which are identical with each other and metacen-tric, can be easily derived by a simple pericentric inversion from the Bornean chromosome 2. Derivation from the Sumatran chromosome 2, however, would require a more complex rearrangement involving at least three breaks (Fig. 11).

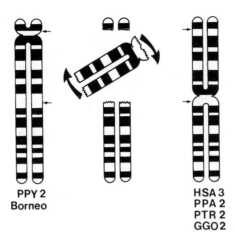

PPY 2
Borneo

HSA 3
PPA 2
PTR 2
GGO 2

Fig. 11. Schematic representation of the G-band patterns of human chromosome 3 (homologous to chimpanzee and gorilla chromosome 2) and chromosome number 2 of the Bornean orang utan, showing that these chromosomes can be derived from each other by a simple pericentric inversion. Derivation of human chromosome number 3 from the Sumatran orang utan chromosome number 2 would require a more complex rearrangement. (PPY = *Pongo pygmaeus*, HSA = *Homo sapiens*, PPA = *Pan paniscus*, PTR = *Pan troglodytes*, GGO = *Gorilla gorilla*.)

Thus, for reasons of parsimony, it is tempting to accept that the Bornean type of chromosome 2 is the original one in *Pongo pygmaeus*, while the Sumatran type is the derived one.

In contradiction to these findings regarding the second chromosome pair of *Pongo pygmaeus*, the occurrence of a variant chromosome 9, which originated from a complex rearrangement, such as an inversion within an inversion, suggests that this variant chromosome has been maintained in both populations as a chromosome polymorphism (Seuánez et al. 1976b), because in both Bornean and Sumatran animals this variant occurs with a higher incidence than would be expected in the case of recurrent mutation. It also suggests that the initial rearrangement which gave rise to this variant chromosome must have occurred prior to the geographic separation of the orang utan into two subpopulations, since the independent occurrence of such a complex rearrangement on two occasions would seem highly improbable. Since that time the variant chromosome 9 must have been maintained in both populations polymorphically without fixation. It is notable that in both Borneo and Sumatra the frequency of the variant chromosome 9 is similar.

There are two possible explanations for the present existence of the chromosome 9 polymorphism. First is that we are dealing with a transient polymorphism which may finally lead to fixation of the variant 9 chromosome and total disappearance of normal 9 in both populations. (It may be assumed that the normal 9 is the original chromosome of the species and the variant 9 is the derived one, since the normal 9 has an identical homologue in man [human chromosome 12]; in chimpanzee and gorilla the homologous element can be derived by a simple pericentric inversion from normal 9 in the orang utan or human number 12.) In this case there would be a selective advantage of the

157

derived chromosome, slowly leading to an increase of its frequency in the population. This selective advantage, however, would have to be clearly lower than that of the inverted (Sumatran) chromosome 2, which apparently reached fixation much more rapidly, having originated later in history and having already been fixed some time ago. The second possible explanation for maintenance of the variant chromosome 9 at a high frequency would be the existence of some kind of selective advantage in favour of heterozygous carriers. In that case, after the initial appearance of the variant, its frequency in the population would steadily increase until an equilibrium would have been reached, after which (under constant environmental conditions) the frequencies of both the normal and the variant chromosomes would remain constant, even after geographic disconnection of the distribution area. With the data currently available it is impossible to choose which one of these explanations is more likely.

It is obvious that in order to maintain a variant chromosome with a high frequency as a (balanced) polymorphism over such a long period (encompassing perhaps 800 to 1,000 generations), heterozygous carriers of this variant chromosome must be as fertile or almost as fertile as either type of homozygote (unless there was a very strong heterozygous advantage). In the heterozygous condition, however, pericentric inversion chromosomes may give rise to a considerable risk of the production of unbalanced gametes (i.e. spermatozoa or eggs either with absent or supernumerary chromosome segments) when, during meiosis crossing-over occurs within the inverted section. These unbalanced gametes could lead to gross abnormalities in the offspring or to foetal death, and thus reduce the fertility of heterozygous carriers. Since the inverted segment in variant chromosome 9 covers about half of the length of the chromosome, the probability of cross-overs within this region is certainly not negligible. During the first meiotic division, pairing of a normal and a variant chromosome 9 possibly involves the formation of two inversion loops (Seuánez et al. 1976b; Seuánez 1979). Meiotic pairing through the formation of inversion loops has been reported in many groups of lower animals, although so far they have not been shown in mammals. In order to reduce the risk of the production of unbalanced gametes there may be some mechanism preventing the occurrence of crossing-over within the inversion loops of heterozygous orang utans (Fig. 12), thus confining cross-overs to the terminal regions of both chromosome 9 arms. In the C-group chromosomes of man, which include the

---

Fig. 12. Meiotic configuration in an animal heterozygous for the normal and the variant chromosome 9, showing the formation of a double inversion loop. The products of meiotic disjunction, following the occurrence of a cross-over within one of the inversion loops, are shown on the right. Two of the resultant gametes are balanced, one containing a normal chromosome 9, the other containing a variant chromosome 9. The other two are unbalanced, one containing a dicentric chromosome, the other an acentric fragment. This diagram should be considered as purely theoretical, because inversion loops so far have not been reported in mammals.

Fig. 13. Phase-contrast photomicrograph of spermatozoa of an orang utan heterozygous for the variant chromosome number 9 (specimen SI15 in the pedigrees of Fig. 7) (left). On the right, *ibidem*, stained with Papanicolaou. Note regular size and shape of the spermatozoa heads. The same results were obtained in an animal homozygous for the variant chromosome number 9 (specimen SI27). From Seuánez et al. (1977).

158

Fig. 12.

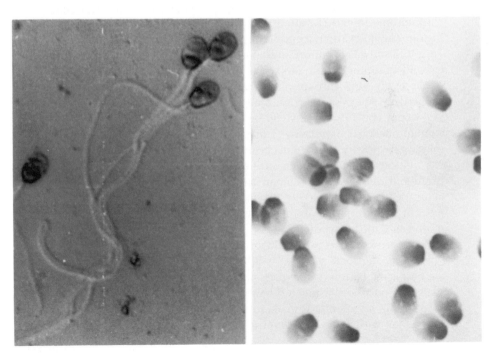

Fig. 13.

159

homologue of chromosome 9 of *Pongo pygmaeus* (human chromosome 12), most meiotic chiasmata actually occur in the terminal regions (Hulten 1974). Thus, if chiasmata reflect crossing-over sites, such a preferential occurrence of crossing-overs may be an important factor in the maintenance of the chromosome number 9 polymorphism in the orang utan.

There is already good evidence that animals heterozygous for both types of chromosome 9 are fertile. Firstly, Turleau et al. (1975, 1976) have already reported a heterozygous carrier of a variant chromosome 9 with proven fertility (animal SI31 in the pedigrees of Fig. 7, which produced four young). In addition, one hybrid animal studied by us (HI23), heterozygous for both Bornean and Sumatran chromosome 2 as well as for the normal and variant chromosome 9, recently started reproducing. Secondly, the morphology and DNA quantification of mature spermatozoa in semen samples of two adult orang utans (SI15 and SI27 in Fig. 7) have been analysed (Seuánez et al. 1977). One of these animals is a heterozygous carrier of variant chromosome 9, the other is homozygous for this element. Both animals, however, produced a similar proportion (98%) of regular sized spermatozoa (Fig. 13) and in neither of them the distribution of individual-cell-DNA-content suggested the presence of aneuploid spermatozoa as would have been the case if unbalanced gametes had been produced with any significant frequency.

Little can be said on the nature of the variant chromosomes 14 and 22 from the point of view of population genetics. As indicated above, the variant chromosome 14 may be characteristic of the wild Sumatran population. The variant chromosome 22 probably occurs in only a single wild-caught animal from Borneo and may have originated from a *de novo* mutation. In both the chromosome 14 and 22 variants, we are dealing with the loss of the short arm region. The short arm regions of the orang utan chromosomes 11, 12, 13, 14, 15, 16, 17, 22 and 23 are completely heterochromatic and positively stained with the C-banding technique, but unlike the centromeric C-band regions found in most of the orang utan chromosomes they also exhibit a number of other specific properties. They specifically bind the fluorescent dyes distamicyn/DAPI (Schweizer et al. 1979); they can be positively stained with silver because they contain nucleolar organizers (Tantravahi et al. 1976; Anderle et al. 1979) and they are frequently involved in chromosome associations. They contain 18S and 28S ribosomal DNA (Henderson et al. 1976; Gosden et al. 1978) and DNA that is homologous to various human satellite DNAs (Gosden et al. 1977; Mitchell et al. 1977; Seuánez 1979). In man, these kinds of chromosome regions tend to be the most variable spots in the karyotype, showing many polymorphisms including duplications and deletions. As such, the discovery of variant chromosome types in the orang utan, involving deletion of this specific type of chromosome material is not surprising. More detailed study of these regions, using various techniques, may even reveal the existence of more minor structural polymorphisms in the orang utan karyotype.

In Fig. 14 an attempt is made to reconstruct the phylogenetic history of the orang utan with respect to its intraspecific karyotypic variation.

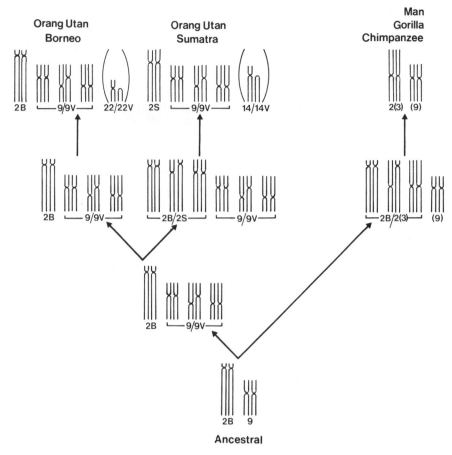

Fig. 14. Possible evolutionary pathways with regard to chromosome pairs 2, 9, 14 and 22 of the orang utan. 9/9v, 2B/2S and 2B/2(3) indicate stages of polymorphisms, during which three possible combinations of the original and the derived chromosome of the chromosome pair in question are found in the same population. (Note that chromosome 2 of the apes is homologous to chromosome 3 in man, and that orang utan chromosome 9 is homologous to chromosome 12 of man, and chromosome 10 of chimpanzee and gorilla.)

Relevance of karyotypic data to captive breeding and conservation of the orang utan

For captive breeding in zoological gardens and primate centres the structural difference in the second chromosome pair between Bornean and Sumatran orang utans is of special interest. In spite of various reports pointing to the external differences between both subspecies (mainly concerning hair length and colour, beard growth, skin colour, relative positions of the eyes and nostrils and the shape of the facial mask; Chasen 1940; Van Bemmel 1969; Jones 1969; MacKinnon 1974) in practice subspecific identification on the basis of external features sometimes causes considerable problems. As a consequence, Sumatran and Bornean animals have been accidentically confused on many occasions.

161

Moreover, some institutions have not even attempted to separate the two forms because of ignorance of the existence of valid subspecific distinction. It goes without saying that the resultant hybrid animals were even more difficult to discern from pure Bornean and Sumatran orang utans.

Since reliable data on the exact geographic origin for many of the captive orang utans are missing, the study of their karyotypes now offers a relatively simple and conclusive possibility for subspecific identification. This method has the great advantage that the results are not debatable; chromosome number 2 structure is either Bornean or Sumatran, without the variability range that is always found with conventional morphological characteristics. For any well-equipped routine human cytogenetics laboratory the methodology of analysing orang utan karyotypes is comparatively simple, so that proper identification should be unproblematic for most institutions maintaining these animals. The only difficulty lies in the fact that a small volume (a few millilitres) of aseptically drawn venous blood is required for the test, because of which sub-adult and adult animals need to be anaesthetised. Modern anaesthetics, however, minimise risks, and medical care of captive apes currently requires more frequent anaesthetisation, so that blood samples can be obtained from the majority of captive animals within a relatively short time.

The importance of chromosomal identification is clearly indicated by our study of 72 animals. Among these there are as many as 12 hybrid orang utans, most of which were bred accidentally. Apart from that, some eight animals were found whose subspecific status was wrongly indicated in the original orang utan studbook data (see Jones, 1980). Using chromosomal identification it should be possible to determine exact origins of all captive orang utans within the next few years. This study must indeed be carried out quickly as is illustrated by the limitations of this method: it only enables proper identification of wild-caught animals and their pure-bred offspring and of first generation hybrids. It does not allow the identification of the offspring of hybrid animals (see Fig. 15). The offspring of a pair of hybrid orang utans, for instance, may either show the chromosome pair 2 structure of a hybrid (chance $\frac{1}{2}$), of a pure Bornean (chance $\frac{1}{4}$) or of a pure Sumatran animal (chance $\frac{1}{4}$). For this reason, all captive animals should be identified before too many second and third generation captive born orang utans are produced.

For captive breeding policy, the question as to what to do with the existing hybrid orang utans (a total of some 95 animals living in captivity are registered in the studbook, see Jones 1980 and 1982) is of crucial importance. It goes without saying that, whenever possible, the breeding of pure Bornean and Sumatran animals is preferable to producing hybrids and breeding with them. The present birth rates and numbers of pure specimens of both forms in captivity (approximately 350 for the Bornean and 250 for the Sumatran subspecies) seem to be amply sufficient to undertake separate, pure Bornean and Sumatran breeding programmes. Mixing of the two forms in order to minimise risks of inbreeding, in future generations, seems not at all necessary (De Boer 1982a). However, should this be reason to totally exclude the presently existing hybrid animals from breeding? Of prime importance in this respect is the question as to the fertility of hybrid orang utans. On the basis of the

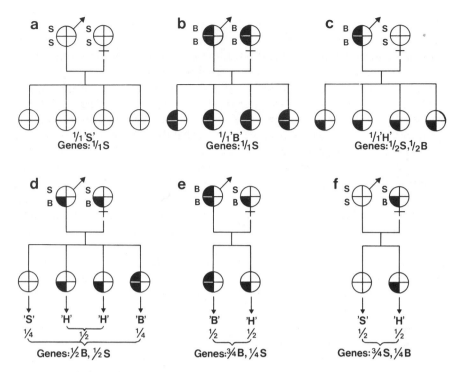

Fig. 15. Morphology of chromosome pair number 2 in the offspring of different crossings in the orang utan. The top row shows crosses between pure Sumatran (a), between pure Bornean (b) and between Bornean × Sumatran (c) animals. In these cases the offspring can always be identified by their chromosome pair 2 morphology. The bottom row gives crosses between two hybrids (d), between a hybrid and a pure Bornean animal (e) and between a hybrid and a pure Sumatran animal (f). In these cases the offspring cannot always be identified by their chromosome pair 2 morphology. Symbols used are the same as those in Fig. 7. 'B', 'S' and 'H' indicate the Bornean, Sumatran and hybrid types of chromosome pair 2 morphology.

structural differences in the second chromosome pair, hybrids, being heterozygous for the pericentric inversion, theoretically would have a certain risk of producing unbalanced gametes (Fig. 16) leading to a risk of embryonic death or malformations in the offspring, and thus to reduced fertility. However, the pericentric inversion in this chromosome covers only approximately one quarter of the chromosome length, which considerably reduces the possibility of the occurrence of crossing-over within this region during first meiotic division. In addition, we do not know whether mechanisms exist to suppress crossing-over in this chromosome segment, like those which possibly act in the inverted segment in chromosome 9 (see above). Finally, during the period of chromosome polymorphism, that supposedly existed after the initial appearance of the derived (Sumatran) chromosome 2, heterozygous animals must have occurred frequently and must have been fertile.

Recently, Seuánez and Fletcher (1978) had the opportunity of examining a hybrid orang utan of seven years of age following hemicastration (animal HI8 in Fig. 7). The study of this animal's testicular histology revealed testicular tubules devoid of spermatogenetic activity (Fig. 17), since the majority of

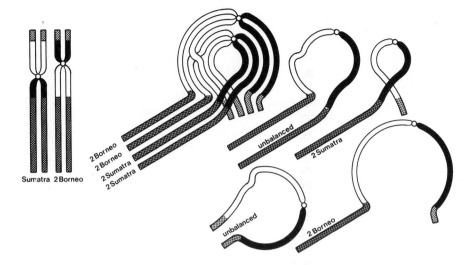

Fig. 16. The theoretical meiotic configuration in bivalent number 2, formed in a Sumatran × Bornean cross-bred orang utan, and the products of meiotic disjunction in the resultant gametes after the occurrence of a cross-over within the inversion loop. For further comments see Fig. 12.

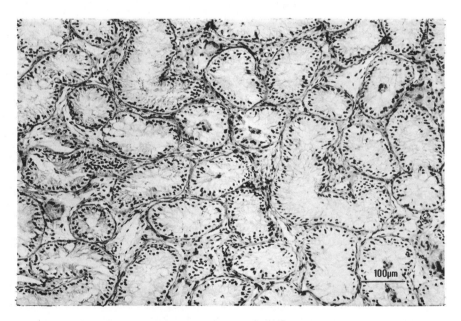

Fig. 17. Testicular histology of the hybrid orang utan Tunku (specimen HI8 in the pedigrees of Fig. 7), showing total absence of spermatogenic activity (from Seuánez and Fletcher 1980).

tubules were lined with Sertoli cells only, and there was no evidence of spermatogenesis in any of them. The testis was reduced in size (weight 5.0 g) and no spermatozoa were found in the epididymis. The reasons for these findings are as yet unknown; they cannot be explained as being solely a consequence of the heterozygozity for the pericentric inversion in the second chromosome pair.

As opposed to these findings, so far nine hybrid orang utans are known to have reproduced successfully. Two of these animals (HI23 and HI24) are included in the pedigrees of Fig. 7; the karyotype of one of them has been studied and the heterozygosity in the second pair of chromosomes was confirmed. The orang utan studbook data (Jones 1980 and 1982) reveal information on seven more hybrids which have bred (one male and six females). Since they all started breeding rather recently (the first birth was recorded in 1971), it is impossible at this moment to assess whether or not these hybrids are as fertile as pure-bred animals. Of the total of 15 conceptions recorded for hybrids (among which there is one case of five and another of four conceptions in a single animal), one ended in an abortion and another in a stillbirth. Again these figures are too low to justify any estimation of abortion and stillbirth frequencies in hybrids as compared to those in pure-breds.

As indicated above, the existence of a chromosomal difference between the two orang utan subspecies gives reason to envisage the possible existence of (many) more as yet undetected differences which might have negative influence on hybrid fertility. Therefore, it would seem of great importance to collect as many data as possible in this connection, including data from breeding experiments. At this moment it would seem appropriate to establish three separate orang utan lines, one for pure Bornean orang utans, another for pure Sumatran orang utans and a small third one for the hybrids. The latter will obviously provide abundant information on hybrid fertility that may be useful for establishing a future breeding policy (see also De Boer 1982a).

Since the chromosome number 9 double inversion found in both Bornean and Sumatran orang utans apparently represents a natural polymorphism there is no reason whatsoever to impose any (unnatural) selection in this respect. Heterozygous carriers should be allowed to breed exactly like the homozygotes, and any animal should be crossed with any other regardless of the fact that they might be homozygous for the normal 9, homozygous for the variant 9 or heterozygous for the double inversion. The same applies for other possible chromosome polymorphisms.

The karyotypic difference between Bornean and Sumatran animals in fact is of less practical importance for the conservation of orang utans in their natural habitats, since here we are dealing with the conservation of two geographically isolated populations between which no exchange of genes takes place. Nevertheless, it is crucial to know that the two populations are actually different, which indicates that they are in no way supplementary: both deserve to be conserved as independent, unique populations.

Natural chromosome polymorphisms are of more importance to the conservation of feral orang utans. As mentioned already in the introduction to this paper, previously various (sub)species of orang utans have been described, suggesting the existence of considerable variability within the two distribution areas. More recent reports also mention this phenomenon. Rijksen (1978), for instance, describes two clearly distinguishable types of orang utans from the same locality in Sumatra (the Gunung Leuser Reserve). It is not impossible that in such cases subpopulations or local races, which were originally kept separate, were driven into each other's distribution area as a result of the reduction of their own home ranges by habitat destruction, thus resulting in a mixed population. In this respect, together with studies of other genetic markers (Bruce and Ayala 1979; Meera Khan et al. 1982; De Boer and Meera Khan 1982; De Boer 1982a), examination of the karyotypes of wild living orang utans may provide an important tool in understanding population structure and population dynamics. Chromosome polymorphisms are extremely valuable in detecting the possible existence of intra-subspecific (and/or local) differences. This type of knowledge may prove to be of crucial importance for a future conservation policy for the orang utan in its natural home land.

Acknowledgements

The authors are greatly indebted to all institutions that supplied them with blood samples of their orang utans. These institutions are listed in the heading of Table 2. We are also greatly obliged to Mr R.H.R. Belterman (Rotterdam Zoo) for technical assistance in karyotyping a large orang utan sample, and to Dr A.R. Glatston (Rotterdam Zoo) for reading and correcting the manuscript. The drawn illustrations were prepared by Mr F.A.C.M. Hollink and Mr G. Looten (both of Rotterdam Zoo). Finally, we want to express our thanks to Mr M.L. Jones (Zoological Society of San Diego, Studbook keeper for the orang utan) for supplying us with the studbook data of the orang utans studied cytologically.

Addendum

While this paper was in press, 11 additional orang utans were studied cytogenetically in the Biological Research Department of the Royal Rotterdam Zoological and Botanical Gardens, bringing the total of animals in this study up to 83.

Three of these additional animals are already included in the pedigrees of Fig. 7 and the list of Table 2. They are:

B_I42 *Sylvia*   *00833B F,*   which is homozygous for the Bornean chromosome 2, and homozygous for the variant chromosome 9.

$S_{II}34,$ *Sexta,*   *01251S F,*   which is homozygous for the Sumatran chromosome 2, and homozygous for the normal chromosome 9.

$H_{II}1,$ *Klaasje,*   *-------X M,*   which is homozygous for the Bornean chromosome 2, and homozygous for the normal chromosome 9.

A further three of these animals are recently born offspring of parents already shown in the pedigrees of Fig. 7 and in Table 2. They are (having been consecutively numbered):

$B_{II}21$,	Leonie,	------B F,	see pedigree below.
$B_{II}22$,	Indra,	------B F,	see pedigree below.
$B_{II}23$,	Sandra,	------B F,	offspring of B_I16 and B_I17, being homozygous for the Bornean chromosome 2, and homozygous for the normal chromosome 9.

The remaining five additional animals belong to a family of orang utans from the Basle Zoological Gardens, which is unrelated to any of the other individuals mentioned in this paper. To this family belong the following animals (the numbers are given consecutively to those of Fig. 7 and Table 2; animals printed in italics have been studied cytologically):

B_I45,	Niko,	00153B M		H_I26,	Xempaka,	01458B M
S_I37,	*Elsi,*	*00446B F*		H_I27,	Bulu,	------B M
S_I38,	*Kiki,*	*00154S F*		H_I28,	*Kasi,*	*00485X F*
H_I25,	*Suma,*	*01077B M*		$H_{II}3$,	*Atjeh,*	*01646X F*

The results of the karotype studies of these animals are shown in the pedigree below. Because of the apparent misidentification of S_I37 as Bornean while the chromosome pair 2 structure determines it as Sumatran, the animals H_I25, 26 and 27 are not pure Borneans as indicated in the Studbook of the Orang utan, but hybrids. The latter was proven in one of them (H_I25). It is noteworthy that $H_{II}3$ is a half second generation hybrid orang utan with the chromosome pair 2 structure of a pure Bornean animal (compare Fig. 15). Chromosome 9 structure was not tested in the Basle animals.

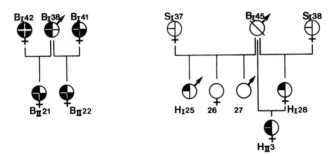

References

Anderle, M., W. Fiedler, A. Rett, P. Ambros and D. Schweizer (1979). A case of trisomy 22 in *Pongo pygmaeus*. Cytogenet. Cell Genet., 24: 1–6.

Bemmel, A.C.V. van (1968). Contribution to the knowledge of the geographical races of *Pongo pygmaeus* (Hoppius). Bijdragen tot de Dierkunde, 38: 13–15.

Bender, M.A. and E.H.Y. Chu (1963). The chromosomes of primates. In: J. Beuttner-Janusch (ed.), Evolutionary and genetic biology of the primates. Academic Press, New York.

Beneviste, R.E. and G.J. Todaro (1976). Evolution of type C viral genes: evidence for an Asian origin of man. Nature (London), 261: 107–108.

Bobrow, M. and K. Madan (1973). A comparison of chimpanzee and human chromosomes using the Giemsa 11 and other chromosome banding techniques. Cytogenet. Cell Genet., 12: 107–116.

Boer, L.E.M. de (1974). Cytotaxonomy of the Platyrrhini (Primates). Genen Phaenen, 17: 1–115.

Boer, L.E.M. de (1982a). Genetics and conservation of the orang utan. In: L.E.M. de Boer (ed.), The orang utan. Its biology and conservation. Junk, The Hague.

Boer, L.E.M. de (1982b). Karyotaxonomy. In: A. Coimbra-Filho and R.A. Mittermeier (eds.), An introduction to New World primatology (in press).

Boer, L.E.M. de and R.H.R. Belterman (1982). A karyotypic study in 53 Bornean, Sumatran and cross-bred orang utans. (In preparation.)

Boer, L.E.M. de and P. Meera Khan (1982). Haemoglobin polymorphisms in Bornean and Sumatran orang utans. In: L.E.M. de Boer (ed.), The orang utan. Its biology and conservation. Junk, The Hague.

Boer, L.E.M. de and C. van Oostrum-van der Horst (1975). A note on the karyotypes and chromosome associations of the Hylobatidae, with special reference to *Hylobates* (*Nomascus*) *concolor*. J. Hum. Evol., 4: 559–564.

Boer, L.E.M. de and J.W.F. Reumer (1979). Reinvestigation of the chromosomes of a male owl monkey (*Aotus trivirgatus*) and its hybrid son. J. Hum. Evol., 8: 479–489.

Bruce, E.J. and F.J. Ayala (1979). Phylogenetic relationships between man and the apes: electrophoretic evidence. Evolution, 33: 1040–1056.

Chasen, F.N. (1940). The mammals of the Netherlands Indian Mt. Leuser expedition 1937 to North Sumatra. Treubia, 17: 479–502.

Chiarelli, B. (1961). Chromosomes of the orang utan (*Pongo pygmaeus*). Nature (London), 192: 285.

Chiarelli, B. (1962a). Comparative morphometric analysis of primate chromosomes. I. The chromosomes of the anthropoid apes and of man. Caryologia, 15: 99–121.

Chiarelli, B. (1962b). Some new data on the chromosomes of Catarrhina. Experientia, 18: 405–407.

Chiarelli, B. (1963). Comparative morphometic analysis of primate chromosomes. III. The chromosomes of the genera *Hylobates*, *Colobus* and *Presbytis*. Caryologia, 16: 637–648.

Chu, E.H.Y. & M.A. Bender (1961). Chromosome cytology and primate evolution. Science, 133: 1399–1405.

Dutrillaux, B., M.O. Rethoré and J. Lejeune (1975). Comparaison du caryotype de l'orang utan (*Pongo pygmaeus*) à celui de l'homme, du chimpanzé et du gorille. Ann. Génét., 18: 153–161.

Goodman, M. (1975). Protein sequence and immunological specificity. Their role in phylogenetic studies of the primates. In: W.P. Luckett and J.S. Szalay (eds.), Phylogeny of the primates. Plenum, New York.

Gosden, J.R., A.R. Mitchell, H.N. Seuánez and C.M. Gosden (1977). The distribution of sequences complementary to human satellite DNAs I, II and IV in the chromosomes of chimpanzee (*Pan troglodytes*), gorilla (*Gorilla gorilla*) and orang utan (*Pongo pygmaeus*). Chromosoma (Berl.), 63: 253–271.

Gosden, J.R., S. Laurie & H. Seuánez (1978). Ribosomal and human-homologous DNA distribution in the orangutan (*Pongo pygmaeus*). Cytogenet. Cell Genet., 21: 1–10.

Grouchy, J. de, C. Turleau, M. Roubin and F. Chavin-Colin (1973). Chromosomal evolution of man and the primates. In: T. Caspersson and L. Zech (eds.), Chromosome identification, techniques and applications in biology and medicine. Academic Press, New York.

Hamerton, J.L., M. Fraccaro, L. de Carli, F. Nuzzo, H.P. Klinger, L. Hullinger, A. Taylor and E.M. Lang (1961). Somatic chromosomes of the gorilla. Nature (London), 192: 225–228.

Hamerton, J., H.P. Klinger, D.E. Mutton and E.M. Lang (1963). The somatic chromosomes of the Hominoidea. Cytogenetics, 2: 240–263.

Henderson, A.S., K.C. Atwood and D. Warburton (1976). Chromosomal distribution of rDNA in *Pan paniscus*, *Gorilla gorilla beringei* and *Symphalangus syndactylus*: comparison to related primates. Chromosoma (Berl.), 59: 147–155.

Hösli, P. and E.M. Lang (1970). Die chromosoman des Zwergsiamang (*Symphalangus klossi*). Schweizer Archiv für Tierheilkunde, 112: 296–297.

Hoyer, B.H., N.W. van de Velde, M. Goodman and R.B. Roberts (1972). Examination of hominid evolution by DNA sequence homology. J. Hum. Evol., 1: 645–649.

Jones, M.L. (1969). The geographical races of orangutan. Proc. 2nd Int. Congr. Primat., vol. 2: 217–223. Karger, Basel.

Jones, M.L. (1980). Studbook of the orang utan, *Pongo pygmaeus*. Zoological Society of San Diego, San Diego.

Jones, M.L. (1982). The orang utan in captivity. In: L.E.M. de Boer (ed.), The orang utan. Its biology and conservation. Junk, The Hague.

Klinger, H.P., J.L. Hamerton, D. Mutton and E.M. Lang (1963). The chromosomes of the Hominoidea. In: S.L. Washburn (ed.), Classification and human evolution. Aldine, Chicago.

Koenigswald, G.H.R. von (1982). Distribution and evolution of the orang utan, *Pongo pygmaeus* (Hoppius). In: L.E.M. de Boer (ed.), The orang utan. Its biology and conservation. Junk, The Hague.

Lucas, M., C. Page and M. Tanmer (1973). Chromosomes of the orangutan. Jersey Wildlife Preservation Trust Ann. Rep., 1973: 57–58.

Ma, N.S.F., T.C. Jones, R.W. Thorington and R.W. Cooper (1974). Chromosome banding patterns in squirrel monkey (*Saimiri sciureus*). J. Med. Primat., 3: 120–137.

168

MacKinnon, J.R. (1974). The behaviour and ecology of wild orang utans (*Pongo pygmaeus*). Anim. Behav., 22: 3–74.

Markvong, A. (1973). Karyotypes of three species of gibbons. Mamm. Chrom. Newsl., 14: 149.

McClure, H.M., K.H. Belden, W.A. Pieper and C.B. Jacobson (1969). Autosomal trisomy in a chimpanzee: resemblance to Down's syndrome. Science, 165: 1010–1012.

McClure, H.M., K.H. Belden, W.A. Pieper, C.B. Jacobson and D. Picciano (1970). Cytogenetic studies and observations in the Yerkes great ape colony. Med. Primat. (Proc. 2nd Conf. exp. Med. Surg. Primates, New York 1969): 281–296.

McClure, H.M., W.A. Pieper, M.E. Keeling and C.B. Jacobson (1971). Mongoloid-like condition in a chimpanzee. Proc. 3rd int. Congr. Primat., Zürich 1970, vol. 2: 110–115. Karger, Basel.

Meera Khan, P., H. Rijken, J. Wijnen, L.M.M. Wijnen and L.E.M. de Boer (1982). Red cell enzyme variation in the orang utan; Electrophoretic characterization of 45 enzyme systems in Cellogel. In: L.E.M. de Boer (ed.), The orang utan. Its biology and conservation. Junk, The Hague.

Mitchell, A.R., H.N. Seuánez, S.S. Lawrie, D.E. Martin and J.R. Gosden (1977). The location of DNA homologous to human satellite III DNA in the chromosomes of chimpanzee (*Pan troglodytes*), gorilla (*Gorilla gorilla*) and orang utan (*Pongo pygmaeus*). Chromosoma (Berl.), 61: 345–358.

Moorhead, P.S., P.C. Nowell, W.J. Mellman, D.M. Battips and D.A. Hungerford (1960). Chromosome preparations of leucocytes cultured from human peripheral blood. Exp. Cell Res., 20: 613–616.

Paris Conference (1971), Supplement 1975. Standardization in human cytogenetics. Birth Defects, Original Article Series, vol. XI, 9. The National Foundation, New York.

Pearson, P.L., M. Bobrow, C.G. Vosa and P.N. Barlow (1971). Quinacrine fluorescence in mammalian chromosomes. Nature (London), 231: 326–329.

Reumer, J.W.F. and L.E.M. de Boer (1980). Standardization of *Aotus* chromosome nomenclature. J. Hum. Evol., 9: 461–482.

Rijksen, H.D. (1978). A fieldstudy on Sumatran orang utans (*Pongo pygmaeus abelii* Lesson 1827); ecology, behaviour and conservation. Meded. Landbouwhogeschool, Wageningen, 78–2: 1–420.

Sarich, V. and J.E. Cronin (1976). Molecular systematics of the primates. In: M. Goodman, R.E. Tashian and J. Tashian (eds.), Molecular anthropology, genes and proteins in the evolutionary ascent of the primates. Plenum Press, London.

Schweizer, D., P. Ambros, M. Anderle, A. Rett and W. Fiedler (1979). Demonstration of specific heterochromatic segments in the orangutan (*Pongo pygmaeus*) by a distamycin/DAPI double staining technique. Cytogenet. Cell Genet., 24: 1–14.

Selenka, E. (1896). Die Rassen und der Zahnwechsel des Orang-utan. Sitzungsberichte der K. Preuss. Akad. der Wissensch. zu Berlin, 16: 381–393.

Selenka, E. (1898). Menschenaffen: Rassen, Schädel und Bezahnung des Orang-utan. Kreidel, Wiesbaden.

Seuánez, H.N. (1977). Chromosomes and spermatozoa of the great apes and man. Ph.D. Thesis, University of Edinburgh.

Seuánez, H.N. (1979). The phylogeny of human chromosomes. Springer, Heidelberg.

Seuánez, H.N. and J. Fletcher (1978). Chromosomes of the orang utan. Dodo, J. Jersey Wildlife Preservation Trust, 15: 77–80.

Seuánez, H., J. Fletcher, H.J. Evans and D.E. Martin (1976a). A chromosome rearrangement in an orang utan studied with Q-, C- and G-banding techniques. Cytogenet. Cell Genet., 17: 26–34.

Seuánez, H., J. Fletcher, H.J. Evans and D.E. Martin (1976b). A polymorphic structural rearrangement in the chromosomes of two populations of orangutans. Cytogenet. Cell Genet., 17: 327–337.

Seuánez, H., A.C. Carothers, D.E. Martin and R.V. Short (1977). Morphological abnormalities in the spermatozoa of man and the great apes. Nature (Lond.), 270: 345–347.

Seuánez, H., H.J. Evans, D.E. Martin and J.A. Fletcher (1978). A chromosomal difference between Bornean and Sumatran orangutans. 186th Meeting Genetical Society of Great Britain (Abst.) Heredity, 41: 127.

Seuánez, H., H.J. Evans, D.E. Martin and J. Fletcher (1979). An inversion in chromosome 2 that distinguishes between Bornean and Sumatran orang utans. Cytogenet. Cell Genet., 23: 137–140.

Stockholm Conference (1977). An international system for human cytogenetic nomenclature, 1978. I.S.C.N. (1978). Cytogenet. Cell Genet., 21: 313–409 (1979).

Tantravahi, R., V.G. Dev, I.L. Firschein, D.A. Miller and O.J. Miller (1975). Karyotype of the gibbons *Hylobates lar* and *H. moloch*: inversion in chromosome 7. Cytogenet. Cell Genet., 15: 92–102.

Tantravahi, R., D.A. Miller, V.G. Dev and O.J. Miller (1976). Detection of nucleolus organizer regions in chromosomes of human, chimpanzee, gorilla, orang utan and gibbon. Chromosoma (Berl.), 56: 15–27.

Turleau, C., J. de Grouchy and M. Klein (1972). Phylogénie chromosomique de l'homme et des primates hominiens (*Pan troglodytes, Gorilla gorilla* et *Pongo pygmaeus*). Essai de réconstitution de l'ancêtre comun. Ann. Génét., 15: 225–240.

Turleau, C., J. de Grouchy, F. Chavin-Colin, J. Mortelmans and W. van den Bergh (1975). Inversion péricentrique du 3, homozygote et héterozygote, et translation centromérique du 12 dans une famille d'orangutans: implications évolutives. Ann. Génét., 18: 227–233.

Turleau, C., J. de Grouchy, F. Chavin-Colin and J. Mortelmans (1976). Inversion péricentrique du 3, homozygote et héterozygote et translation centromérique du 12 dans une famille d'orangs-outangs. Implications évolutives. Acta Zool. Pathol Antverpiensa, 64: 69–79.

Warburton, D., A.S. Henderson and K.C. Atwood (1975). Localization of rDNA and Giemsa-banded chromosome complement of white-handed gibbon, *Hylobates lar*. Chromosoma (Berl.), 51: 35–40.

White, M.J.D. (1977). Modes of speciation. Freeman, San Francisco.

Wijnen, J. Th., H. Rijken, L.E.M. de Boer and P. Meera Khan (1982). Glucose-6-phosphate dehydrogenase (G6PD) variation in the orang utan. In: L.E.M. de Boer (ed.), The orang utan. Its biology and conservation. Junk, The Hague.

Wurster, D. and K. Benrischke (1969). Chromosomes of some primates. Memm. Chrom. Newsl., 10: 3–5.

Yeager, C.H., T.S. Painter and R.M. Yerkes (1940). The chromosomes of the chimpanzee. Science, 91: 74–75.

Young, W.J., T. Mergt, M.A. Ferguson-Smith and A.W. Johnston (1960). Chromosome number of the chimpanzee (*Pan troglodytes*). Science, 131: 1672–1673.

Authors' addresses:
L.E.M. de Boer
Biological Research Department
Royal Rotterdam Zoological and Botanical Gardens
Rotterdam
The Netherlands

H.N. Seuánez
Laboratorio de Citogenetica Humana
Universidade Federal do Rio de Janeiro
Rio de Janeiro
Brazil

9. Veterinary aspects of the maintenance of orang utans in captivity

David M. Jones

Introduction

Although the orang utan is kept by a large number of zoological collections, very little information has been published on the veterinary aspects of its management. Because the chimpanzee is being used extensively in biomedical research, more is known of the physiology and medicine of this species and much of that information is of relevance to the care of the other apes. The attention of the reader is drawn to the six volume series of books on the chimpanzee (Bourne 1969–73) which, apart from summarising most of the published data on that species, also provides a wealth of comparative data on the gorilla and orang utan.

In this paper, the author has concentrated specifically on the available information on the orang utan, based on published work, a survey of the results of post-mortem examinations from many collections around the world, and on the management and research data collected by staff of the Zoological Society of London.

General management

The orang utan is the least gregarious of the apes. Although strong bonds may exist between adult females and territorial males in the wild, close associations will only form for mating and are therefore temporary. After leaving their mothers, juveniles often form loosely attached groups which last until the animals are adolescent. Groups of young animals can also be set up in captivity, but adult pairs cannot usually be kept together for any length of time, and they often have to be separated once the female is pregnant and during the period that she is nursing an infant. This means that even a relatively small breeding collection of orang utans takes up a considerable amount of room and most zoos must be prepared to manage this species jointly with others in order to make the best of their animals and facilities.

Orang utans very quickly become bored and lethargic if sensory stimulation is not provided in a number of ways and if there is insufficient room to exercise. An outdoor area of not less than 15 m long, 10 m wide and at least 4 m high can

The orang utan. Its biology and conservation, edited by L.E.M. de Boer

be used in turn by six adults, either individually or in pairs. Each animal should have its own heated den of at least 6 m² floor area and 3 m in height with, in addition, access for part of every day to an indoor exercise area of at least 15 m² floor area and 3 m in height. Ten gauge weldmesh of 4 × 10 cm rectangles is used in London and is strongly wired or welded to a steel frame. Poor quality workmanship in cage construction will soon be found by these animals which are capable of undoing machine-tightened nuts and bolts. They will also use levers and rocks to break mesh and glass if such 'tools' are available. Laminated glass of 3.2 cm in thickness has also been successfully used at London. All floors and walls should be constructed of materials which are easy to wash.

It is most important with this species to provide possibilities for climbing and moving horizontally at a high level in the cage. Resting platforms should also be placed well above the ground. As many playthings as possible should be provided. Robust playground toys for children are ideal but expensive and have to be replaced frequently. Only stiff ropes and chains must be used because younger animals in particular have a tendency to hang themselves in flexible ropes. Water moats or pools of more than 30 cm in depth are dangerous to orang utans and may lead to deaths from drowning. All corners and edges of the enclosure should be rounded and smooth to prevent injuries especially to the female when she is chased by the male. Orang utan hair is very brittle and easily damaged by abrasive surfaces. Even young animals of 3–4 years can be very strong and facilities should ideally be provided to catch animals for hand injection or at least to allow the use of a projectile dart. Orang utans, like other apes, become adept at hiding in corners where they are impossible to hit with a dart and this must be taken into account when designing accommodation. Young animals of up to 3 years of age are usually safe to handle when fully conscious, but beyond this age orang utans should not be trusted. The young of all the great ape species can be kept together up to adolescence but the risks of injury increase beyond this stage.

In the wild, orang utans do not experience marked changes in the ambient temperature. Usually these would vary from 20° to 30°C although the extreme ranges might be from 17° to 35°C. The relative humidity of the Asian tropical rain forests remains very high at from 65% to 90%. Orang utans do not acclimatise to temperate conditions as well as chimpanzees and although they can tolerate temperatures as low as 5°C for short periods they quickly become chilled. They must therefore always have access to a warm den and a relative humidity of at least 60% is recommended as levels below this probably predispose to respiratory disease particularly in young animals.

Nutrition

From the now considerable volume of observational data from the wild, orang utans have been confirmed as being mainly vegetarian. Between July and

---

Fig. 1 (*opposite page*). Orang utans need plenty of exercise. Adult female orang utan using the space frame in the top of her cage at London.

November they feed principally on fruit, mainly durians, figs and rambutans. At other times of the year they feed on new leaves and shoots, epiphytes and liana bark. Most of the primates including the orang utan also take some animal protein when the opportunity arises. Insects, young birds, birds' eggs and small mammals are regularly eaten and soil is also occasionally ingested. The animal items of the diet are trapped manually and transferred to the mouth, but branches are usually bent towards the mouth and the fruit or leaves plucked off with the lips. Larger fruits are either bitten open or broken up with the nails.

The intestinal tract is very long with all the cellulose digestion being undertaken in the capacious caecum. The stomach is that of a typical monogastric but is very capacious. The teeth are deeply rooted and the permanent dentition consists of 8 incisors, 4 canines, 8 premolars and 12 molars.

Although orang utans will drink from puddles and off branches and leaves after a rainstorm, they also rely on the water content of the leaves and fruit consumed for part of their water supply. The faeces are usually relatively soft and poorly formed. Although generally they defecate at random, they will never soil their nests or beds.

There is often a tendency to believe that, because animal protein forms a relatively small part of the diet of orang utans in the wild, it is unnecessary to include it in the design of the diet for captive animals. It has to be remembered that in the wild the animal is able to select and eat 'growing' foods which almost certainly include proteins of a high biological value. Foods bought for an urban zoo may have been stored for long periods, and often deteriorate in nutritional value. The variety of foods given to a captive ape will be far less than that available to it in the wild and it is therefore more likely that a deficiency of some important item will result in a zoo. For this reason, the addition of a carefully balanced supplement of animal proteins, vitamins and minerals to the more obvious items of the diet is essential. Most of the energy should be provided as carbohydrate with a moderate amount of protein and fibre together with low levels of fats. The diet should be based on a commercial primate pellet made by a company with a good record of quality control. On a dry weight basis the ratios of the pellet to vegetables and fruit should be 3:2:1. When introducing the pellet to animals which have not eaten it before, it may be necessary to feed it alone in the morning and not offer vegetables and fruit until the pellet has been finished.

As an approximate guide, a 50 kg, moderately active adult female requires about 2,700 kcal of energy every day. This should be provided in about 2 kg of bulk (about 1 kg of dry matter). About 20% of the dry weight of the diet should be digestible crude protein and 15–20% should be fibre. The fat content should be kept down to about 5%. The diet for such an animal should also be calculated to provide a daily level of not less than 8 g of calcium, 7 g of phosphorus, 20,000 iu of vitamin A, 4,000 iu of vitamin D_3 and 50 iu of vitamin E. Double these levels would be required by a large male and a female nursing an infant. It needs to be stressed that these are guidelines only and that in the final analysis it is the manager of the animals who has to decide the actual quantities of food needed by each individual at any given time.

Food can be used to stimulate activity in captive orang utans and it is preferable to give the bulk of the items in two to three separate meals a day, supplementing this by scattering nuts and seeds in the enclosure for the animals to search out. Water should be provided at all times from a fitting which is not easily damaged by the animals.

Reproduction and rearing

It is generally thought that male orang utans are mature at 10 years of age and females at about 9 years, although records exist of fertile matings occurring in 7-year-old females. In captivity, orang utans will breed throughout the year. The male has relatively large scrotal testes and a small penis with a slender rudimentary baculum. The female has a simple uterus and one pair of thoracic mammary glands. The menstrual cycle is of approximately 29 days duration and a slight menstrual flow lasting 3–4 days is occasionally noticeable. There is no periodic vulval swelling. Contraception, using steroid 'pills' suitable for humans can be used and intra-uterine devices have been fitted (Florence et al. 1977). In London, it was found to be impossible to fit a coil in a young adult female as the device could not be passed through the cervix.

Animals will copulate at almost any time, even during pregnancy. The male usually mounts from behind but in the wild he will often mate while suspended by his arms and with the female lying on her back.

The gestation period is about 275 days and the course of pregnancy can be followed by assay of the steroids in the urine (Hobson 1976). The vulva begins to swell during pregnancy and becomes very pronounced near the time of birth. The mammary glands become prominent during the last third of the gestation period. Only one young is usually born, but twins have been recorded. No wild births have been observed but observations on captive animals suggest that the young of apes are usually born in anterior presentation. The birth weight of orang utans is usually in the range of 1.4–1.6 kg. The mother bites through the umbilical cord and sometimes eats all or part of the foetal membranes. Births usually take place at night. The female may hold the infant in both hands to nurse it or it may suck while clinging to the fur on the mother's chest with the mother giving it additional support when necessary. The sucking infant often squeezes the mother's breast with one hand. Older infants may be carried sideways but usually in a position where they can reach the nipple. Suckling usually takes place frequently and for short periods but there is considerable individual variation in this. The newborn orang utan usually takes milk within 6 hours of birth but occasionally up to 2 days will elapse before this happens. Young animals should not therefore be taken away for hand-rearing unless they are obviously ill or 2 days have passed without them feeding. The infant takes its first solid foods 3–4 months after birth, although it will mouth solids before this time. By 9 months of age, the animal is taking significant quantities of solid food and it is usually weaned at 2 years of age.

Where pairs of adult orang utans cannot be kept together, the male should be introduced at about days 13–15 of the oestrous cycle. A successful mating is

Fig. 2. Doppler foetometer being used to detect a foetal heartbeat in a pregnant, lightly sedated orang utan.

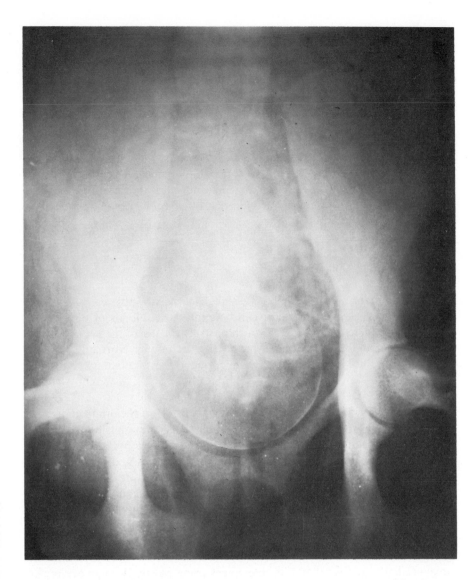

Fig. 3. Radiograph showing the head of a full-term foetus in the birth canal of the same female orang utan on which the foetometer was being used (Fig. 2).

probably more likely if the animals are already familiar with one another. As far as possible only unrelated animals should be allowed to mate and the two subspecies should not be cross bred. It is probably wise to separate the male from the female in late pregnancy as deaths in infants are often related to injuries inflicted by their fathers. Clinical problems involving the reproductive tract often occur in orang utans just before and while giving birth and these are considered later.

In order to prevent boredom, nursing females should be kept within sight and sound of familiar animals but given the option to hide away from them when she so desires. She should be given plenty of bedding and encouraged to move about to gather food. Maternal behaviour usually improves with the second and subsequent infants and it is often unwise in the long run to remove the first or even the second youngster for hand-rearing until it is quite evident that the infant has no chance of survival with the mother. Young animals are probably best left with the mother for $2\frac{1}{2}$ years and if possible allowed to watch the nursing of younger animals. This is thought to be important for the development of maternal behaviour in apes.

Where hand-rearing is absolutely necessary, milk substitutes suitable for human babies can be used. The most suitable (but the most expensive) are those that come prepared in sterilised containers and ready for use after heating. Hand-reared animals should be introduced to other young orang utans as soon as possible after 6 months of age.

Handling

Young orang utans up to 5 kg in body weight can easily be physically restrained in a light net. The same technique can be applied to animals of up to 15 kg providing that there are no objects which obstruct the use of the net in the cage and that the handlers are experienced in this technique. Normally, juvenile orang utans are not aggressive, but when cornered they may attack, especially when the captors are strangers. For animals above 15 kg in weight, stronger equipment such as a mechanical crushing device is necessary. The animal has to be enticed into the device and this is virtually impossible if the orang utan is not used to passing through it daily. Once a mobile crush crate has been used it is unlikely that the animal will enter it again. It is therefore important to build the cage into a passageway through which the animal passes and where on most occasions that it passes through, the crush is not used.

In practice, orang utans over 10 kg are most effectively handled for anything more than a brief examination by sedating them. Often, quiet juveniles can be quickly hand injected if an arm is presented to take food. More usually the use of a projectile syringe is necessary and although metal alloy syringes propelled by compressed gas or an explosive charge are sometimes necessary where the range needed is greater than 7 m, blowpipes using plastic syringes are preferable. They are relatively silent in use and have a low impact velocity, as a result of which the animal does not usually become very excited. Orang utans are generally much easier to hit with such projectile darts than other apes as they tend to try to hide under bedding or cage furniture rather than rush about the enclosure. The dart should be aimed at the upper arm, shoulder or hindquarters. When necessary the operator has to be very patient in order to ensure that the dart hits one of the correct body sites.

Marsboom et al. (1963) and Field et al. (1966) advocated the use of a potent narcotic, in their case Fentanyl citrate, in combination with the butyrophenone sedative Droperidol for primates. The dose rates suggested for apes were 0.02 mg/kg of Fentanyl with 2.0 mg/kg of Droperidol. The author has used this and other mixtures of drugs incorporating a narcotic in a number of primate species.

An immobilising effect is achieved but muscular relaxation is often inadequate for the purposes of clinical examination. Respiration sometimes becomes depressed and the safety margin is poor. Such mixtures do have the advantage that the narcotic effect can be largely abolished with a specific antagonist.

During the last two decades it has become evident that the drugs of choice for primates are the cyclohexamines. Chen and Weston (1960), Rutty and Thurley (1962), Spalding and Heymann (1962) and Melby and Baker (1965) used Phencyclidine hydrochloride. Although this drug is still being used by many workers it is not now commercially available. The dose rate of Phencyclidine in the orang utan is 0.7 to 1.0 mg/kg when given intramuscularly.

Bree (1972) described the use of Ketamine hydrochloride in the rhesus macaque and this agent, a closely related drug to Phencyclidine, is now used extensively in primates. Bonner et al. (1972) describe its use in chimps and other great apes. The dose rate for the orang utan is 8–12 mg/kg. Some workers advocate the use of a sedative such as Azaperone or Acepromazine in combination with Ketamine but this is not necessary in apes as the degree of relaxation is usually adequate for clinical examination. Adult orang utans, especially large males, often salivate profusely when under the influence of cyclohexamines but this seems to be a problem of particular individuals. In these cases, 5–8 mg of atropine sulphate should be added to the Ketamine to reduce salivation. Ketamine is not as potent as Phencyclidine but is shorter acting. Although commercially prepared solutions of Ketamine usually contain 50 or 100 mg/ml of the hydrochloride it is soluble in warm water at a concentration of 200 mg/ml. This means that a 50 kg orang utan can be immobilised using a single 2.5 ml capacity blowpipe dart but that a second dart or a metal alloy syringe with a greater capacity has to be used for larger animals. Phencyclidine with a potency of 10 times that of Ketamine is therefore a more practical drug to use because the volume required to immobilise the animal is much smaller. Anyone who still holds stocks of Phencyclidine should keep this drug available for use on their apes. The drug's effective shelf life when kept in a cool cupboard is probably 10 years.

A number of other cyclohexamines, notably Tiletamine, have been used on primates, but for various reasons they have not become generally available. A mixture of 5 mg/kg of Ketamine and 0.5 mg/kg of Xylazine has been shown to have a similar effect to the larger dose suggested for Ketamine alone (10 mg/kg) in smaller primates and should prove in time to be a practical mixture for apes as well.

When under sedation, adult male orang utans especially are at risk from asphyxia, not only by inhalation of saliva, but also because the flaccid soft palate and walls of the pharynx tend to occlude the laryngeal opening during inspiration. It is often wise to apply a gag to these animals and insert a cuffed endotracheal tube where this is practical. Adult orang utans will take endotracheal tubes of from 9 to 14 mm in diameter depending on their body weight. Anaesthesia for surgery can be induced following initial sedation with Ketamine by giving Sodium Thiopentone intravenously at a dose rate of 3–5 mg/kg. If intubation has not already been carried out, the addition of the barbiturate will facilitate this. This dose of Thiopentone after Ketamine will give 10 min of anaesthesia which may be prolonged by giving additional Thiopentone or by connection of the endotracheal tube to a conventional closed circuit system for administration of a volatile anaesthetic. Halothane in oxygen at a concentration of 2% to 4% is usually used. Orang utans have a tendency to become obese in captivity and this may lead to fatty infiltration of the liver in adults. Where liver function might be impaired, recovery from anaesthesia will be prolonged and this risk must be considered when a decision is made as to the depth of anaesthesia and the length of time that the animal is to be maintained on Halothane.

The clinical approach to orang utans

Specific clinical problems will be considered in the discussion of the disease survey and the published work on orang utan pathology, but a few comments on the general clinical approach to this species are worth emphasising. Orang utans do not make good patients and quickly learn to distrust a particular individual. As much of the nursing and daily drug administration to a sick orang utan is likely to be carried out by the animal's usual keepers, do not ask such a keeper to be involved for example in efforts to immobilise the animal by projectile dart or by hand injection using a mechanical crushing device. It is important that the animal can trust at least one of its attendants. Orang utans are very easily put off their food when sick and, as far as possible, should not be moved from the accommodation with which they are familiar, unless other animals are at risk from infection. If an animal is particularly nervous and especially if it is difficult to fire an immobilising dart at it quickly, try giving it oral Valium (0.5–1.0 mg/kg) or Phencyclidine (2.0–3.0 mg/kg) in fruit or a drink after removing all food and water overnight. Although this does not always have a reliable effect, it will often produce a degree of sedation which makes it easier to give further drugs later by injection. It is not a good idea to try to achieve deep sedation solely by using drugs given orally (e.g. by doubling or trebling the dose) as some individuals become very markedly affected and then need emergency support of respiration.

Many adult orang utans in captivity are overweight and may already have some degree of fatty change in the liver. This should be borne in mind when calculating the dose rate of drugs like the cyclohexamines which are detoxified by the liver. The speed of recovery is often a good indication of liver function.

Apart from the data obtained at Yerkes Primate Center (Bourne 1969–73), there is very little published information on the normal biochemical and haematological values in apes. It is always worthwhile establishing normal values for one's own animals because they often vary from those found elsewhere. Tables 1 and 2 give the values for clinically normal juvenile and adult orang utans maintained at London. Blood samples can usually be obtained from the prominent cephalic vein which runs under the surface of the forearm with the arm held in the supine position. In large males, this vessel is often not visible and in such a case blood is most easily taken from the cubital vein, the main branch of the cephalic, where it crosses the elbow joint just lateral to the midline in a similar position to that in man.

Table 1. Plasma biochemistry of 5 normal adult orang utans kept at Regent's Park.

	Range	Mean
Urea (mmol/l)	3.3–4.0	3.6
Creatinine (μmol/l)	78–113	90.0
Bicarbonate (mmol/l)	23–28	25.6
Chloride (mmol/l)	94–100	98.2
Sodium (mmol/l)	137–140	138.4
Potassium (mmol/l)	3.3–3.8	3.5
Ions difference (mmol/l)	13–23	18.0
Total protein (g/l)	75–96	82.0
Albumin (g/l)	53–59	56.0
Globulin (g/l)	18–30	22.5
Calcium (mmol/l)	2.5–2.7	2.6
Inorganic phosphate (mmol/l)	0.9–1.5	1.1
Alkaline phosphatase (iu/l)	145–323	241.8
Total bilirubin (μmol/l)	5–14	8.0
Conjugated bilirubin (μmol/l)	—	1.0
Alanine transaminase (iu/l)	11–48	29.8
Aspartate transaminase (iu/l)	3–22	13.4
Urate (μmol/l)	116–194	145.2
Cholesterol (mmol/l)	4.0–6.9	5.7
Iron (μmol/l)	33.9–66.3	43.2

Table 2. Haematology of 21 normal adult orang utans kept at Regent's Park.

	Range	Mean
RBC ($\times 10^{12}$/l)	4.1–6.0	5.0
WBC ($\times 10^9$/l)	6.1–13.6	8.9
Hb (g/dl)	10.0–14.1	12.0
PCV (%)	34–51	40.1
Platelets ($\times 10^9$/l)	95–260	186.0
Reticulocytes (%)	0–0.3	0.1
Neutrophils (%)	17–87	44.3
Lymphocytes (%)	13–76	51.0
Monocytes (%)	0–3	1.1
Eosinophils (%)	0–10	3.3
Basophils (%)	—	—
MCV (fl)	71.6–92.3	80.6
MCH (pg)	20.9–27.9	24.2
MCHC (g/dl)	27.6–31.7	30.0
ESR (Wintrobes)	0–3	0.3

Routine administration of drugs to orang utans often presents a considerable problem. It is usually impractical to inject such animals daily and can be very distressing for them. Preparations of drugs designed for human children are usually taken most readily as they tend to be suitably flavoured. Failing this, powders and liquids can often be hidden in fruit yoghurts, particularly strongly flavoured ones such as cherry or blackcurrant, and it is a good idea to offer apes such delicacies occasionally (without drugs) to eliminate any suspicions when the food has to be used to provide treatment.

The disease survey

Most of the published information on diseases of orang utans tends to concentrate on the unusual or spectacular cases. A survey of this literature does not indicate the important problems of relevance to the captive management of this species. In order to discover the more common causes of death, the managers of most of the world's larger collections of orang utans were asked to supply details of animals which they had had examined post-mortem. Information was received from 19 collections and these cases together with a number of others which were found in the literature but known not to have been included in the private replies from the zoos totalled 170. The survey was by no means exhaustive and it is evident that some collections have quite different problems to others. Nevertheless, analysis of the data reveals what are probably the main management difficulties with this species and the fact that some of the histories included are of animals examined 50 years ago also reveals some of the improvements in health care that have taken place over that period.

The animals included in the survey have been divided into four age groups: Infants (less than 1 month), Juveniles (1 month to 2 years), Sub-adults (2–8 years) and Adults (9 years and over). This division has been arrived at because, as is clear from the data, each of these age groups tends to have its own particular problems. The data are summarised in Table 3. Where more than one problem (e.g. a respiratory tract infection and malnutrition) contributed to the death of an animal, both diagnoses are indicated in the table in order to establish the relative importance of particular problems to each age group.

The upper digestive tract

Although two reports of gastritis in juveniles appear amongst the replies from the post-mortem survey, no specific cause was given for the lesions seen and the problem in both cases appeared to be part of a generalised inflammatory response of the gastrointestinal tract. Ulcers of the stomach were not reported.

Dental problems were not mentioned in the survey, probably because they are unlikely to have contributed to the cause of death. Nevertheless, alveolar abscesses, gingivitis, periodontitis and caries regularly occur in older animals, especially if soft, sugar-rich diets have been fed. An increasing awareness of the need for carefully balanced diets in most collections of apes should reduce the incidence of dental disorders. If any orang utan is sedated for some reason, examination of its teeth should always be part of the routine health check. Colyer (1936) examined 255 orang utan skulls from animals shot in the wild and found that only six animals had evidence of caries. Schultz (1941) looked at 217 skulls from wild orang utans and found occasional dental caries in young adults. Caries were relatively frequently seen in the skulls from old animals. No evidence of oral pathology was found in 65 young animals, but of 97 young adults 80 had no evidence of any oral lesions, four had teeth with caries and 13 had one or more teeth with alveolar abscesses. Of 45 older adults, five had lost teeth, 27 had evidence of alveolar abscesses and six had fresh carious lesions.

Table 3. Principal post-mortem findings and the systems involved in the examination of 170 orang utans.

System/Cause	Infants (32) (less than 1 month)		Juveniles (38) (1 month–2 years)		Sub-adults (25) (2–8 years)		Adults (75) (9 years and over)	
	Cases affected	%	Cases affected	%	Cases affected	%	Cases affected	%
Enteric parasites	0	0	12	31.6	11	44.0	4	5.3
Enteric (non parasitic)	1	3.1	4	10.5	6	24.0	6	8.0
Liver	3	9.4	3	7.9	6	24.0	11	14.7
Upper respiratory	0	0	0	0	0	0	7	9.3
Lower respiratory	5	15.6	17	44.7	4	16.0	16	21.3
Stillborn	8	25.0	—	—	—	—	—	—
Neoplasms	0	0	1	2.6	0	0	3	4.0
Injuries	5	15.6	3	7.9	0	0	4	5.3
Nutritional	3	9.4	10	26.3	2	8.0	5	6.7
Central nervous	2	6.2	6	15.8	1	4.0	4	5.3
Renal	0	0	0	0	4	16.0	7	9.3
Cardiovascular	2	6.2	1	2.6	4	16.0	15	20.0
Bacterial infection	4	12.5	2	5.3	6	24.0	23	30.7
Viral infection	0	0	1	2.6	4	16.0	5	6.7
Reproductive	0	0	0	0	0	0	8	10.7
No diagnosis	0	0	0	0	2	8.0	2	2.7

Supernumary molar teeth were seen in nine of the 217 animals. Treatment of a fistulated molar abscess has been described by Fagan and Robinson (1978).

Parasites of the digestive system

One of the most important causes of death in younger orang utans is strongyloidosis. Judging from the replies of the survey, several major collections appear to have been very severely affected by this problem. The disease mainly affects animals of between 6 months and 5 years, but adults can also be clinically affected. The status of a *Strongyloides* infestation in a group of orang utans is usually difficult to monitor. Adult animals may carry large numbers of the nematode and the eggs or larvae are never detected when the faeces are examined. Some zoos find that almost every mother-reared infant becomes affected. The presence of *Strongyloides* should always be suspected when young orang utans develop a persistent diarrhoea. This nematode is fairly resistant to most anthelmintics but the best method of treatment is probably to give Levamisole or one of the more recent benzimidazole derivatives orally in relatively low doses for a 3–4 day period, repeating the procedure two weeks later (Levamisole 7.5 mg/kg, Fenbendazole 5 mg/kg, Albendazole 10 mg/kg).

Another nematode which occasionally causes problems is an *Oesophagostomum* sp. (Weinberg 1909). As with *Strongyloides*, a generalised enteritis may result, or specific haemorrhagic caecal and colonic ulcers may be caused by the parasites. Secondary bacterial infections and peritonitis often follow. Less frequently encountered helminths which are probably of doubtful clinical significance include an ascarid which is indistinguishable from *Ascaris lumbricoides* of man and the cestode *Bertiella studeri* (Stiles and Hassal 1929). The human pinworm *Enterobius vermicularis* although reported frequently in chimpanzees is less commonly encountered in the orang utan, probably because relatively fewer orang utans have come into close contact with man.

The hydatid cysts of *Echinococcus granulosus* caused the deaths of a number of young orang utans at one American zoo (Bernstein 1972). Of eight juvenile animals, six showed clinical signs and three died. Large multiple cysts attached to the mesentery and in the liver were described. Surgery was of little value unless the cysts were discrete and could be removed intact. The severely affected animals became lethargic and began to lose weight. One died post-operatively after a large cyst had been drained and flushed with formalin. Another died of internal haemorrhage. The source of the infection was probably a wild coyote or a carnivore imported into the zoo and carrying *Echinococcus*. A single report exists of *Cysticercus tenuicollis* in the liver of an old female orang utan at London, but the parasite was probably of no clinical significance.

Entamoeba histolytica and *Balantidium coli* are both relatively frequent findings in the faeces of clinically normal captive orang utans, but occasionally they become pathogenic, especially in animals of the juvenile and sub-adult age groups. Thirty per cent of the parasitic problems leading to death in these two age groups were probably caused by amoebiasis and balantidiosis. Cases have also been reported in the literature (Patten 1939). Acute and chronic cases are known where the principal changes may vary from a profuse dysentery to intermittent diarrhoea with occasional masses of mucous, necrotic mucosal cells and a little blood. A generalised inflammation of the mucosa of the large intestine or specific areas of necrotic ulceration may be found on post-mortem examination. The parasites probably become pathogenic in young animals when they are stressed for some reason and if the number of parasitic cysts being passed by carriers and picked up by the younger animals of the group becomes excessive. Regular faecal examinations will reveal whether the numbers of cysts being passed are increasing and all carriers should then be treated. As is the case with the management of most parasitic infections it is not good practice simply to use drugs just because the parasite is there, unless the manager can be certain that he is going to be able to clear the organism completely and ensure that it does not return. Healthy animals will usually develop and maintain an equilibrium with their parasites which is only broken down if their general management deteriorates.

Treatment of affected animals should be aimed at destroying the pathogenic protozoa, preventing further diarrhoea and providing fluid therapy where this is applicable and practical. Drugs are now available to promote more rapid healing of intestinal ulcers. Metronidazole at a dose rate of 20 mg/kg for 10 to 20 days is recommended for its specific antiprotozoal activity.

Other problems of the intestinal tract

A number of cases of enteritis were reported for all age groups in the survey where no specific diagnosis was reached. In many cases symptomatic treatment was sufficient to arrest the problem. Some orang utans appear to need a considerable quantity of fibre in their diets to maintain relatively firm faeces. This can easily be given by providing a proportion of wheat bran in their daily ration ('All Bran' — Kelloggs) or by giving fresh, clean leaves and twigs from trees and shrubs which are not known to be toxic. One case of volvulus of the colon involving a 20-month-old animal which had been treated with Ampicillin was reported in the survey and a case of an intussusception in an 8-year-old animal has been recorded in the literature (Hood et al. 1973). This was successfully treated by surgical resection of the bowel and there was a strong suspicion that enteritis caused by *Balantidium coli* had been the predisposing cause.

Colitis was reported as contributing to the deaths of eight animals in the survey, four of them between 2 and 4 years of age. Ulceration of the mucus membrane was noted in three cases but no specific cause was found. A single case is also reported in the literature (Cragg and Scott 1974). In all the cases in the survey there were a number of other pathological findings of significance and therefore the colitis might have been a secondary stress effect in animals which were already in poor condition. Appendicitis has been reported in the literature (Vervat 1961). The animal showed symptoms of vomiting and a painful abdomen. It had a raised temperature and had not defecated for some days. An inflamed appendix of abnormal length was removed.

Bacteria and viruses have occasionally been identified as causing enteritis in orang utans and these will be covered in the sections dealing with specific causes of infection.

The liver

Obstructive jaundice has been reported in two newborn orang utans which were being hand-reared. One case has been published (Boever et al. 1977). Signs of liver failure in the published case developed from birth and despite treatment, the animal died 13 days later. The main bile duct was not obstructed, but there was an impression that there were fewer small bile ducts than usual in the liver. There were large quantities of bile in the hepatic canaliculi. Bilirubin and cholebilirubin were present in the urine.

Fibrosis of the liver was reported in three older animals in the survey and varying degrees of fatty infiltration of the liver were recorded in many adults. Diets which are too high in their energy content are frequently fed to orang utans which may also have little opportunity to exercise. Hepatic necrosis and degeneration was recorded in seven animals, two of them sub-adults. In general no cause was evident but in one case, severe amoebiasis was a concurrent problem. Bacterial, viral and parasitic diseases of the liver are occasionally reported and these are covered in more detail in other sections.

Cholecystitis was recorded in two adults in the survey. Fox (1930) reports a case where a gallstone had destroyed the gallbladder and penetrated the large intestine where it had lodged in the caecum. Peritonitis and haemorrhage from the wall of the colon had led to the animal's death.

The respiratory tract

Of the 13 deaths in animals included in the survey which were under one month of age, seven were stillborn, one inhaled amniotic fluid and one died of *Klebsiella* pneumonia. The cause of the stillbirths was not determined although dystokia was suspected in two cases. Inhalation of milk being given by a human foster parent led to a fatal pneumonia in two of the remaining four cases.

The common cold and influenza viruses of man will also affect orang utans and occasionally such infections are severe, particularly in young animals. If affected animals are lethargic and fail to eat, they should be given broad spectrum antibiotics because the risk of a secondary pneumonia occurring is considerable. Providing this is done and the animals are carefully nursed, mortality from this cause is very low.

Complement fixing antibodies to *Mycoplasma pneumoniae* have been found in orang utans (Fiennes 1972) but the clinical significance of this is not known. *Klebsiella* sp., *Haemophilus* sp., staphylococci and streptococci have all caused pneumonia, especially in younger animals. Tuberculosis which was a frequent cause of death is now uncommon. A single case of pulmonary nocardiosis has been reported (McClure et al. 1976).

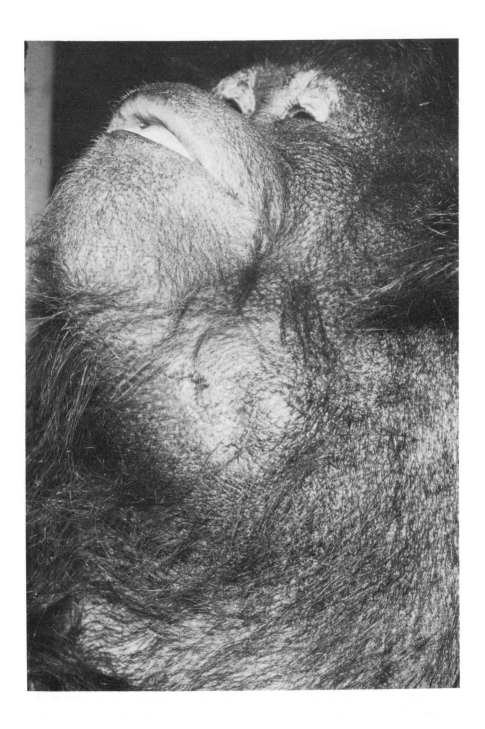

Fig. 4. Submandibular swelling caused by laryngeal sac infection. Note the point of rupture on the surface of what is effectively a large abscess.

Infection of the large laryngeal air sac of orang utans appears to be a severe problem in one of the collections surveyed and is seen occasionally in some of the others (Guilloud and McClure 1969; Gucwinski and Michalaska 1964; Hassko 1929). Treatment of the condition has been described (Clifford et al. 1977). The first signs of the infection are usually a nasal discharge, anorexia and a distended laryngeal sac. It is probable that the localised infection of the sac is secondary to a rhinitis or laryngitis. Various bacteria including *Pseudomonas aeruginosa*, *Escherichia coli*, haemolytic streptococci, *Aerobacter cloacae* and *Proteus* spp. have been isolated from samples of the pus in these cases. The clinical approach should be the same as for a large abscess or infected cyst. Surgical drainage of the sac is required, followed by flushing of the cavity with a mild antiseptic. The cavity should then be packed with swabs soaked in a broad spectrum antibiotic to which the bacteria are sensitive. It may also be necessary to give antibiotic cover by the oral route or by intramuscular injection, particularly if the animal is becoming septicaemic.

Pulmonary osteoarthropathy (Marie's disease) occurred in an adult male orang utan at London (Hime et al. 1972). This animal had an air sac infection with bronchitis and chronic pneumonia. Histological examination revealed the presence of discrete foci consisting of basophilic material which appeared to consist of colonies of an actinomycete. Only *E. coli* was isolated from the lungs and no anaerobes were found. There was marked deposition of periosteal new bone over almost all the long bones. It is probable that the skeletal lesions followed the chronic respiratory infection as has been reported in similar cases in man and domestic animals (Jubb and Kennedy 1963; Smith and Jones 1966).

Another unusual case affecting the respiratory tract of an orang utan was that of a pulmonary alveolar microlithiasis in a 19-year-old male (Kelly 1976). Multiple calcified foci were found throughout the lungs. Death followed signs of respiratory distress which lasted for about a week.

Neoplasms

A squamous cell carcinoma of the oesophagus was reported in a 35-year-old female, and a uterine leiomyoma was seen in another female which was 45 years old. Both were from the Philadelphia collection. A mammary carcinoma with metastases was seen in a 15-year-old female at Frankfurt (Brack 1966) and a testicular seminoma is described in another report by Voronoff (1949). Gardner et al. (1978) have reported on a case of myelomonocytic leukaemia in an orang utan.

Injuries

The survey revealed that multiple injuries had led to the deaths of eight orang utans under the age of 2 years. Five of these animals were less than a month old. All the deaths were probably caused by adult animals and in four cases the father was to blame. It was evident from the response to the survey that many collections found that a particular female could not be trusted with her own infant and that despite giving the nursing mother every opportunity to look after the young animal properly, it had ultimately been removed for hand-rearing. There were no fatal injuries reported in the 2 to 8-year-old age group, although a single case of drowning was seen. Four deaths from injuries were recorded amongst the adult males and all these cases were probably caused by fighting with other males. Sonntag (1924), surveying the skeletons of wild orang utans, commented that there was a high incidence of injuries probably caused by falling from trees and Fox (1939) found evidence of arthritis in two of 41 orang utan skeletons that he examined.

Nutrition

There are three stages in the development of a young orang utan where inadequate nutrition often causes problems. The first difficulties arise if the animal fails to suckle properly after birth. Usually, a healthy baby will easily survive 2 days without milk, but failure to suckle after this time will probably mean that the infant will have to be hand-reared. Many collections remove infant apes at the slightest sign of a problem developing and because of this many inexperienced mothers are never given a chance to develop the technique of rearing their own young successfully. It is usually wiser to take a risk by leaving the first two infants born to each mother with her in order to ensure that she gains experience. Some female orang utans never become good mothers and their young will always have to be removed. Hand-reared apes are often said to be difficult to return to the breeding population although the evidence for this is less convincing for the orang utan than for the other two species. It may be possible to take an infant orang utan away temporarily and reintroduce it to its mother at a later date (Mehren and Rapley 1978).

The second stage of growth where problems arise often originates from the first. The mother may feed the infant but for various reasons her milk provides insufficient nutriment for optimum growth of the young animal. This may occur if the dam is not producing enough milk, as is sometimes the case in a young or very old animal. More often, the problem is caused by inadequate nutrition of the mother, notably through a lack of sufficient protein of a high biological value and a relative imbalance of calcium, phosphorus and vitamin D_3. Juveniles that fail to thrive sometimes have to be removed at from 4 to 18 months for hand-rearing. At about 2 years, the young animal is eating a high proportion of solids in its diet. If its daily ration is not correctly balanced, problems may then occur at this third stage of development.

In the survey, three animals died as a direct result of the mother failing to suckle them properly. Seventeen animals, all juveniles, died as a result of an inadequate diet. Poor growth rates and nutritional bone disease were the main reported findings. It is probable that many of the deaths ascribed to infections with bacteria and endoparasites were secondary to nutritional problems. There is strong evidence from the survey that with better standards of feeding in the

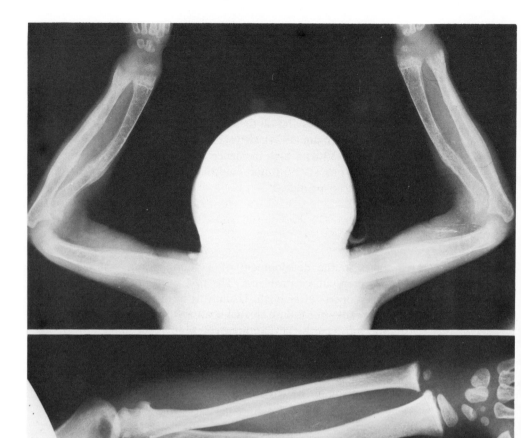

Fig. 5 (top). Radiograph of the forearms of the orang utan shown in Fig. 2 when she was 16 months old. The severe nutritional osteodystrophy responded rapidly to correct management of the diet. *Fig. 6.* Radiograph of the forearm of the young orang utan shown just prior to birth in Fig. 3. Normal bones. This animal is the daughter of the orang utan in Figs. 2 and 5.

major zoos, problems relating to inadequate nutrition are now much less frequently encountered. Smaller zoos in the developed world and those in the 'third' world regrettably still have to learn many of these lessons.

The nervous system

The replies from the survey indicate that signs of involvement of the central nervous system were frequently seen prior to death. It is not clear in many cases whether these were secondary to other problems or whether a primary neural lesion was the cause, principally because the nervous system was not examined in detail.

190

Meningitis was described in five infants and juveniles and cerebral oedema in a sixth, but no cause was found. Other pathological signs in these animals suggested a septicaemia and three of the cases were associated with concurrent nephritis. Measles virus was probably responsible for encephalitis in one of the adults in the survey and Hamerton (1937) records that after two keepers contracted measles, an orang utan in their care developed meningo-encephalitis. Paralysis due to polio virus type 1 has been seen in an orang utan (Allmond et al. 1967). Nine other orang utans in the affected colony showed no signs of the disease but were excreting the virus.

Scherer (1944) records a case of demyelination of the white matter of the brain in an orang utan but was unable to comment on the likely cause. Van Bogaert and Innes (1962) in a review of the neurological diseases of apes and monkeys described meningo-encephalitis and degenerative changes in the brain of an orang utan with dysentery.

The renal system

From the results of the survey, problems with the renal system appear to be largely confined to older animals. Glomerulonephritis was the main finding. This was usually associated either with other specific lesions, principally of the meninges or upper respiratory tract which suggested a septicaemia, or as part of a 'degenerative syndrome' usually involving the liver or heart as well. In one case (Fiennes 1967) pyelonephritis was caused by a massive infection with *Pseudomonas aeruginosa*.

Cardiovascular

One collection reported a newborn animal with signs suggestive of 'haemolytic disease of the newborn' but was not able to confirm this. A persistent foramen ovale led to the death of another young orang utan in the same zoo. Congenital abnormalities are reported in the literature. Degaris (1934) described a newborn orang utan where the pericardium was incomplete over the left ventricle. Chase and Degaris (1938) found an abnormality of the left superior vena cava which allowed some venous drainage into the pulmonary venous return.

The nematode *Dirofilaria pongoi* has been found in the orang utan heart (Vogel and Vogelsang 1930). Sandosham (1951), who found *Dirofilaria immitis* in the peritoneal cavity of another orang utan, considered that *D. immitis* and *D. pongoi* were actually the same species.

The natural malarial parasite of the orang utan is *Plasmodium pitheci* and this has been found on a number of occasions in wild and newly-caught animals (Voller 1972; Killick-Kendrick et al. 1973; Stafford et al. 1978). Stafford et al. also found a *Hepatocystis* sp. in blood smears from one of 26 animals examined. The clinical significance of these blood protozoa is not known.

Degenerative changes of the myocardium and major arteries were reported in many older animals covered by the survey. Degenerative changes of the liver and kidneys were also found in many of these cases. Focal myocarditis associated with a streptococcal infection was seen in one juvenile. Not all the cases of myocardial degeneration occurred in old animals and degenerative lesions were found in the hearts of three young adults, all of which were obese. Sudden death caused by coronary arteriosclerosis and myocardial infarction was seen on two occasions in old males and death from this cause can be a danger when anaesthetising older orang utans. In the literature, Finlayson (1965) and Ratcliffe et al. (1960) have reported on arteriosclerosis in apes.

The reproductive system

Three of the adult females in the survey died during or just after parturition. A fourth died of a septicaemia one month after the birth, probably as a result of an incompletely involuted uterus, which had become infected. Incomplete expulsion of the foetal membranes with subsequent septicaemia has also been the cause of death of a number of females in collections not covered by the survey according to M. Jones (pers. comm., 1979). As the female often eats the foetal membranes, it is difficult for the keeper to know whether the placenta has been totally expelled. Occasional instances of slight bleeding from the vagina of a pregnant female during the last month of gestation is fairly common and may not be clinically significant, but if it is regular and profuse, the animal should be examined in case this indicates an impending abortion or early detachment of the placenta. A case of 'placenta praevia' where the placenta becomes detached prematurely and effectively blocks the passage of the foetus through the birth canal has been reported recently at London (Kingsley and Martin 1979).

Although there is no recorded case of a Caesarian section being carried out on an orang utan, there have been reports of this procedure in gorillas and chimpanzees. Although theoretically surgical interference should be considered if dystokia or 'placenta praevia' is suspected, orang utans are usually relatively more obese than other apes in captivity and thus present a considerable surgical problem. The risk involved is worth taking if an accurate diagnosis can be made, but the surgeon should be wary about undertaking an exploratory laparotomy in a heavily pregnant orang utan if there are no obvious reasons for interference.

Congenital scrotal hernias have been seen occasionally in apes. One case is reported by Straus (1936) in the orang utan. It is probable that some hernias go unnoticed in young apes and are of no clinical significance.

Bacterial causes of disease

As many bacteria affect a number of body systems concurrently, it was decided to discuss the more important of these in a separate section.

Diseases of the digestive tract which have definitely been shown to be caused by bacteria are not common. Apart from the occasional mention of coliforms being associated with neonatal diarrhoea, especially in hand-reared animals, only one confirmed case of shigellosis and two of salmonellosis were noted in the survey, all in young animals. *Salmonella munchen* was isolated from an 8-day-old animal in one case and the mother was shown to have been a carrier. Occasional cases of infection with *Shigella* sp. and *Salmonella* sp. have been mentioned in the literature (Ruch 1959; Fiennes 1967).

By far the most important bacterial disease of orang utans in recent years has been tuberculosis. Fourteen confirmed cases were included in the survey, mostly in adult animals. Better standards of husbandry and nutrition combined with radical efforts to control the disease have meant that most major collections are now free of the problem. It still persists in some of the smaller zoos, especially in the 'third' world. Before a collection which is free of the disease introduces a new animal, especially from a source with poor medical facilities, that animal should be thoroughly examined and isolated for at least 3 months before being introduced to other members of the colony. Human tuberculosis is the usual type found in orang utans, although *Mycobacterium bovis* has occurred in apes and they are probably susceptible to other mycobacteria including *M. kansasii* and *M. avium*. The course of the disease in orang utans is chronic and progressive. Clinical signs may only develop slowly over a considerable period of time, often years. The digestive, respiratory and lymphatic systems are usually all involved although occasionally, in cases where the disease has not progressed, isolated caseous lesions are found in just one of these systems. The post-mortem picture is fairly typical of tuberculosis in large mammals.

Treatment of affected individuals has been tried with variable success. It is doubtful that drugs are of much use in controlling a well-established case and usually these animals have to be destroyed. It is probably worthwhile treating animals which have been in contact with a confirmed case whether or not they react to a tuberculin test. A combination of 10 mg of Isoniazid (INH) per kilogram of body weight with 60 mg of Paraminosalicylic acid (PAS) per kilogram of body weight is usually used. Streptomycin and, more recently, Rifampicin, have also been tried. The drugs are often given continuously for some years and the dose rate gradually decreased. This may not be good medical practice because of the danger of producing a drug-resistant strain of *Mycobacterium*. The intra-palpebral skin test for tuberculosis using standard tuberculin is applied at regular intervals.

The drugs probably only hold the infection in check and cases have been recorded when the disease flares up again if the treatment is stopped. Some dealers have been in the habit of giving INH to all imported primates but this practice should not be encouraged, again because of the danger of producing drug resistance.

There is no absolutely reliable test for tuberculosis in the living animal. Although new immunological techniques of diagnosis are being developed at the present time, the intra-palpebral skin test is still the only one that has some current use. Usually a purified mammalian tuberculin, human or bovine, is injected subcutaneously into one upper eyelid and avian tuberculin is injected into the other. A more accurate indication of the status of the animal with regard to mycobacterial infection is achieved by repeating the test every 60 days. False positives and false negatives occur and complicate the picture. Many

German zoos vaccinate their apes with BCG while zoos in the United States and Britain do not. This means that apes from German zoos will react to the skin tests whether they are infected or not. No serological tests are applicable because mycobacteria produce cell-mediated immune response.

Apart from carrying out tuberculin skin tests it is often helpful, where practical, to take a series of radiographs of the thorax of all newly arrived primates under anaesthesia. The chest should be inflated mechanically via an endotracheal tube as each plate is exposed. The plates are then examined for evidence of lesions of pulmonary tuberculosis.

Apart from the cases described in the survey, various descriptions of the pathology and treatment of the disease have appeared in the literature (Vertruyen 1956; Fiennes 1972; Ruch 1959; Harberle 1970).

Other bacterial infections of the respiratory tract sometimes run concurrently with chronic tuberculosis and may be the actual cause of death. Air sac infections in adults and pneumonia in younger animals often involve a number of bacteria and these have already been listed elsewhere in the paper.

Viral infections

Although evidence of infection with viruses has been shown in a number of surveys of sera from orang utans, the clinical significance of many of these viruses is not known. Harrison (1967) showed that 26% of 126 sera from six genera of primates including orang utans which had recently been imported into the United States contained serum neutralising antibodies to *Chikungunya* and related arboviruses. Fiennes (1972) describes the occasional finding of anti-bodies to the picornaviruses of poliomyelitis and Coxsackie virus. Paralysis due to polio virus type 1 in a captive orang utan has already been described and Fowler (1978) advocates the use of the oral vaccine designed for humans in apes. Kalter (1972) found rubella (German measles) antibodies in orang utan serum and also reported that antibodies to Reoviruses types 1, 2 and 3 were seen in over 50% of the sera he examined from this species.

Although chimpanzees have been shown to be carriers of viral hepatitis type A, examination of orang utans so far has not shown this. The hepatitis A antibody though is seen occasionally in orang utan sera. The antibody of hepatitis type B has also been recorded in sera from this species. It is possible therefore that like chimpanzees, orang utans may also be carriers of the antigen of both types of hepatitis and could therefore be potentially infective for man (Zuckerman et al. 1978; Shulman et al. 1969; Prince 1969). Measles virus causes encephalitis and occasional deaths in orang utans and one case of mumps was recorded in the survey. The animal with mumps developed a severe parotitis with erosions of the soft palate and pharynx. The same collection also had a hand-reared orang utan with persistent sore lips attributable to *Herpes simplex* virus. The disease was of little significance to the animal but it was probably the source of the virus which caused deaths in five of a large group of young gibbons with which it had contact. No case of herpes virus B has ever been reported in an orang utan.

Herpes zoster (chicken pox) has been described in orang utans by Heuschele (1960) but the disease is mild in this species. Peters (1965) reported an outbreak of monkey pox in a group of orang utans. The infection was probably introduced by newly imported anteaters (*Myrmecophaga tridactyla*) in which the disease was also diagnosed. The lesions in the orang utans were almost identical with those of man. There was a severe rhinitis with ulcers on the mucosae of the lips and buccal cavity. Cutaneous pustules were also present on the inside of the thighs and on the hands and feet. Seven orang utans died and four recovered. In most cases, secondary bacterial infection with staphylococci, streptococci, *Klebsiella pneumoniae* and *Escherichia coli* was probably of more clinical significance than the virus. Other great apes were also affected but survived. Arita and Henderson (1968) advocate the use of human smallpox vaccine to protect against monkey pox in primates.

Kalter et al. (1967) found antibodies to SV19 and SV49 type enteroviruses in orang utan serum but the clinical significance of this is not known.

Miscellaneous problems

Skin lesions are not often seen in the orang utan, but a few cases have been reported. Weidman (1923) described a case of sarcoptic mange and the occurrence of mycotic dermatosis with a *Blastomyces* sp. and also with *Tinea captitis* and *T. circinata*. Blastomycosis was also suspected in an orang utan with a vesicular dermatitis at the Seattle Zoo (Binkley 1959). The lesions coalesced and ultimately formed extensive dry areas eroded of epithelium. At London, the author has seen a moist dermatitis affecting the axillary and femoral areas in an orang utan from which a *Tinea* sp. was isolated. A single case of an allergy to 'wood wool' bedding has also been seen at London.

Conclusions and summary

It is evident from the survey that many zoological collections appear to have their own particular problems with orang utans such as air sac infections or strongyloidosis. Different pathologists and clinicians place different emphasis on the importance of a particular finding and whereas one may feel that the presence of *Strongyloides* in any carcass is significant, another may virtually ignore it. Nevertheless a pattern of the more frequently encountered difficulties with orang utans emerged from the survey and as a form of summary these are listed below.

1. The period around the birth of an orang utan is critical to both mother and baby. Any premature bleeding or straining on the part of the mother should be treated with suspicion. Where possible after the birth she should be observed to see whether the placenta is passed. The neonate should feed within 48 hours of birth, but keepers should not be in too much of a hurry to remove it, especially if it is the first or second youngster;

2. Neonatal animals are prone to injury from either parent, but especially from the father, so he should be separated from the nursing mother. The next most frequently encountered causes of death in this age group are: being neglected by the dam (starvation); septicaemia (*E. coli* or *Klebsiella pneumoniae*) and jaundice;

3. In the juvenile age group, pneumonia, malnutrition and a heavy infestation with *Strongyloides* nematodes are the most frequent problems. Malnutrition probably predisposes to the other two and should now be avoidable. If the adults are known to be carriers of *Strongyloides*, all the animals should be regularly treated, even if evidence of the parasite is not found on examination of the faeces;

4. In the sub-adult age group, enteritis is still caused by *Strongyloides* but *Balantidium coli* and *Entamoeba coli* appear to become important pathogens and their presence, although not always significant, should be looked for in regular faecal checks. The incidence of colitis, which sometimes involves ulceration of the mucous membrane, increases in this age group as also does renal and cardiovascular disease;

5. In the adults, degenerative changes of the cardiovascular system, liver and kidneys are more frequent, as might be expected, but in some cases these changes have occurred in relatively young animals (10–15 years of age). There is a generally held opinion that overweight, underexercised, isolated and often bored orang utans, especially males in this age group, are particularly prone to premature degenerative disease and an early death.

Acknowledgements

The author would like to acknowledge the considerable assistance given to him by many people in the compilation of this review. Dr Michael Brambell and Miss Susan Mathews were responsible for preparing the data sheet on the orang utan which is used as the basis for management of this species at Regent's Park. Miss Veronica Worth and Miss Linda Juul assisted in the preparation of the manuscript. The following individuals and collections provided much information on their orang utans, especially from their post-mortem data: Mr Marvin Jones (studbook keeper), Dr W. de Meurichy (Antwerp), Dr Wilbur Amand (Philadelphia), Dr Richard Montali (Washington), Dr Gerald Cosgrove (San Diego), Dr Richard Faust (Frankfurt), Dr Howell Hood (Phoenix), Dr Dieter Ruedi (Basle), Dr William Boever (St. Louis), Dr William Rapley (Toronto), Dr Maarten Frankenhuis (Rotterdam), Dr Christian Schmidt (Zürich), Dr Michael Hughes (Brownsville), Dr Lee Simmons (Omaha), Dr Tom Spence (Perth), Dr Harold McClure (Yerkes Primate Center), Mr Geoffrey Greed (Bristol) and Mr Jeremy Mallinson (Jersey).

The haematological data resulted from the work of Dr Christine Hawkey of the Zoological Society's Institute of Zoology, and the biochemical data from the work of Dr David Gardner and his colleagues at University College Hospital, London.

References

Allmond, B.W., J.E. Froeschle and N.D. Guilloud (1967). Paralytic poliomyelitis in large laboratory primates, virologic investigation and report on the use of oral poliomyelitis virus vaccine. Am. J. Epidem., 85: 229–239.

Arita, I. and D.A. Henderson (1968). Smallpox and monkeypox in non-human primates. Bull. W.H.O., 39: 277.

Bernstein, J.J. (1972). An epizootic of hydatid disease in captive apes. J. Zoo. Anim. Med., 3: 16.

Binkley, K.L. (1959). Treatment of skin disease in an orang-utan at Woodland Park Zoo, U.S.A. Int. Zoo Yearb., 1: 52.

Boever, W.J., D. Dietsch and K.K. Kane (1977). Obstructive jaundice in a newborn orangutan. Vet. Med. Sm. An. Clin., 72: 1227–1229.

Bonner, W.B., M.E. Keeling, E.T. Van Ormer and J.E. Hayrie (1972). Ketamine anaesthesia in chimps and other great ape species. In: G.H. Bourne (ed.), The chimpanzee, vol. 5, Karger, Basel.

Bourne, G.H. (ed.) (1969–1973). The chimpanzee, 6 vols. Karger, London.

Brack, M. (1966). Carcinoma solidum simplex mammae bei einem Orang-utan (*Pongo pygmaeus*). Zbl. Allg. Path. Anat., 109: 474–480.

Bree, M.M. (1972). Dissociative anaesthesia in *Macaca mulatta*. Clinical evaluation of CI744. J. Med. Prim., 1: 256.

Chare, R.E. and D.F. de Garis (1938). Anomalies of vena cavae superiores in an orang. Am. J. Phys. Anthrop., 24: 61–65.

Chen, G.M. and J.K. Weston (1960). The analgesic and anaesthetic effect of 1-(1-phenylcyclohexyl) piperidine HCl on the monkey. Anaesthesia and Analgesia (Cleveland), 39: 132.

Clifford, D.H., S.Y. Yoo, S. Fazekas and C.J. Hardin (1977). Surgical drainage of a submandibular air sac in an orang utan. J. Am. Vet. Med. Ass., 171: 862–865.

Colyer, F. (1936). Variations and diseases of the teeth of animals. John Bale, Sons and Danielsson, London.

Cragg, J. and G.B.D. Scott (1974). Ulcerated colitis in an adult male Bornean orang utan (*Pongo p. pygmaeus*). Jersey Wildlife Preservation Trust Annual Report No. 11: 100–101.

Fagan, D.A. and P.T. Robinson (1978). Endodontic surgery for treatment of a fistulated molar abscess in an orang-utan. J. Am. Vet. Med. Ass., 173: 1141–1144.

Field, W.E., J. Yelnowsky, J. Mundy and J. Mitchell (1966). The use of Droperidol and Fentanyl for analgesia and sedation in primates. J. Am. Vet. Med. Ass., 149: 896.

Fiennes, R.N.T.-W. (1967). Zoonoses of Primates. Cornell University Press, Ithaca.

Fiennes, R.N.T.-W. (ed.) (1972). The pathology of simian primates, parts 1 and 2. Karger, Basel.

Finlayson, R. (1965). Spontaneous arterial disease in exotic animals. J. Zool., 147: 239–343.

Florence, B.D., P.J. Taylor and J.M. Busheikin (1977). Contraception for a female Borneo orangutan. J. Am. Vet. Med. Ass., 171: 974–975.

Fowler, M.E. (ed.) (1978). Zoo and wild animal medicine. W.B. Saunders, Philadelphia.

Fox, H. (1930). Animals of special importance. Rep. Lab. comp. Path. Philad., 12–22.

Fox, H. (1939). Chronic arthritis in wild mammals. Trans. Am. Phil. Soc., 31: 73–148.

Gardner, M.B., G. Esra, M.J. Cain, S. Rossman and C. Johnson (1978). Myelomonocytic leukaemia in an orang utan. Vet. Path., 15: 667–670.

Garis, D.F. de (1934). Pericardial patency and partial ectocardia in a newborn orang-utan. Anat. Rec., 59: 69–82.

Gucwinski, A. and Z. Michalaska (1964). Eitrige Kehlsackentzündung beim Orang-utan (*P. pygmaeus*). Internat. Symp. Erkrankungen Zootiere 6, 62–65.

Guilloud, N.G. and H.M. McClure (1969). Air sac infection in the orang utan, *Pongo pygmaeus*. In: D.J. Chivers and K.A. Joysey (eds.), Recent advances in primatology, vol. 3. Karger, Basel.

Haberle, A.J. (1970). Tuberculosis in an orang utan. J. Zoo. Anim. Med., 1: 10.

Hamerton, A.E. (1937). Report on the deaths occurring in the Society's gardens during the year 1936. Proc. Zool. Soc. Lond., B. 107: 443–474.

Harrison, V.R. (1967). The presence of antibody to Chikungunya and other serologically related viruses in the sera of sub-human primate imports to the U.S. J. Immunol., 98: 979–981.

Hassko, A. (1929). Uber die Kehlsäcke des Orang-utans und eine todbringende Erkrankung im Kehlsacke eines jungen Orang. Z. Hals-Nas-Ohrenheilk., 23: 258–262.

Heuschele, W.P. (1960). Varicella (chickenpox) in three young anthropoid apes. J. Am. Vet. Med. Ass., 136: 256–257.

Hime, J.M., K.F. Keymer and E.C. Appleby (1972). Hypertrophic pulmonary osteoarthropathy in an orang-utan. Vet. Rec., 91: 334–337.

Hobson, B.M. (1976). Evaluation of the sub-human primate tube test for pregnancy in primates. Lab. Animals, 10: 87–91.

197

Hood, H.B., N. Rokey and A.M. Smith (1973). Intussusception in an orangutan. J. Zoo. An. Med., 4: 15–16.

Jubb, K.V.F. and P.C. Kennedy (1963). Pathology of domestic animals, Vol. 1. (1st ed.) Academic Press, New York.

Kalter, S.S., J. Ratner, G.V. Kalter, A.R. Rodriguez and C.S. Kim (1967). A survey of primate sera for antibodies to viruses of human and simian origin. Am. J. Epidem., 86: 552–568.

Kalter, S.S. (1972). Identification and study of viruses. In: R.N.T.-W. Fiennes (ed.), Pathology of simian primates, part 2. Karger, Basel.

Kelly, D.F. (1976). Pulmonary alveolar microlithiasis in the orang-utan (*Pongo pygmaeus*). Acta Zool. et Path. Antverpiensis, 66: 53–57.

Killick-Kendrick, R., N. Rajapaksa, W. Peters, P.C.C. Garnham, W.H. Cheong and F.C. Cadigan (1973). Malaria parasites of the orang utan. Roy. Soc. Trop. Med. Hyg., 67: 1–2.

Kingsley, S.R. and R.D. Martin (1979). A case of placenta praevia in an orang utan. Vet. Rec., 104: 56–57.

Marsboom, R., J. Mortelmans and J. Vercrysse (1963). Neuroleptanalgesia in monkeys. Vet. Rec., 75: 132.

McClure, H.M., J. Chang, W. Kaplan and J.M. Brown (1976). Pulmonary nocardiosis in an orangutan. J. Am. Vet. Med. Ass., 169: 943–945.

Mehren, K.G. and W.A. Rapley (1978). Reintroduction of a rejected orangutan infant to its mother. Proc. Annual Meeting American Association Zoo Veterinarians, Knoxville 1978.

Melby, E.C. and H.J. Baher (1965). Phencyclidine for analgesia and anaesthesia in simian primates. J. Am. Vet. Med. Ass., 147: 1068.

Patten, R.A. (1939). Amoebic dysentery in orang utans (*Simia satyrus*). Aust. Vet. J., 15: 68–71.

Peters, J.C. (1965). Eine 'Monkey-Pox' Enzootie im Affenhaus des Tiergartens 'Blijdorp'. Int. Symp. Diseases Zoo Animals, 7: 197–201.

Prince, A.M. (1969). Studies on human serum hepatitis in primates. Proc. 2nd Conf. exp. Med. Surg. Prim. Karger, Basel.

Ratcliffe, H.L., T.G. Yerasimides and G.A. Elliot (1960). Changes in the character and location of arterial lesions in mammals and birds in the Philadelphia Zoological Gardens. Circulation, 21: 730–738.

Ruch, T.C. (1959). Diseases of laboratory primates. W.B. Saunders, Philadelphia.

Rutty, D.A. and D.A. Thurley (1962). Anaesthesia of small primates. Vet. Rec., 74: 883.

Sandosham, A.A. (1951). On 2 helminths from the orang utan, *Luptertrema rewelli* and *Dirofilaria immitis* (Leidy 1856). J. Helm., 25: 19–26.

Scherer, H.J. (1944). Vergleichende Pathologie des Nervensystems der Säugetiere. Georg. Thieme, Leipzig.

Schuetz, A.H. (1941). Growth and development of the orang-utan. Contr. Embryol. Carneg. Instn., 29: 57–110.

Shulman, N., R. Raphael, J. Hirschman and L.F. Barker (1969). Identification of the virus-like antigen of human hepatitis in non-human primates. Proc. 2nd Conf. Exp. Med. Surg. Prim. Karger, Basel.

Smith, H.A. and Jones T.C. (1966). Veterinary pathology (3rd ed.). Lea and Febiger, Philadelphia.

Sonntag, C.F. (1924). On the anatomy, physiology and pathology of the orang utan. Proc. Zool. Soc. Lond., 1924: 349–450.

Spalding, V.T. and C.S. Heymann (1962). The value of phencyclidine in the anaesthesia of monkeys. Vet. Rec., 74: 158.

Stafford, E.E., B. Galdikas-Brindamour and R.L. Beaudoin (1978). Hepatocystis in the orang utan, *Pongo pygmaeus*. Trans. Roy. Soc. Trop. Med. Hyg., 72: 106–107.

Stiles, C.W. and A. Hassal (1929). Key catalogue of parasites reported for primates with their possible public health importance. Bull. No. 152 Hyg. Labs. U.S. Pub. Hlth. Service: 409–601.

Straus, W.L. (1936). The thoracic and abdominal viscera of primates with special reference to the orang-utan. Proc. Am. Phil. Soc., 76: 1–85.

Bogaery, L. van and J.R.M. Innes (1962). Neurological disease of apes and monkeys. In: Innes and Saunders (eds.), Comparative neuropathology. Academic Press, New York.

Vertruyen, H. (1956). The diagnosis and treatment of tuberculosis in anthropoids. Société Royale de Zoologie d'Anvers Antwerp.

Vervat, D. (1961). Appendicitis in an orang-utan (*Pongo pygmaeus*) at Rotterdam Zoo. Int. Zoo Yb., 3: 112.

Vogel, H. and E.G. Vogelsang (1930). Neue Filarien aus dem Orang Utan und der Ratte. Zbl. Bakt. Abst., 1, 117: 480–485.

Voller, A. (1972). Plasmodium and hepatocystis. In: R.N.T.-W. Fiennes (ed.), Pathology of simian primates, part 2. Karger, Basel.

198

Voronoff, S. (1949). Tumeurs spontanées chez les singes. In: Groupes sanguines chez les singes. La greffe du cancer humain aux singes. Doin, Paris.

Weidman, F.D. (1923). Certain dermatoses of monkeys and an ape. Arch. Derm. Syph. N.Y., 7: 289–302.

Weinberg, M. (1908). Oesophagostomose des anthropoïdes et des singes interieurs. Arch. Parasit., 13: 161–203.

Zuckerman, A.J., C.R. Howard, A. Thornton, K.N. Tsiquaye, D.M. Jones and M.R. Brambell (1978). Hepatitis B outbreak among the chimpanzees at the London Zoo. Lancet, 2: 652–654.

Author's address:
D.M. Jones
The Zoological Society of London
Regent's Park
London NW1 4RY
United Kingdom

10. The physiology of reproduction of the orang utan

Jacob J. van der Werff ten Bosch

Reproduction comprises the full circle from the birth of an individual to the birth of its offspring. The physiology of reproduction should be concerned with parental aspects such as the care of the newborn — lactation, cleaning and rearing — and the growth and development of the infant, the latter's somatic and behavioural maturation and the establishment of sexually dimorphic characters in appearance and behaviour, the development of social behaviour resulting in copulation, fertilization and nidation, and the growth and sexual differentiation of the conceptus and its delivery.

For many of these aspects there is a dearth of information with regard to the orang utan. The data that are available derive from widely differing sources which may yield conflicting evidence, i.e. from animals born and reared in a zoo as opposed to animals born in the wild and grown up in the wild or in a zoo. Behavioural data in particular are likely to suffer, but purely morphological characteristics, such as growth in body weight, may of course also be strongly affected by such early environmental variables. For example, in captive animals great differences have been noted in the rate of growth depending on whether the mother nursed the young or whether the young was hand-reared (Brandes 1931; Lippert 1977). In the latter condition growth was much faster.

Pregnancy

About 2–4 weeks after conception there is an abrupt appearance of oedema of the labia majora; the oedema develops in 1–2 days' time and later disappears again immediately after parturition (Lippert 1974). About a month after conception, swelling occurs of the nipples and the surrounding mammary region (Fox 1929). Another early sign of pregnancy that has been reported is a change in the pigmentation of the skin around the eyes (Van Doorn 1964). Three months after conception the vulvar region begins to swell (Fox 1929), while further swelling of the peri-vulval tissue (Fig. 1) and enlargement of the clitoris can be seen during the last trimester of pregnancy (Fox 1929; Graham-Jones and Hill 1962).

Pregnancy can be ascertained by examination of the female's urine for the presence of chorionic gonadotrophins with the aid of the Sub-human (Non-

The orang utan. Its biology and conservation, edited by L.E.M. de Boer
© *1982, Dr W. Junk Publishers, The Hague, ISBN 90 6193 702 7*

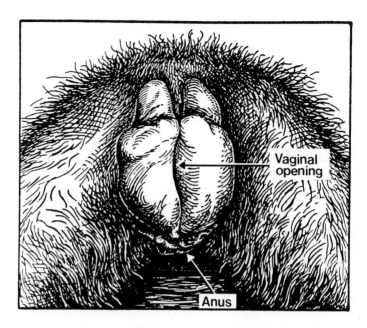

Fig. 1. Genital swelling in an adult, pregnant, wild-shot orang utan. The animal is lying on her back with her legs hanging over the edge of the supporting box (after a photograph, taken in the field, by S.L. Washburn) (from Schultz 1938).

human) Primate Pregnancy test. This test is a haemagglutination inhibition test in which an antiserum to the beta subunit of ovine luteinizing hormone is used. In one study the test was negative 248 days before parturition and positive nine days later, 239 days prior to parturition (Davis 1977). Since pregnancy lasts about 270 days, the pregnancy test apparently becomes positive between days 22 and 31 of pregnancy. From another study it is known that positive results can be obtained with this test until the end of pregnancy, including the day after delivery (Hodgen 1977).

Parturition

Parturition has been witnessed in several zoos and in the rehabilitation camp of the Sepilok Forest Reserve in East Sabah on Borneo. In the latter case the semi-wild *Joan* became pregnant by a wild male on one of her trips into the forest and gave birth to a daughter at the camp. She licked and ate the afterbirth and bit through the umbilical cord (De Silva, quoted by MacKinnon 1974). The first orang utan born in captivity was *Buschi* (the later subject of a monograph by Brandes [1939]), whose mother *Suma* gave birth on board a ship in the Red Sea on April 18, 1927. Suma nursed her son for 6.5 years in the Dresden Zoo.

The first births of orang utans which conceived in captivity took place in 1928, in zoos of Berlin, Nuremberg and Philadelphia. All three young died soon after birth, presumably because of dietary deficiency of their mothers (Harrisson 1967). Between 1928 and 1978 612 orang utans were born in captivity; this figure includes stillbirths but not abortions (Jones 1982). One young has been reported to die during birth in breech position. This was caused by placenta praevia, which had given rise to vaginal blood loss 5 and 3.5 weeks earlier (Kingsley and Martin 1979). Another orang utan has been delivered through forcipal extraction. Although this animal was born with gross damage to the skull, its growth and development were uneventful during the first two years of life (Hochbaum 1975).

Parturition may be preceded by refusal to eat on the day before. When labour begins the animal is restless and she may change her posture between uterine contractions: lateral and dorsal decubitus, the arms extended as though reaching for support, indicating pains. Delivery may occur with great speed. In one case (Graham-Jones and Hill 1962) 'abdominal contractions commenced as a series of 3 or 4 contractions, of 10 seconds' duration at intervals of seconds only. The vulval aperture widened, to display a mucilaginous secretion on the infant's head. Two and a half minutes later delivery of the head was effected by three rapid abdominal contractions over, approximately, a one-minute period. Immediately after head delivery, the female crouched, resting mainly on right knee and forearm, the left forelimb being left free to grasp the infant's head and to synchronise manual traction with a further and final abdominal contraction, which resulted in delivery of the infant's body. The mother was extremely gentle in her manipulative movements; she appeared to be fully predetermined and never at a loss as to procedure'.

Manual assistance in delivering the infant has also been observed by others (Van Bemmel 1963; Lippert 1974). Occasionally a mother has been noted to be less proficient and to have let the infant drop on the floor of the cage (Gorgas 1972). There are several accounts of a very curious posture adopted by the parturient female, who stands on her head, as it were, leaning on lower arm and elbow with extended legs so that the pelvis reaches upward (see Ullrich 1970). The latter author allowed the father to be present during delivery, and vividly describes the behaviour of that male before, during and after the birth of his son, *Djambi*. Initially the father, *Buschi-II*, attempted to copulate with the delivering mother, *Suma-II*, but as soon as the infant's head became visible in the vulva his behaviour changed completely. He took a position behind the female who was standing with pelvis extended upward, opened his mouth and closed it round the infant's head which disappeared fully in its father's mouth (Fig. 2). Slowly he thus pulled the infant out of the birth canal until the legs

Fig. 2. Obstetric aid by orang utan father (from Ullrich 1970).

appeared in the vulva. He then laid an open hand under the infant's body, let go of the head and took the infant in his lap and started to remove the foetal membranes from the head by sucking with pouted lips. Prompted by this successful cooperation between a mother and a father others have attempted to repeat the procedure, with less success however (Gorgas 1972).

Rearing

In the wild very young infants are continuously carried around, usually in side-ventral position. 'One hand clings on to the hair of the mother's shoulder, the foot holds the hair on her back, the opposite hand and foot hold on to the ventral hair so that the infant is on the mother's side behind her arm where it does not get in the way during tree travel' (MacKinnon 1974). Suckling takes place during short periods of rest on the part of the mother, which last a few minutes only and occur at intervals of, on average, 40 min in the first year of life (Rijksen 1978). As the animals grow older their relationship with the mother and their feeding pattern changes gradually (see Table 1).

Lactation is always associated with amenorrhoea. Menses return about 70 days after weaning. If a male is present only one or two menses occur before the female becomes pregnant again (Lippert 1974). Second births have been reported at intervals of just under 3 (Lippert 1977) and 4 (Chaffee 1967) years after the first young had been born.

Many captive-born young have been hand-reared by humans from birth or some time after birth. In the past it has often proved difficult to feed infants properly (Van Bemmel 1963; Klös 1966; Gorgas 1972), but it now appears that

Table 1. Life stages of the orang utan (from Rijksen 1978).

Life stage	Estimated age (in yr)	Estimated weight (in kg)	Physical characteristics	Behavioural characteristics
Infant	0–2½	2–6	Light pigmented zones around eyes and muzzle contrast with darker facial pigmentation; hair surrounding face long and standing out.	Always carried by the mother during travel; largely dependent on mother for food; sleeps in nest with mother.
Juvenile	2½–5	6–15	Facial characteristics as in infant.	Still mostly carried by mother, but makes short exploratory trips alone within her field of vision; plays often alone or with peers; initially still sleeps with mother, but later it builds own nest close to mother; towards the end of this stage mother may bear a new infant and attention for juvenile weakens.

Table 1. (Continued)

Life stage	Estimated age (in yr)	Estimated weight (in kg)	Physical characteristics	Behavioural characteristics
Adolescent	5–8	15–30	Hair surrounding face still long and standing out; initially facial colouration has the obvious light patches, but changes to completely dark; changes teeth; male and female are difficult to distinguish unless there is full view of anogenital region.	Frequent contact with mother; seeks contact with peers, plays with them and moves about with them in adolescent groups; cautious during contacts with adults, especially adult males; sometimes still traveling with mother; starts showing sexual behaviour; females are sexually mature at about 7 yr.
Sub-adult male	8–13/15	30–50	Facial colouration completely dark, with hard rims of undeveloped cheek flanges; beard starts to develop; hair surrounding face short; not standing out but flattened against the skull; testicles completely descended.	This stage commences with sexual maturity and continues until individual is socially mature; avoids contacts with adult males.
Adult female	8+	30–50	Old females may develop a beard, and are difficult to distinguish from sub-adult males, if not accompanied by offspring; nipples enlarged.	Usually accompanied by offspring.
Adult male	13/15+	50–90	Extremely large animals; maximal development of secondary sexual characteristics: cheek callosities, beard, throat pouches and long hair.	Sexually and socially mature; travels alone, moving cautiously; characteristic vocalisation is loud calling: 'long call' (MacKinnon 1971).

artificially fed youngsters may, in fact, grow faster than mother-fed peers (Lippert 1977). In the Berlin Tierpark *Biggy* and *Dunja* were born in the spring of 1973. Their ages differed three weeks. *Dunja* was reared by her mother, and at one year she weighed 3.5 kg. *Biggy* had to be taken from her mother when she was one week old, and was reared in the home of an animal caretaker; at one year she weighed 5.5 kg. However, when she was subsequently returned to full-time life in the zoo she appeared socially disturbed and grew poorly, so that at 2.5 years *Dunja*, who was socially well-adapted, weighed more than *Biggy* (Dathe et al. 1976).

Somatic growth and development

At birth orang utans weigh about 1.5 kg. In Figure 3 the birth weights of a number of species are plotted against the duration of gestation. McCance and Widdowson (1978) distinguish three types of prenatal growth, indicated by the dotted curves in the figure. The orang utan clearly belongs to the class of species shared by man, macaque, ox and elephant, which grow relatively slowly. From data summarised by Seitz (1969) the following median values can be calculated for male young at different postnatal ages (there were five to nine animals per age group): weight at birth, 1,750 g; at 6 months, 5.1 kg; at 12 months, 6.8 kg; and at 2 years, 10.4 kg. Females weighed less, the difference being about 100 g at birth and 400 g at one year. Adult weights are reported to average about 70 kg for males and about 40 kg for females for both Sumatran and Bornean orang utans (Eckhardt 1975) but higher weights have also been recorded. Many measurements have been taken of dead animals or skeletal material (Schultz 1941), but only few are available from longitudinal studies. The weight curve of a particularly fast growing male, *Viko*, is depicted in Fig. 4. *Viko* was the first born son of *Vicki* and *Hummel*, both of whom were about 8 years old. *Vicki* had no milk so that *Viko* was taken from her at the age of 48 hours (Seitz 1969). His weight has been plotted against a background which depicts the 3rd and 97th (per-)centiles for the bodyweights of Dutch boys (Van Wieringen 1972). The figure shows that *Viko* grew at about the same rate as boys from birth until 6 months, then grew very rapidly, and slowed down to almost human speed of

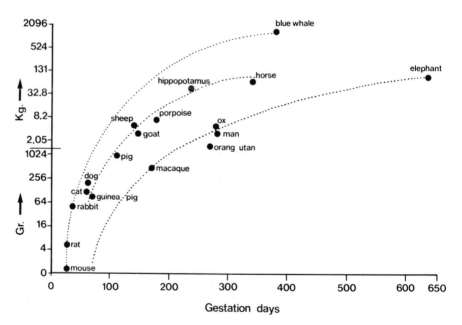

Fig. 3. Birth weights of 18 mammalian species and their lengths of gestation. The dotted curves show that the species can be classified roughly into those that grow (1) very fast from conception whether they are destined to become small or large animals; (2) fast; and (3) comparatively slowly. (After a figure of McCance and Widdowson 1978, to which datum for orang utan has been added.)

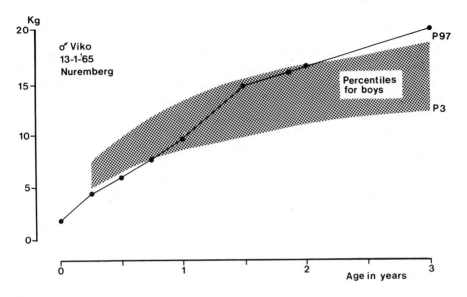

Fig. 4. Growth in weight of the male orang utan *Viko* from birth until the age of three years. Stippled area delimits the 94% range of weights at various ages for boys of the 1964–1966 national Dutch survey (Van Wieringen 1972).

growth again a year later. At ages of 6, 12 and 24 months *Viko* was twice as heavy as *Buschi-I* had been in the 1920s. It would be extremely interesting to have data on skeletal dimensions to see how much of the difference in weight can be attributed to faster skeletal growth and development, distinct from the accumulation of fat.

Sitting height at birth is about 30–40 cm, and in adults 70–90 cm (Schultz 1941). In one study length data were obtained on day 3 and at one year of age (Gorgas 1972): *Lotti* had a crown–rump length of 39.8 cm and a leg length of 11.9 cm at birth and at one year these figures were 61.0 and 21.0 cm. When the two measurements are added we get an estimate of 'stature', which increases from 51.7 at birth to 82 cm at one year, not differing very much from the human infant's growth from 50 to 75 cm over the same period. Adult stature may reach 150 cm (De la Fuente 1974). The proportions of the adult orang utan are very different from those of man. From data published by Schultz (1941) it can be calculated (using his Tables 1 and 6) that adult male orang utans with an average sitting height of 82.4 cm, had legs with an average length of 58.6 cm and thus a 'stature' of 141 cm, the legs constituting only 42% of stature as against 50% in man. In contrast, the arms are very long, about 175% of the leg length (Schultz 1941) which results in a very wide span of, in one instance, about 225 cm (Harrisson 1967). The development of the difference in bodily proportions can be gauged by calculating the ratio of the upper segment (sitting height) over the lower segment (leg length). The above figures yield the following ratios for the orang utan: at birth 3.34, at one year 2.90 and in adulthood 1.41. Human data at corresponding ages are (see Wilkins 1957) 1.69, 1.54 and 0.98; at the age of two years the ratio is 1.44, the same as in the adult orang utan.

Sexual maturation

Gross somatic sexually dimorphic characters other than the external genitalia do not arise until adulthood (see Table 1). Great body size, cheek flanges, beard and long hair on arms and back are the most obvious signs of male adulthood (Rijksen 1978). However, the cheek flanges or pads (Backenwülster) may not have reached full size by the time that the animal becomes fertile (Van Doorn 1964; Kingsley 1981).

In captive females the first menstruation, menarche, takes place when they are about 7 years old (Van Doorn 1964; Asano 1967; Lippert 1977). First mating has been recorded within a year after menarche (Asano 1967; Seifert 1970). First conception is known to have occurred some months after first mating (Chaffee 1967). Figure 6 illustrates the gentle awakening of motherhood in *Gypsy* of the Tama Zoo of Tokyo. *Gypsy* started to menstruate in January 1963, was first seen copulating a year later and conceived about 7 months after; in May 1965 she gave birth to a female baby (Asano 1967). From the few accounts available it would seem that the period of adolescent sterility, if there is any, must be very short indeed in this species.

Urinary steroid excretion has been studied for two orang utans during three menstrual cycles (Collins et al. 1975). The result of the findings during one cycle are shown in Fig. 7. It was found that values for oestrone and oestradiol-17β were similar to those found in the human and chimpanzee menstrual cycle. Oestriol values were between those for human and chimpanzee. Levels of pregnanediol and androsterone were significantly lower than in the human. Oestrone showed a mid-cycle and a luteal peak, whereas pregnanediol and androsterone excretion rose during the luteal phase only.

Earliest recorded male age at fertilisation was 10.5 years, 1.5 years before the appearance of cheek pads (Van Doorn 1964). The transition from sub-adult to adult male is not clearly defined (see Table 1), and seems mostly based on behavioural aspects of the animals. Obviously there is interaction between somatic and behavioural characters, as the following quotation may illustrate (MacKinnon 1974): 'If dominant males are less active sexually than the sub-adult males it is not easy to see any advantage in becoming dominant. It is possible that dominant males are sexually more potent when they do mate. Male orang utans in zoos do not usually develop full cheek flanges and large size if they are dominated by a larger male: even a dominating zoo-keeper can retard these developments. If being dominated can have such marked somatic effects on sexual secondary characteristics then it is possible that reduced potency also results.'

Mating behaviour

The earliest description of mating by orang utans in captivity was rendered by Fox in 1929. 'The copulatory act of our orangs is worthy of description because

---

Fig. 5 (opposite page). Orang utan baby Geert (Sumatran, born Rotterdam Zoo 13 December, 1963) at the age of 10 months.

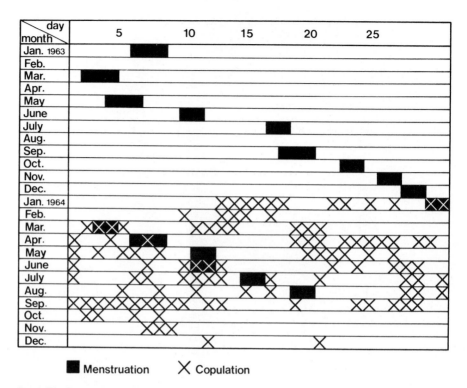

day month	5	10	15	20	25

Menstruation   X Copulation

Fig. 6. The first and subsequent menstruations, and copulatory activity of orang utan female, *Gypsy* (from Asano 1967).

of its dissimilarity from that described for the chimpanzee. When the desire animates the male and is reciprocated by the female, he pushes and mauls her a little, whereupon she lies upon her back on the floor. The male then approaches her and separates her legs. During the act he remains in a sitting or crouching attitude and though they are face to face, he does not lie upon the abdomen of the female. The male will sometimes grasp a leg of the female and hold it up and to the side during the conjugation. During the act, there is no fondling, nor do they mouth each other either before or after the act. The female lies passive and often has a hand over her face. The act is practiced daily, without relation to the sexual cycle. ... At no time has the keeper observed ... that copulation has occurred *more canum.*' Other observers have noticed different mating positions. One report states that during copulation the female stands on all four limbs and the male mounts her from the rear, usually wrapping his arms round her torso (Chaffee 1967). Vocal accompaniments of copulation are cited by several authors. According to De la Fuente (1974), 'Copulation is preceded by a noisy courtship ritual. The male begins by emitting soft purring sounds which are gradually transformed into loud roars, soon dying away. He then approaches the female and the animals mate, both uttering deep growls.' A slightly different account has been given by Harrisson (1967). She also reported an *andante* beginning ('mit leisem, vibrierendem Brummen hebt es an'), but during coitus both partners were silent.

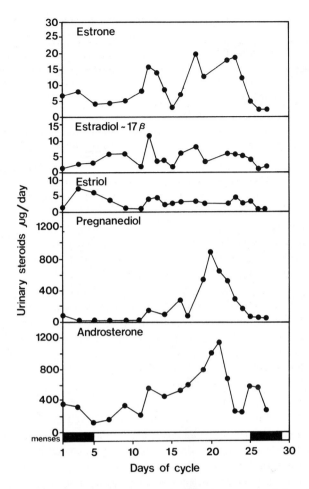

Fig. 7. Urinary excretion of oestrone, oestradiol-17β, oestriol, pregnanediol and androsterone during a menstrual cycle of an orang utan (from Collins et al. 1975).

The observation made by Fox (1929) that copulations occur independently of the ovarian cycle, has since been confirmed (Asano 1967, see, Fig. 6). Others have noted periods of 7–10 days per cycle during which copulations took place in a permanent pair in a zoo (Chaffee 1967). When daily tests were carried out through the menstrual cycle, copulations were found to take place daily, but 'female resistance to sexual advances by the male was lower during the midcycle phase of the cycle and multiple copulations occurred more frequently during that period' (Nadler 1977). More recently Nadler (1982) has carried out the following investigation. A male and a female orang utan were housed in adjacent rooms which were separated by a wall with an opening through which the female could, but the male with its much larger body size could not, pass to seek the other animal's company. It appeared that females would only visit the male neighbour in the course of the pre-ovulatory week; the male would then present, lying on his back, and be mounted by the female while he remained passive. The active enterprising role of the adult female in establishing sexual intercourse has also received special mention by several other investigators (MacKinnon 1974; Rijksen 1978).

Copulatory activity may continue throughout pregnancy (Asano 1967; Ullrich 1970; Gorgas 1972) and has even been observed during parturition (Ullrich 1970); on the other hand, pregnant females have also been reported to resist mating attempts (Lippert 1974).

Sometimes couples in a zoo never copulate, while change of a partner may be quickly followed by fruitful mating (Gorgas 1972). When they had a choice, males were found to have a preferred partner (Van Doorn 1964). In the wild, pair formation occurs, which is often of a non-sexual nature and frequently involves a sub-adult male and an adolescent female (Galdikas-Brindamour 1975) (see also Schürmann 1982). Two patterns of male–female sexual interactions have been distinguished, of which Rijksen (1978) provides the following observations and interpretations. 'One sequence of behaviour elements I called "rape" after MacKinnon (1971), the other sequence of elements I called "cooperative mating". In its extreme form, the "rape" is a straightforward copulatory act, initiated by sub-adult males, in which females usually display distress and a lack of cooperation. In its most extreme form, "cooperative mating" is an interaction, usually initiated by the female, in which both partners show considerable coordination of movements before, during and after the copulation, while copulation itself is performed cooperatively. Inevitably there were a number of cases that were difficult to classify. Even in "rapes" some coordination and some form of cooperation was discernible. It is noteworthy for example that females subjected to "rape" behaviour usually waited for a sub-adult male to approach or they moved away using a "hesitant avoid" locomotor pattern, which enabled the male to catch up. Contact would certainly not have been achieved if the female really avoided or even fled. On the other hand, some matings were observed that showed close coordination and cooperation of both partners in the initial stages, but turned into "struggling" and "screaming" on the part of the female participant, followed by fierce "positioning" on the part of the male.' It is of interest that Rijksen (1978) observed 'rape' behaviour being carried out by sub-adult males only, and never by adult males. This author also notes that 'rape' behaviour typically occurred during encounters, when both partners had not been in contact for at least one day; overt social behaviour between the partners before the male's initiating approach, and after the copulation was absent in most of the observed 'rape' interactions. Newly arriving adolescent females were always raped, usually by the highest ranking sub-adult male. Cooperative matings were commonly seen in partners who cooperated in many aspects of social behaviour, and who stayed and travelled together for days or months, in so-called consortship (Rijksen 1978; see also MacKinnon 1974).

Reproduction of the wild-living orang utan seems largely dependent on the behaviour of adult females. When she chooses an adult male and provokes him into sexual activity by presenting behaviour, she may soon become pregnant and through most of her adult life she will rear offspring. Sub-adult males are much more active sexually than adults, but most of their activity consists of rape; they may be infertile for much of their sub-adulthood, although they are known to have successfully inseminated females in zoos (see Van Doorn 1964). It is not clear whether rape behaviour includes intromission and intravaginal ejaculation. It has been stated that for intromission to occur, the active cooperation of the female is required in view of the small size of the penis and the difficulty of arboreal suspended copulation (MacKinnon 1974).

Acknowledgement

It is a very great pleasure to thank Mr A.E. Hijlkema, librarian of the Royal Rotterdam Zoological and Botanical Gardens, for his invaluable assistance in collecting the material on which this review is based.

References

Asano, M. (1967). A note on the birth and rearing of an orang-utan *Pongo pygmaeus* at Tama Zoo, Tokyo. Intern. Zoo Yearb. 7: 95–96.

Bemmel, A.C.V. van (1963). Das Züchten des Orang-Utan im Zoo. Z. Morph. Anthrop., 53: 65–71.

Brandes, G. (1931). Das Wachstum der Menschenaffen im Vergleich zu dem des Menschen in Kurven dargestellt. Zool. Garten (NF), 4: 339–347.

Brandes, G. (1939). 'Buschi', von Orang Säugling zum Backenwülster. Quelle und Meyer, Leipzig.

Chaffee, P.S. (1967). A note on the breeding of orang-utans *Pongo pygmaeus* at Fresno Zoo. Intern. Zoo Yearb., 7: 94–95.

Collins, D.C., C.E. Graham and J.R.K. Preedy (1975). Identification and measurement of urinary estrone, estradiol-17β, estriol, pregnanediol and androsterone during the menstrual cycle of the orangutan. Endocrinology, 96: 93–101.

Dathe, R., H.H. Dathe and R. Nagel (1976). Beobachtungen zur Mutter-Kind-Beziehung beim Orang-Utan (*Pongo pygmaeus*). Zool. Garten (NF), 46: 39–51.

Davis, R.R. (1977). Pregnancy diagnosis in an orangutan using two prepared test kits. J. Med. Primatol., 6: 315–318.

Doorn, C. van (1964). Orang Utans. Freunde des Kölner Zoo, 7: 3–9.

Eckhardt, R.B. (1975). The relative body weights of Bornean and Sumatran orangutans. Am. J. Phys. Anthrop., 42: 349–350.

Fuente, F.R. de la (1974). World of wildlife: The East. Animals of the jungle. Orbis, London.

Fox, H., (1929). The birth of two anthropoid apes. J. Mamm., 10: 37–51.

Galdikas-Brindamour, B. (1975). Orangutans, Indonesia's 'people of the forest'. National Geographic, 148: 444–472.

Gorgas, M. (1972). Zur Problematik der Aufzucht von Orang-Utans im Zoo. Z. Kölner Zoo, 15: 79–89.

Graham-Jones, O. and W.C.O. Hill (1962). Pregnancy and parturition in a Bornean orang. Proc. Zool. Soc. London, 139: 503–510.

Harrisson, B. (1967). Der Orang-Utan. In: Grzimeks Tierleben vol. 10. Kindler, Zürich.

Hochbaum, M. (1975). Bericht über die Aufzucht eines durch Zange entbundenen Orang-Utans, *Pongo pygmaeus*. Zool. Garten (NF), 45: 427–435.

Hodgen, G.D., C.K. Turner, E.E. Smith and R.M. Bush (1977). Pregnancy diagnosis in the orangutan (*Pongo pygmaeus*) using the subhuman primate pregnancy test kit. Lab. Anim. Sci., 27: 99–101.

Jones, M.L. (1982). The orang utan in captivity. In: L.E.M. de Boer (ed.), The orang utan. Its biology and conservation. Junk, The Hague.

Kingsley, S.R. (1982). Causes of non-breeding and the development of the secondary sexual characteristics in the male orang utan — a hormonal study. In: L.E.M. de Boer (ed.), The orang utan. Its biology and conservation. Junk, The Hague.

Kingsley, S.R. and R.D. Martin (1979). A case of placenta praevia in an orang-utan. Vet. Rec., 104: 56–57.

Klös, H.G. and U. Klös (1966). Bemerkungen zur künstlichen Aufzucht von Orang-Utans (*Pongo pygmaeus*). Sber. Ges. Naturf. Freunde (NF), 6: 66–76.

Lippert, W. (1974). Beobachtungen zum Schwangerschafts- und Geburtsverhalten beim Orang-Utan (*Pongo pygmaeus*) im Tierpark Berlin. Folia Primat., 21: 108–134.

Lippert, W. (1977). Erfahrungen bei der Aufzucht von Orang-Utans (*Pongo pygmaeus*) im Tierpark Berlin. Zool. Garten (NF), 47: 209–225.

MacKinnon, J. (1974). The behaviour and ecology of wild orang-utans (*Pongo pygmaeus*). Anim. Behav., 22: 3–74.

MacKinnon, J.R. (1971). The orang-utan in Sabah today. Oryx, 11: 141–191.

McCance, R.A. and E.M. Widdowson (1978). Glimpses of comparative growth and development. In: F. Falkner and J.M. Tanner (eds.), Human growth, vol. 1. Baillière Tindall, London.

Nadler, R.D. (1977). Sexual behavior of captive orangutans. Arch. Sex. Behav., 6: 457–475.

Nadler, R.D. (1982). Reproductive behavior and endocrinology of orang utans. In: L.E.M. de Boer (ed.), The orang utan. Its biology and conservation. Junk, The Hague.

Rijksen, H.D. (1978). A field study on Sumatran orang utans (*Pongo pygmaeus* abelii Lesson 1827). Ecology, behaviour and conservation. Mededel. Landbouwhogeschool Wageningen, 78–2: 1–420.

Schultz, A.H. (1938). Genital swelling in the female orang-utan. J. Mammal., 19: 363–366.

Schultz, A.H. (1941). Growth and development of the orang-utan. (Carnegie Inst.) Contrib. Embryol. no. 182: 59–110.

Schürmann, C. (1982). Mating behaviour of wild orang utans. In L.E.M. de Boer (ed.), The orang utan. Its biology and conservation. Junk, The Hague.

Seifert, S. (1970). Uber eine Schwangerschaft und die Frühgeburt eines toten Jungen bei einem Orang-Utan (*Pongo pygmaeus* Hoppius 1763) im Leipziger Zoo. Zool. Garten (NF), 38: 310–321.

Seitz, A. (1969). Notes on the body weights of new-born and young orang-utans *Pongo pygmaeus*. Intern. Zoo Yearb., 9: 81–84.

Ullrich, W. (1970). Geburt und natürliche Geburtshilfe beim Orang-Utan. Zool. Garten (NF), 39: 284–289.

Wieringen, J.C. van (1972). Secular changes of growth, 1964–1966 height and weight surveys in the Netherlands in historical perspective. Ned. Inst. Proev. Geneesk., Leiden.

Wilkins, L. (1957). The diagnosis and treatment of endocrine disorders in childhood and adolescence, 2nd ed. Blackwell, Oxford.

Author's address:
J.J. van der Werff ten Bosch
Department of Endocrinology, Growth and Reproduction
Faculty of Medicine
Erasmus University
P.O. Box 1738
3000 DR Rotterdam
The Netherlands

11. Causes of non-breeding and the development of the secondary sexual characteristics in the male orang utan: a hormonal study

Susan Kingsley

Introduction

Although there is a high incidence of reproductive failure in captive male orang utans, there are as yet no studies on the hormonal correlates of reproduction and development in this species.

The orang utan is unusual among primates in that, in the male, physical development is slowed for a period of several years before the final stages of physical maturation are completed. The animals become sexually mature and capable of siring young between the ages of 7 and 10 years (MacKinnon 1974; Rijksen 1978), sometimes earlier (e.g. a captive born male, *Schorsch*, at Hanover Zoo sired at $6\frac{1}{2}$ years old, Everts pers. comm.) but do not begin to develop their most striking secondary sexual characteristics until up to 7 years later (Ulmer 1958). Males which have completed the first stage of development retain many of the physical characteristics of the adolescent such as short hair, hard rims on the side of the face and a flat chest (Fig. 1a). After remaining in this 'sub-adult' (MacKinnon 1974) or 'non-flanged' stage for a variable number of years, at least in certain captive conditions, the male starts to develop the fatty cheek flanges and the large, muscle-covered, throat pouch of the 'mature' (MacKinnon 1974) or 'flanged' male (Fig. 1b). The larynx also develops, the hair lengthens and thickens, there is general body growth and a musty scent becomes apparent. Bornean orang utans also develop a fatty crown. Once these characteristics begin to develop the animal grows rapidly and full flange expression is often reached within a year.

In captivity the retardation of development is thought to have a social cause and may also be socially controlled in the wild. Zoo keepers have long held that the presence of a flanged male suppresses the development of flanges in a non-flanged male (Callard, Carmen, Smith and Wheeler pers. comm.). In the present study, this hypothesis was tested by studying a pair of adult males of similar age, one flanged and the other non-flanged. A second aspect of the study was concerned with examining the hormonal basis of flange development, and oestrogen and androgens were measured in adolescents, non-flanged adults and flanged adults as hormonal differences between the three groups might be expected. This study also included the hormonal monitoring of a male during

The orang utan. Its biology and conservation, edited by L.E.M. de Boer
© *1982, Dr W. Junk Publishers, The Hague, ISBN 90 6193 702 7*

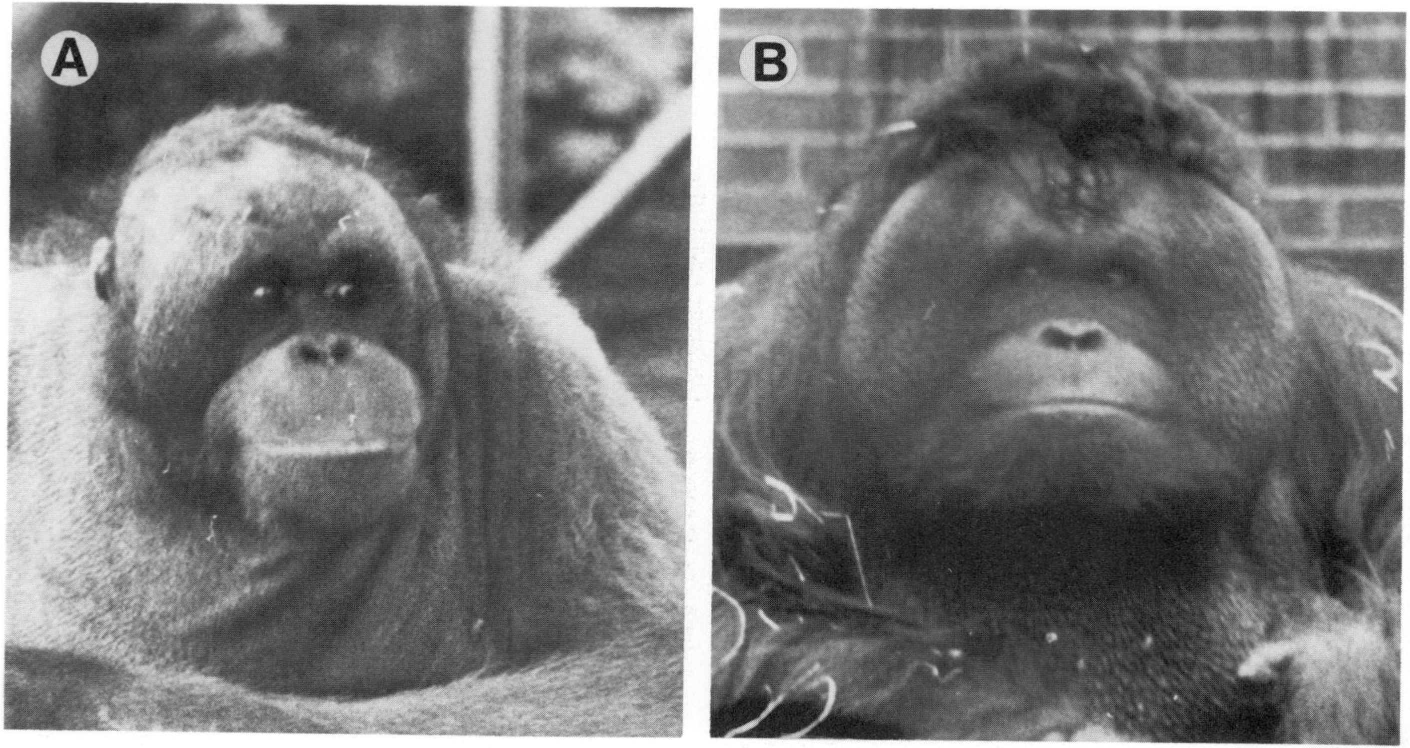

Fig. 1. A: Non-breeding or sub-adult sexually mature male showing typical facial characteristics (see text); B: flanged or mature adult male showing typical facial characteristics (see text).

216

flange development. The main androgen measured was testosterone and, as such, the term testosterone is used throughout when referring to androgens.

A third related study was undertaken to discover the reasons for the high incidence of non-breeding found among captive orang utans, in order to improve the chances of establishing viable breeding populations. Over 50% of adult captive males in the U.K. have not sired young (nine out of 19 are proven breeders, see Table 1), despite the fact that all have at some time been caged with proven breeding females. The situation seems to be similar among captive orang utan males in general. The cause may be hormonal, behavioural or both, but the present research was designed to investigate the hormonal aspect of the problem.

In the human, hormonal abnormalities such as elevated levels of FSH and LH and decreased levels of testosterone have been associated with reproductive failure (e.g. Legros et al. 1973; Hunter et al. 1974; Lawrence and Swyer 1974; Aafjes et al. 1977; Nieschlag et al. 1978; Spera et al. 1978). Although decreased levels of testosterone are less commonly associated with infertility in humans than are elevated LH levels (Lawrence and Swyer 1974; Aafjes et al. 1977; Corker et al. 1978), low testosterone levels are particularly associated with gross abnormalities of the testes (Lawrence and Swyer 1974; Nieschlag et al. 1978). A remarkable number of captive male gorillas have been reported to possess atrophied testes (Koch 1937; McKenney et al. 1944; Steiner 1954; Steiner et al. 1955; Antonius et al. 1971; Dixson et al. 1980) and measurement of urinary

Table 1. Adult male orang utans in zoological collections in the U.K. (correct to Jan. 1980)

Name	Birthdate	Where maintained	Breeding	Flanged
Adam	1955	Dublin	−ve	
Coco	1971	Weybridge	−ve	−ve
Joe	1959	Dudley	+ve	+ve
Adam	1960	Colchester	+ve	+ve
Jod	1961	Twycross	+ve	+ve
Gambar (Gamb or Ga)	1962	Jersey	+ve	+ve
Harry (Har or Ha)	1962(59?)	Weybridge	+ve	+ve
Toby	1963	Twycross	?	+ve
Dodo	1964	London	+ve	+ve
Tommy	1964	Dublin	−ve	
Kabir	1965	Edinburgh	−ve	+ve
Saleh	1965	London	+ve	−/+ve
Giles (Gi)	1965	Jersey	+ve	+ve
Louis (Lo)	1965	Weybridge	−ve	+ve
Tuan	1966	Chessington	−ve	
Dennis	1966	Chester	+ve	−/+ve
David (Dav)	1966	Chester	−ve	−ve
Barry	1967	Chester	−ve	−ve
Rajang (Raj)	1968 (c)	Chester	−ve	−ve
Sabah	1968	Blackpool	−ve	
Total	20	Breeding	9	
		Non-breeding	10	

c = captive-born
Name abbreviations used in Figs. 3 and 4 are in brackets

testosterone levels in two of the males subsequently studied at their death by Dixson et al. (1980) showed testosterone levels well below that of normal male gorillas (Kingsley, unpublished data). Both males had never bred as is the case with a large percentage of captive male gorillas (only four out of 13 adult males in the U.K. have sired, personal observation). Lack of breeding success may therefore be due to a high incidence of testicular abnormality and the high rate of non-breeding in male orang utans may have a similar explanation. As yet, no known testicular tissue from non-breeding orang utans has been studied and biopsy of the testes of living zoo-kept males is not permissible. In order to establish an easily identifiable basis for non-breeding in these males, urinary testosterone levels were measured and compared with those of breeding males.

Materials and methods

Animals

A total of 20 males from four age-status classes (juvenile and adolescent; non-flanged adult; adults with flanges growing; flanged adult) were used in the three aspects of the study, their ages ranging from 1 year to 16 years (see Fig. 3 for names and ages). They were maintained in six zoological collections in the U.K. and Germany, and all were in visual, olfactory and auditory contact with other animals. All caging had indoor and outdoor areas, and bars for climbing. The animals were fed several times a day with dried pellet food (e.g. Mazuri Primate Diets) and fresh fruit and vegetables of the season, and most had water *ad libitum*.

The study undertaken to determine hormonal differences between the different age-status classes involved all 20 animals, and one of these, *Saleh*, was monitored throughout the process of flange development. The second aspect of the study concerned the testing of the flange growth suppression hypothesis and involved two of the 20 males, *Saleh* (originally non-flanged) and *Dodo* (flanged). Both males had sired several infants and were therefore sexually mature, and both were originally caged in visual, olfactory and auditory contact with each other and with three adult females. After one year they were separated, *Dodo* remaining in contact with the three females. *Saleh* was removed from the flanged male (which may have inhibited flange development) and was caged only in contact with juveniles and adolescents. The non-breeding aspect of the study involved only 13 adults, all of which were, or had been, caged with adult breeding females. Of these males, seven had sired and six had not.

Collection of samples

Urine was collected between late 1977 and early 1979. It was either collected in a special trap fitted into the cage drain or syringed from the floor immediately after excretion. It was not possible to collect 24-hour samples but the routine collection of early morning samples minimised variation in hormone levels due to the circadian nature of testosterone excretion (e.g. Resko and Eik-Nes 1966; Goodman et al. 1974). Normally six or more samples were collected from each male within a short time period. Weekly samples, when possible, were collected from the two males (*Saleh* and *Dodo*) involved in the flange growth and separation studies.

Samples were stored at $-20°C$ for determination of salt concentration and analysis of testosterone and oestrogen by radioimmunoassay.

Radioimmunoassay of urinary steroids

Urinary oestrogen and testosterone were measured by a procedure used successfully in this laboratory for a number of other primate species.

Urine (50 μl) was hydrolysed with 0.4 i.u. β-glucuronidase 'Pasteur' (Uniscience) in 50 μl phosphate-buffered saline (pH 7.0, 0.02 M phosphate buffer, containing 0.2% NaCl, 0.02% NaN₃) for 1 h at 37°C and then diluted with an appropriate volume (between 1:2 and 1:150) of Tris-buffered saline (pH 8.0, containing 0.1 M NaCl, 0.05 M Tris, 0.1% NaN₃, 0.1% gelatin) prior to direct radioimmunoassay without extraction or chromatography. A biological standard was included in each assay. All samples were assayed in duplicate in 12×75 mm plastic tubes (Luckham Ltd.).

Oestrogens. Each assay tube contained 100 μl diluted hydrolysed urine, 100 μl (~ 50 pg) $[2,4,6,7(n)-^3H]$ oestradiol (specific activity 86 Ci/mmol: Radiochemical Centre) and 100 μl antiserum 461/9 (raised in a goat to oestradiol-17β-succinyl-bovine serum albumin by Dr Furr, ICI Pharmaceutical Division, Macclesfield, U.K.) diluted to give approximately 20% binding of ^3H oestradiol-17β in the absence of unlabelled steroid. Antiserum cross-reactivities with the other oestrogens were: oestradiol-17α, 50.3%; oestrone, 27.9%; oestriol, 2.8%. Cross-reactivity with androstenedione, testosterone, progesterone and pregnanediol was less than 0.1%.

Each tube was mixed briefly, incubated for 1 h at 37°C, then cooled to 4°C for 30 min. Separation of antibody bound and free steroid was achieved by the addition of 200 μl of dextran-coated charcoal suspension (containing 1.0% charcoal: Sigma; and 0.1% Dextran T-70: Pharmacia, in assay buffer). Each tube was then briefly mixed, incubated for 3 min at 4°C followed by centrifugation at 4°C and 1,700 g for 4 min.

Supernatant (300 μl) was removed to a mini scintillating vial containing 1 ml of scintillant (Pico-Fluor 15: Packard Instruments Ltd., Caversham, U.K.) and the vial was counted for 4 min in a Packard Tri-Carb scintillation spectrometer (Model 3255). A standard curve ranging from 0.019 to 5.0 ng per tube was processed similarly for each set of samples analysed. The 50% inhibition occurred at 0.604 ± 0.118 ng per tube (n = 13).

Intra- and inter-assay coefficients of variation were 4.7% (n = 10) and 9.4% (n = 7), respectively. These were calculated by repeated assay of a hydrolysed urine pool in a single assay and in separate assays. Accuracy was determined by measuring added amounts of oestradiol (1.25, 0.625, 0.312, 0.156 ng/tube) and

219

hydrolysed urine (diluted to 1:10 and treated with charcoal to remove endogenous hormone). A linear regression was found for estimated oestradiol against added oestradiol (y = 0.93x − 0.023), the slope of which was not significantly different from the expected value of 1.

Testosterone. The same procedure was followed as for oestrogens, using 100 μl (∼60 pg) [1,2,6,7,(n)−^{3}H] testosterone (specific activity 81 Ci/mmol: Radiochemical Centre) and 100 μl antiserum (raised in a rabbit to testosterone-3-bovine serum albumin, by Dr Horth, G.D. Searle and Co. Ltd., High Wycombe, U.K.) diluted to give approximately 27% binding of ^{3}H testosterone in the absence of cold testosterone, in each tube.

Anti-serum cross-reactivities were: 5α-dehydrotestosterone (DHT), 35.3%; 19-nortestosterone, 5.1%; androstenedione, 3.4%; dehydroepiandrosterone, 0.3%. Cross-reactivity with 5β-DHT, oestrone, oestradiol-β, oestradiol-α, oestriol and progesterone was less than 0.1%. The 50% level of inhibition occurred at 0.351 ± 0.060 ng per tube (n = 8).

Intra- and inter-assay coefficients of variation, 6.5% (n = 10) and 10.2% (n = 9), respectively, were calculated by the same procedure as for oestrogens. Accuracy was also determined as above, by adding known amounts of testosterone (1.0, 0.5, 0.25, 0.125 ng/tube) to the treated hydrolysed urine (diluted 1:20). A linear regression was found for estimated testosterone against added testosterone (y = 1.013x − 0.029), the slope of which was not significantly different from the expected value of 1.

Assay for NaCl

Since creatinine could not be detected in any samples from one of the zoos (due to faulty refrigeration), the normal procedure of relating hormone levels of the isolated samples to urinary creatinine (to take into account the concentration of the sample) could not be followed. Hormone levels were therefore related to the pure NaCl equivalent of urinary salts concentration, the salts level reflecting the concentration of the urine. The results obtained using NaCl were of the same pattern as those using creatinine (calculated in samples suffering no creatinine breakdown (Kingsley 1981).

The freezing point depression of each sample (200 μl) was measured using an Osmometer (Advanced Instruments, Inc., Mass., U.S.A.), allowing the osmolarity (the concentration of salts) of the sample to be determined. NaCl forms a major part of urinary salts and therefore the osmolarity was converted totally to NaCl concentration (by means of pure salt standards) for ease of calculation.

Intra- and inter-assay coefficients of variation were 2.0% (n = 8) and 2.3% (n = 7), respectively as established by repeated measurement of an untreated urine pool in a single assay, and in separate assays. Hormone levels were related to urinary salt concentration and the results averaged for each male and used in subsequent analysis. When less than three urine samples were collected from any one male the results were excluded from the analysis. The results from *Saleh* and *Dodo* were pooled into three monthly time periods and the means

220

calculated. Results between the four age-status groups or between three monthly time periods (in the case of *Saleh* and *Dodo*) were compared by multiple comparison statistics (a protected t-test) and those between the breeding and non-breeding groups by Student's t-test. The level of significance chosen was $p < 0.01$.

Measurement of flange growth rate

Two orang utans, *Saleh* and *Dodo*, were involved in the experiment to check whether flange growth is suppressed in a non-flanged male by the presence of a flanged male. Urine was collected weekly from both males for a year before and six months after separation. Once the non-flanged male, *Saleh*, began to develop flanges, full face photographs were taken at monthly intervals, where possible. Photographs were also taken from *Dodo* to check that flange development was complete. Since it was not possible to photograph the animals from a fixed distance (it was difficult to get full face photographs at all) the photographs were of different sizes. The rate of flange growth was therefore calculated by dividing the total width of the face (TF) by the biorbital breadth (BB, Fig. 2) to obtain a ratio of flange size. The total width of the face increased as the flanges developed whereas the distance between the eyes remained essentially constant. The validity of this method was checked by measuring facial photographs from 49 animals from zoological collections throughout the world.

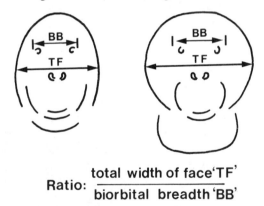

$$\text{Ratio:} \quad \frac{\text{total width of face 'TF'}}{\text{biorbital breadth 'BB'}}$$

Fig. 2. Diagram to illustrate the calculation of flange ratio.

Results

Urinary concentrations of testosterone and oestrogen (mean $\pm$ SD) of all the males in the study are shown in Fig. 3. Non-flanged adult males had generally higher urinary testosterone levels than the juveniles and adolescents, while flanged males had higher testosterone levels than non-flanged adult males ($p < 0.01$). Adults in the process of flange development had levels intermediate between non-flanged and flanged males. Similarly, the flanged males had significantly higher levels of oestrogens than the non-flanged adults ($p < 0.01$)

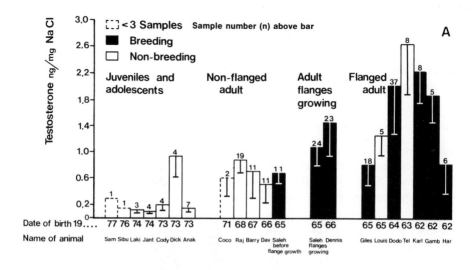

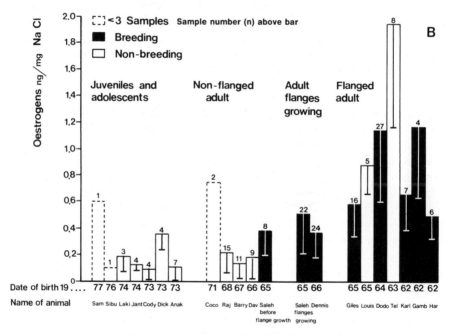

Fig. 3. Mean (±SD) urinary testosterone (a) and oestrogen (b) levels (ng/mg NaCl) in male orang utans of all ages and breeding status (n = 20) (Tel = *Telok*; Jant = *Jantan*; Dav = *David*; Raj = *Rajang*; Har = *Harry*).

and the males growing their flanges had intermediate levels. Similar concentrations of oestrogen were present in the urine of non-flanged males, and juveniles and adolescents.

The breeding status of all the adults is also illustrated in Fig. 3. Testosterone and oestrogen levels in both breeding and non-breeding adults were not

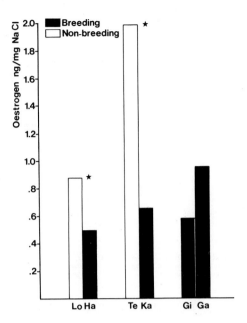

Fig. 4. Mean urinary oestrogen levels (ng/mg NaCl) in pairs of flanged orang utans, caged in visual, auditory and olfactory contact. The pairs are: *Louis* (Lo) and *Harry* (Ha), Weybridge; *Telok* (Te) and *Karl* (Ka), Frankfurt; *Giles* (Gi) and *Gambar* (Ga), Jersey.

statistically different but, taking the results of the flanged males only (so that there is no effect of lower levels of hormone due to stage of development), non-breeding males had significantly higher levels of oestrogen ($p < 0.01$) than breeding males, although testosterone levels were the same in both groups. The oestrogen results from six (which are housed in three pairs) of these seven flanged males are presented in Fig. 4. The non-breeding member of the two pairs comprised of both breeding and non-breeding males had significantly higher ($p < 0.01$) levels of oestrogen than their breeding neighbour, whereas the levels were statistically the same in the pair which had bred.

The method for measuring flange growth was validated by the results illustrated in Fig. 5. Flanged males had a greater facial ratio than non-flanged males (which had the same ratio as adult females) and the increasing flange ratio shown in *Saleh*'s results (Fig. 6) is therefore indicative of flange growth. Figure 6 also illustrates that testosterone levels increased during the three-monthly period before commencement of flange growth, and continued to increase in conjunction with flange development for the next six months. Testosterone secretion then remained at a constant level while flange ratio continued to increase. Oestrogen levels also increased markedly at first, but then decreased significantly after the separation ($p < 0.01$). *Saleh*'s results (Fig. 6) also show that flange development commenced before separation from *Dodo*. *Dodo*'s levels of oestrogen and testosterone remained statistically constant throughout the study, there being no difference in levels either before or after separation from *Saleh* (Fig. 7).

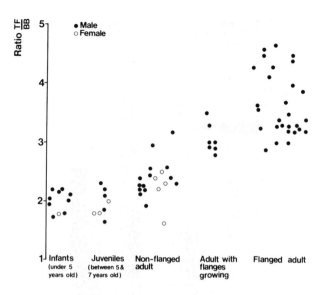

Fig. 5. Face proportions calculated as a ratio TF/BB (depicting flange size in males) in orang utans of all ages (n = 49).

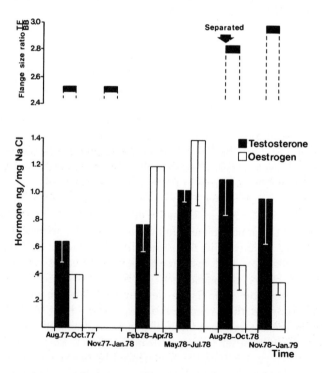

Fig. 6. Relationship between flange size and mean (±SD) 3 monthly urinary oestrogen and testosterone levels in adult male orang utan, *Saleh*, before and after separation from a flanged male (*Dodo*).

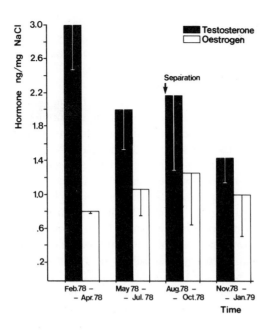

Fig. 7. Mean (±SD) 3 monthly oestrogen and testosterone levels in flanged male, *Dodo*, before and after separation from *Saleh* (originally non-flanged).

Discussion

Non-flanged adult orang utan males are all sexually mature and, as such, have higher testosterone levels than the adolescent males, since spermatogenesis is only achieved by increased testosterone secretion. The higher levels of testosterone observed for the adolescent male, *Dick* (see Fig. 3), may be associated with an earlier onset of pubertal development than the two other 5-year-olds in the study, and he may have been approaching the stage of development of the non-flanged male. The sexually mature males with fully developed secondary sexual characteristics have even higher testosterone and oestrogen levels than non-flanged males, which suggests that flange development requires an increase in testosterone and oestrogen secretion above that necessary for the attainment of breeding alone. This suggestion is supported by the data from *Dodo*, a fully developed male, which shows that the levels of testosterone and oestrogen remained high throughout the study and is partly supported by the data from *Saleh*, which shows that testosterone levels increased as the flanges developed. In fact, it seemed that an increase in the levels of testosterone was a prerequisite for flange development, flange growth commencing a few months after the initial rise in levels. The situation with the oestrogens was slightly different, as the levels increased at first, but then decreased once *Saleh* was separated from *Dodo*. The reason for this is as yet unexplained. As far as the attainment of breeding status is concerned, there seems to be no associated increase in levels of oestrogen, adolescents and non-flanged males having similar urinary concentrations.

The results from the study on suppression of flange growth indicate that the presence of a flanged male does inhibit flange development in a non-flanged male for a time, but not permanently so. One male at Philadelphia Zoo remained at the non-flanged stage of development for about seven years, but eventually developed flanges whilst still in the presence of the flanged male (Ulmer 1958). Most males, however, remain non-flanged for two or three years when housed with a flanged male, finally developing flanges at a later age than their already flanged cage neighbour. In the present study *Saleh* commenced flange development before being separated from *Dodo*, although at a later age, and a similar situation has been found in a number of other zoos (e.g. *Telok* and *Karl* in Frankfurt, Germany [Jantschke pers. comm.]; *Toby* and *Jod* in Twycross, England pers. obs.). This slowing down of development, which is so unusual among mammals, does not always occur, however, at least in captivity. Males caged away from contact with other adult males seem to develop their flanges soon after reaching sexual maturity and the development of the secondary sexual characteristics starts at 5 or 6 years and continues, to culminate in flange growth (and associated characteristics) within four years. This continuum in development occurred with *Buschi* in Dresden Zoo, Germany (Brandes 1939) and *Joe* in Dudley Zoo, England (Masson pers. comm.), both animals being fully developed by the age of ten years.

How the apparent suppression of flange development in a non-flanged male takes place is unknown, but could be related to stress in the presence of a flanged male. In girls, social stress slowed down, or arrested, pubertal development for a time (Tanner 1978) and this might therefore be the cause of flange growth suppression in orang utans. Social stress is also known to decrease testosterone levels in other primates (human: Kreuz et al. 1972; rhesus monkey: Rose et al. 1971, 1975; tree shrews: Von Holst 1977). Assumed stress in the orang utan may therefore cause testosterone secretion to remain at a lower level than that required for flange development to occur. The measurement of cortisol or prolactin (two hormones sensitive to stress) levels in the orang utans may give a clearer picture as to the cause of flange growth inhibition.

Although flange development seems to have a hormonal correlate, non-breeding in adult males as a whole does not. Testosterone and oestrogen levels are similar in both breeding and non-breeding males which suggests that the orang utans under study, and probably captive orang utans in general, do not have the same testicular abnormalities observed in captive gorillas (Dixson et al. 1980) but are more likely to be similar to humans where gross testicular abnormalities only account for a small percentage of infertility (De Kretser 1977; Paulsen 1977), and where peripheral levels of oestrogen and testosterone are similar in both fertile and infertile men (Kelch et al. 1972; Lawrence and Swyer 1974; Aafjes et al. 1977; Corker et al. 1978, Nieschlag et al. 1978). Although the testosterone and oestrogen results suggest grossly normal testes it

226

is possible that some other physiological factor is responsible for the high incidence of non-breeding found. Such a factor might be reflected by elevated LH concentrations as reported in humans (Hunter et al. 1974; Aafjes et al. 1977; De Kretser 1977). Consequently, it would be useful to measure LH levels in the adult orang utans.

Although the overall results of the orang utan study showed no difference in testosterone and oestrogen levels in breeding and non-breeding males, the results from the flanged males alone did show a hormonal correlate, the non-breeding males showing significantly higher oestrogens than the breeding males. The cause of this is uncertain, but behaviourally these non-breeding flanged males appeared to be intimidated by the proximity of their breeding neighbours. The breeding neighbours would show 'chomping' and 'fixed stare' behaviour towards the non-breeding males (Jantschke pers. comm.; Keeling pers. comm.; pers. obs.), as described by MacKinnon (1974) and Rijksen (1978) for aggressive wild males. It can therefore be supposed that the flanged breeding males were acting aggressively towards their non-breeding neighbours; the non-breeding animals were never observed to behave in an aggressive manner towards them. The two flanged males in neighbouring cages who both bred (*Gambar* and *Giles*) were rarely seen to exhibit 'chomping' and 'fixed stare' behaviour, and when they did, neither behaved more aggressively than the other. So it may be that the non-breeding in these flanged males is a socially related phenomenon, since socially subordinate males of other species also fail to reproduce and show little sexual interest towards females (talapoin monkey: Keverne 1979). The high oestrogen levels may be reflecting this subordinate status, although why this should be is as yet unknown.

Finally the source of the urinary oestrogens in the males is uncertain. Over 50% of circulating oestrogen in men is converted from testosterone (Longcope et al. 1969; Lipsett 1974) and a similar conversion may occur in the orang utan. There may be an increase in oestrogen secretion associated with flange development as the oestrogen levels in non-flanged males are similar to those in adolescents and juveniles whereas the testosterone levels are higher.

The main points of the present study are summarised as follows:
1. There is a tendency for higher testosterone levels in non-flanged adults than in adolescents, but the oestrogen levels are similar in both groups;
2. Flanged males have higher testosterone and oestrogen levels than non-flanged males ($p < 0.01$);
3. In the one animal monitored during flange development, an increase in testosterone levels preceded the commencement of flange growth. Testosterone levels continued to increase at first, early in flange development, but later remained constant while the flanges continued to grow;
4. The presence of a flanged male appears to inhibit development in a non-flanged neighbour, although not permanently. Males housed alone tend to show no retardation of development of the secondary sexual characteristics;
5. There is no correlation between urinary testosterone and oestrogen levels and the occurrence of non-breeding in the adult male orang utan;
6. There is, however, a correlation between urinary steroid levels and the

occurrence of non-breeding in flanged males alone, the non-breeding males having significantly higher oestrogens than the breeding males ($p < 0.01$). These high oestrogens may reflect the socially subordinate status of the non-breeding males.

Acknowledgements

I am very grateful to the directors, curators of mammals and ape staff of Chester, Flamingo Park, Frankfurt, Jersey, London and Weybridge zoological collections for their help and cooperation in the study. I would also like to thank Dr Alan Dixson and Dr Rosemary Bonney for their help on various aspects of the study, and Dennis Tanner for preparing the figures. The research was supported by the Medical Research Council, England. This contribution was made possible by funding from the Zoological Society of London.

References

Aafjes, J.H., J.C.M. van der Vijver, R. Docter and P.E. Schenck (1977). Serum gonadotrophins, testosterone and spermatogenesis in subfertile men. Acta Endocrinol. (Kbh.), 86: 651–658.

Antonius, J.I., S.A. Ferrier and L.A. Dillingham (1971). Pulmonary embolus and testicular atrophy in a gorilla. Folia Primat., 15: 277–292.

Brandes, G. (1939). Buschi; Vom Orang-saügling zum Backenwülster. Quelle und Meyer, Leipzig.

Corker, C.S., D.W. Davison and P. Edmond (1978). Plasma testosterone levels in men attending a sub-fertility clinic. J. Steroid Biochem., 9: 375–376.

Dixson, A.F., H.D.M. Moore and W.V. Holt (1980). Testicular atrophy in captive gorillas (Gorilla g. gorilla). J. Zool.,

Goodman, R.L., J. Hotchkiss, F.J. Karsch and E. Knobil (1974). Diurnal variations in serum testosterone concentrations in the adult male rhesus monkey. Biol. Reprod., 11: 624–630.

Holst, D. von (1977). Social stress in tree shrews: problems, results and goals. J. Comp. Physiol., 120: 71–86.

Hunter, W.M., P. Edmond, G.S. Watson and N. McLean (1974). Plasma LH and FSH levels in subfertile men. J. Clin. Endocr. Metab., 39: 740–749.

Kelch, R.P., M.R. Jenner, R. Weinstein, S.L. Kaplan and M.M. Grumbach (1972). Estradiol and testosterone secretion by human, simian and canine testes, in males with hypogonadism and in male pseudohermaphrodites with the feminizing testes syndrome. J. Clin. Invest., 51: 824–830.

Keverne, E.B. (1979). Sexual and aggressive behaviour in social groups of talapoin monkeys. In: Sex, hormones and behaviour (CIBA Fdn. Symposium 62). Excerpta Medica, Amsterdam.

Kingsley, S.R. (1981). The reproductive physiology and behaviour of captive orangutans (Pongo pygmaeus). Ph.D. thesis: University of London.

Koch, W. (1937). Bericht über das Ergebnis der Obduktion des Gorilla 'Bobby' des Zoologischen Gartens zu Berlin. Ein Beitrag zur Vergleichenden Konstitutionspathologie. Veröff. Konst. u. Wehr Path., 9 (3): 36.

Kretser, D.M. de (1977). Evaluation and treatment of infertile men. Paper read to joint meeting of the Society for the Study of Fertility and Blair Bell Research Society, 16 December 1977, at the Zoological Society, London.

Kreuz, L.E., R.M. Rose and J.R. Jennings (1972). Suppression of plasma testosterone levels and psychological stress: a longitudinal study of young men in Officer Candidate School. Arch. General Psychiatry, 26: 479–482.

Lawrence, D.M. and G.I.M. Swyer (1974). Plasma testosterone and testosterone binding affinities in men with impotence, oligospermia, azoospermia and hypogonadism. Br. Med. J., 1: 349–351.

Legros, J.J., M. Palem, J. Servais, M. Margoulies and P. Franchimont (1973). Basal pituitary-gonadal function in impotency evaluated by blood testosterone and LH assays. In: K. Lissak (ed.), Hormones and brain function. Plenum Press, New York.

Lipsett, M.B. (1974). Steroid secretion by the testis in man. In: M. Serio and L. Martini (eds.), The endocrine function of the human testis, Vol. II. Academic Press, New York.

Longcope, C., T. Kato and R. Horton (1969). Conversion of blood androgens to estrogens in normal adult men and women. J. Clin. Invest., 48: 2191–2201.

MacKinnon, J. (1974). The behaviour and ecology of wild orang-utans (*Pongo pygmaeus*). Anim. Behav., 22: 3–74.

McKenney, F.D., J. Traum and A.E. Bonestell (1944). Acute coccidiomycosis in a mountain gorilla with anatomical notes. J. Am. Vet. Med. Assoc., 104: 136–140.

Nieschlag, E., E.J. Wickings and J. Mauss (1978). Endocrine testicular function in infertility. In: A. Fabbrini and E. Steinberger (eds.), Recent progress in andrology. Academic Press, London.

Paulsen, C.A. (1977). Regulation of male fertility. In: R.O. Greep and M.A. Koblinsky (eds.), Frontiers in reproduction and fertility control, Part 2. MIT Press, Cambridge, Mass.

Resko, J.A. and K.B. Eik-Nes (1966). Diurnal testosterone levels in peripheral plasma of human male subjects. J. Clin. Endocr., 26: 573–576.

Rijksen, H.D. (1978). A fieldstudy on Sumatran orang utans (*Pongo pygmaeus abelii* Lesson 1827); ecology, behaviour and conservation. Mededel. Landbouwhogeschool, Wageningen, 78–2: 1–420.

Rose, R.M., I.S. Bernstein and T.P. Gordon (1975). Consequences of social conflict on plasma testosterone levels in rhesus monkeys. Psychosom. Med., 37: 50–61.

Rose, R.M., J.Q. Holaday and I.S. Bernstein (1971). Plasma testosterone, dominance rank and aggressive behaviour in male rhesus monkeys. Nature, 231: 366–368.

Spera, G., M. Re, R. Weisz, G. Montagna, M. Iannitelli and F. Fraioli (1978). Relationship between plasma levels of FSH and impairment of spermatogenesis in male infertility. In: A. Fabbrini and E. Steinberger (eds.), Recent progress in andrology. Academic Press, London.

Steiner, P.E. 1954. Anatomical observations in a *Gorilla gorilla*. Am. J. Phys. Anth., 12: 145–179.

Steiner, P.E., T.B. Rasmussen and L.E. Fisher (1955). Neuropathy, cardiopathy, hemosiderosis and testicular atrophy in *Gorilla gorilla*. Arch. Path., 59: 5–25.

Tanner, J.M. (1978). Foetus into man: physical growth from conception to maturity. Fletcher, Norwich.

Ulmer, F.A., Jr. (1958). Rusty becomes a backenwülster! America's first zoo, 10: 7.

Author's address:
S. Kingsley
Wellcome Laboratories of Comparative Physiology
Institute of Zoology
The Zoological Society of London
Regent's Park
London NW1 4RY
United Kingdom

12. Reproductive behavior and endocrinology of orang utans

Ronald D. Nadler

Laboratory research on reproductive function of orang utans is less advanced than research on the other great apes, chimpanzee and gorilla. For this reason, and because the research on the latter species established the background and impetus for studies with the orang utan, it is appropriate to review the relevant data obtained on these apes before examining data for the orang utan.

Laboratory research on sexual behavior of chimpanzees

The chimpanzee was the first of the great apes to be investigated under controlled laboratory conditions in studies of sexual behavior (Yerkes and Elder 1936; Yerkes 1939; Young and Orbison 1944). The studies with this species, moreover, established the rationale and experimental paradigm for much of the subsequent research on the other apes. The main objective of this research was to determine whether this advanced primate, close taxonomic affiliate of the human, exhibited evidence of estrus, i.e., the periodic increase in female sexual responsiveness at about the time of ovulation, characteristic of most other mammals. Yerkes (1939) proposed that the close taxonomic relationship between the ape and human made the chimpanzee an excellent model for investigating issues of interest to humans that could not be conveniently investigated with humans. The research on sexual behavior and estrus of the chimpanzee was an example of just such an issue.

In these studies, oppositely-sexed pairs were tested daily throughout the sexual cycle of the female and daily measurements were made of the female's genital swelling as a method of monitoring cycle phase. Two major findings resulted from this research: 1) chimpanzees copulated primarily during a 10-day midcycle period of maximal genital swelling, but 2) they also copulated at other phases of the cycle, including the time of menses and during gestation and lactation. Yerkes (1939) reported that the particular pattern of mating that occurred in any given pair was determined to a greater extent by the social relationship of the consorts than by the female's sexual status. When the female was fully mature and well-acquainted with the male, she tended to initiate and control mating, and copulation was restricted to a relatively brief period of the phase of maximal genital swelling. When the female was immature, timid and/or unfamiliar to the male and the male was dominant, he initiated mating

The orang utan. Its biology and conservation, edited by L.E.M. de Boer
© *1982, Dr W. Junk Publishers, The Hague, ISBN 90 6193 702 7*

and copulation could occur without regard to cycle phase. As a result of this research, Yerkes (1939) proposed that display of estrus and the restriction of mating to the periovulatory period (and, presumably, the hormonal regulation of sexual behavior) were a function of the taxonomic status of a species and the differential dominance and sexual initiative of the male and female. Specifically, he proposed that the greater the extent of encephalization of a species and the greater the degree of sexual initiative by the male, the lesser the restriction of mating to the periovulatory phase of the cycle. This general hypothesis was clearly derived from the research on chimpanzees. Evidence was available that ovulation in chimpanzees occurred toward the end of the phase of maximal genital swelling (Elder 1938), the time when copulation was most frequent (Yerkes and Elder 1936; Yerkes 1939). This result did not establish the phenomenon of estrus for the chimpanzee, but was consistent with such an interpretation (Yerkes 1939). On the other hand, the finding that chimpanzees also copulated at phases of the cycle other than the periovulatory phase suggested that their sexual behavior exhibited an emancipation from phys-iological control commensurate with their advanced taxonomic status and 'behavioral adaptiveness (general intelligence)' (Yerkes 1939, p. 79).

Other early research on the chimpanzee suggested, and later studies con-firmed, the hormonal stimulation of female genital swelling by estrogens (Zuckerman and Fulton 1934; Graham et al. 1972) and its inhibition by progestins (Clark and Birch 1948; Graham et al. 1972). Those data and the data on menstrual cycle patterns of hormones in relation to pattern of genital swelling (Graham et al. 1972; Reyes et al. 1975) permit extrapolation to the behavioral studies and provide indirect evidence of the hormonal regulation of sexual behavior in chimpanzees. The increased frequency of copulation at the time of maximal genital swelling, a phase of the cycle when endogenous estrogens are elevated in the female, is consistent with the hypothesis that elevated concentrations of estrogen (estradiol) facilitate mating in this species. The decline in mating frequency during the post-swelling phase, i.e., the luteal phase, despite elevated estradiol concentrations, is consistent with the hy-pothesis that progesterone, also elevated during the luteal phase, exerts an inhibitory influence on mating of chimpanzees. Among the primates, these hypotheses have gained extensive support, of course, through research on the rhesus monkey (Herbert and Trimble 1967; Michael et al. 1967; Johnson and Phoenix 1976; Keverne 1976; Wallen and Goy 1977).

Evidence suggesting that chimpanzees exhibit some emancipation of their sexual behavior from hormonal regulation is derived from the data on mating during phases of the cycle other than maximal genital swelling. Since such mating was accounted for primarily by the sexual initiative of the male, the data suggest that under certain circumstances, the male may act to moderate the hormonal influences on mating which typically act to facilitate mating at a phase of the cycle when conception is likely. The behavioral studies on the chimpanzee, therefore, provided the initial data on menstrual cycle patterns of mating by a great ape species and also proposed a theoretical framework within which to interpret research results on the remaining apes, gorilla and orang utan.

Laboratory research on sexual behavior of gorillas

Laboratory research on menstrual cycle patterns of mating by gorillas was initiated in 1971 (Nadler 1975b, 1976) and provided data relevant to the hypothesis described above (Nadler 1981). When tested under conditions comparable to those used with the chimpanzees, the gorillas exhibited a pattern of mating in relation to female genital swelling that was similar in certain respects to that reported for chimpanzees, but different in others. Mating occurred primarily during the phase of maximal genital swelling, termed maximal labial tumescence (LT) in the gorilla (Nadler 1975a). Maximal LT, however, persists for only 1–2 days, in comparison to approximately 10 days of maximal genital swelling in the chimpanzees. The gorillas, moreover, unlike the chimpanzees, hardly ever copulated when the females were in the detumescent condition. This pattern of mating by gorillas of only 1–2 days per cycle, moreover, was shown to be the species-typical pattern under natural conditions (Harcourt and Stewart 1978).

The pattern of 1–2 days of mating by gorillas does not support the hypothesis that advanced taxonomic status *per se* assures emancipation of sexual be-haviour from hormonal regulation. On the contrary, the relatively infrequent occurrence and circumscribed period of mating, closely associated with a morphological condition (maximal genital swelling) shown to reflect elevated concentration of estrogens in several primate species, e.g., pigtailed macaque (Eaton and Resko 1974) and chimpanzee (Graham et al. 1972; Reyes et al. 1975), suggest that hormones exert a considerable influence on sexual behavior of gorillas. Subsequent research on menstrual cycle hormone patterns in relation to patterns of LT in female gorillas revealed a relationship that was quite similar to that of the chimpanzee (Nadler et al. 1979). LT increased with increasing concentration of estradiol during the follicular phase and reached maximal proportions in association with the midcycle, preovulatory surge of estradiol. LT underwent detumescence in association with the subsequent decline in estradiol and increase in progesterone. The luteal phase elevation in estradiol was not accompanied by increased LT, presumably due to the inhibitory influence of the elevated concentrations of progesterone. Extrapolation of these hormone data to the behavioral study, therefore, suggests a similar facilitation of mating by estradiol and inhibition by progesterone as that proposed for the chimpanzee. The difference between the two species is in the degree to which mating is restricted to the periovulatory period and, presumably, the degree to which the behavior is constrained by endogenous hormonal conditions of the female. This difference in mating patterns, moreover, is related to the difference in sexual initiative between male chimpanzees and male gorillas. Among the chimpanzees, mating dissociated from the periovulatory period resulted pri-marily from the sexual initiative of the males. Among the gorillas, the males were essentially passive in this regard and it was the females that assertively initiated copulation. The data on the gorillas, therefore, support the hypothesis of Yerkes (1939) regarding the role of male dominance and sexual initiative. In the absence of male sexual initiative, the female controlled mating and copulation was restricted primarily to the periovulatory period. The data on both the chimpanzee and the gorilla are consistent with the hypothesis that sexual behavior of these species is influenced to some degree by endogenous hormones of the female. The degree to which hormones are influential, moreover, appears to be related to the differential sexual initiative of the males and females. A relatively closer periovulatory restriction of mating, reflecting a greater influence by hormones, was associated with female sexual initiative, whereas mating dissociated temporally from the periovulatory period, reflecting a lesser influence of hormones, occurred as a result primarily of male sexual initiative.

Laboratory research on sexual behavior of orang utans

The first laboratory study of sexual behavior in the orang utan began an investigation of three related issues in this species that were derived from the research on the other apes: 1) the degree of cyclicity in sexual behavior; 2) the degree to which sexual behavior is restricted to the periovulatory phase of the menstrual cycle; and 3) the degree to which sexual behavior during the cycle is regulated by endogenous hormones of the female (Nadler 1977a). When this study was initiated in 1974, the main sources of information on mating by orang utans were reports on animals living in zoological gardens. The earliest report of this type appeared more than 50 years ago and provided a description that has proved to be characteristic of most captive animals (Fox 1929). 'When the desire animates the male and is reciprocated by the female, he pushes and mauls her a little, whereupon she lies on her back on the floor. The male then approaches her and separates her legs. During the act he remains in a sitting or crouching attitude and though they are face to face, he does not lie on the abdomen of the female. The male will sometimes grasp a leg of the female and hold it up and to the side during conjugation. During the act there is no fondling, nor do they mouth each other either before or after the act. The female lies passive and often has a hand over her face. The act is practiced daily, without relation to the sexual cycle' (Fox 1929, p. 41).

Two aspects of this description are especially noteworthy when considered in relation to Yerkes' (1939) hypothesis, described above: 1) the suggestion that copulation was typically initiated (forcibly) by the male; and 2) the statement that copulation took place daily, irrespective of cycle phase. Further evidence of forcible copulation or rape by male orang utans in zoos was published subsequently (Seitz 1969; Coffey 1972, 1975), as were reports of daily or near-daily mating by these animals (Asano 1967; Heinrichs and Dillingham 1970; Coffey 1972, 1975). Other reports on zoo animals, however, were not entirely consistent with the above characterization. At least two authors reported cyclicity in mating of orang utans (Chaffee 1967; Coffey 1975), in one case, 7–10 days of mating at 25-day intervals (Chaffee 1967). It is not clear from these zoo reports whether or not cyclicity in sexual behavior is characteristic of this species and, if so, whether there is an increased frequency of copulation during the midcycle phase, as was found for the other apes.

The laboratory study of orang utan sexual behaviour was conducted similarly to those on the chimpanzee and gorilla to facilitate a comparative assessment of the data (Nadler 1977a). Four oppositely-sexed pairs were tested daily during the menstrual cycle of the female. Female orang utans differ from chimpanzee and gorilla females in that they develop a genital swelling during pregnancy, but not during the menstrual cycle (Fox 1929; Schultz 1938; Lippert 1974). The menstrual cycle of this species, therefore, was monitored by the periodic occurrence of menses, which was detected by testing urine samples daily with commercially available reagent strips. The behavioral data were analyzed in relation to cycle phase by dividing the intermenstrual period into three approximately equal parts.

Two different types of behavioral tests were administered. Initially, all four pairs were tested through one cycle of the female to a criterion of 30 min without copulation (when copulation occurred, the test was terminated 30 min following completion of copulation; when copulation failed to occur, a relatively rare event, the test was terminated 30 min following its initiation). In a second series of menstrual cycle tests, three of the four original pairs were tested for 5–6 h per day (the remaining pair was not tested because the female became pregnant during the first test cycle). During the first half hour of the 5–6 h test, an observer recorded the behavioral data as in the first series of tests. For the remainder of the test, the observer was absent, but a video camera and tape recorder sampled the test cage periodically throughout the day at intervals sufficiently close together to assure recording of any copulations that occurred.

The length of the menstrual cycle, calculated from the onset of menses in one cycle to the next onset of menses, was determined for eight cycles of three females. The cycles ranged in length from 26 to 32 days, with a median cycle length of 30.5 days. These results confirm those presented earlier for orang utans (Napier and Napier 1967; Heinrichs and Dillingham 1970; Lippert 1974) and suggest that the menstrual cycle in this species is somewhat shorter than those for the chimpanzee (Yerkes and Elder 1936) and gorilla (Nadler 1975a).

The results of the behavioral study, although markedly different from those of the chimpanzee and gorilla in certain respects, revealed some basic similarities as well. Almost every test was initiated in one of two ways. As soon as the door separating the compartments of the male and female was raised at the beginning of the test, either 1) the male rushed into the female's compartment and chased her out into the test cage, or 2) the female rushed out into the test cage and ran toward the front fence of the cage. The male rapidly pursued the female under both circumstances, wrestled her to the floor and roughly positioned her for copulation (Fig. 1). The female initially grimaced, struggled and vocalized in distress (Fig. 2). Once subdued, however, the female became fairly passive and permitted the male to spread her legs and 'mount' her ventro-ventrally (Fig. 3). The male supported himself during copulation by reclining on one elbow, placing his hands on the ground and using his arms as crutches or by grasping the wire mesh of the cage or the cage shelf with his hands (Fig. 4). During copulation, neither sex exhibited any facial expressions suggestive of heightened sexual arousal. The duration of copulation by the orang utans was considerably longer on the average than that reported for the gorilla and chimpanzee, ranging from 1–46 min, with an overall median of 14 min!

Fig. 1. Male orang utan roughly positions female for ventro-ventral copulation.
Fig. 2. Female orang utan grimaces while being pulled across cage floor by male (not shown).

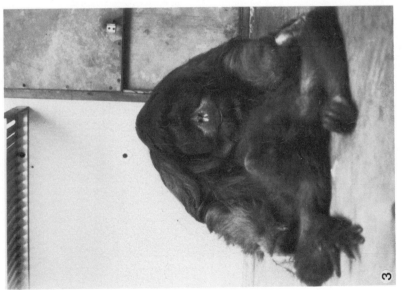

Fig. 3. Male orang utan spreads female's legs prior to initiating ventro-ventral copulation.
Fig. 4. Male orang utan grasps the cage shelf to support himself during ventro-ventral copulation. Female lies passively on her back.

237

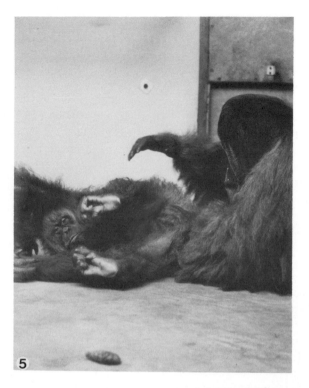

Fig. 5. Male orang utan inspects the genitals of the female during an interruption in copulatory thrusting.
Fig. 6. A form of dorso-ventral copulation.

Chimpanzees were reported to copulate for about 8 sec (Yerkes and Elder 1936; Yerkes 1939) and gorillas for about 53 sec (Hess 1973). The median number of thrusts per copulation by the orang utans was 529, compared to 17 thrusts for the chimpanzee (Yerkes and Elder 1936; Yerkes 1939) and 30 thrusts for the gorilla (Nadler 1976). The rate of thrusting for the orang utans was 1–2 thrusts per sec, somewhat slower than the other apes. Thrusting was not continuous, but was interrupted periodically by the male. During these interruptions, the male 1) repositioned the female, 2) inspected her genitals (Fig. 5), or 3) maintained insertion without any other activity until he resumed thrusting. Ventro-ventral copulation as described by Fox (1929) was the primary position observed (Fig. 4), although considerable variation occurred, including a form of dorse-ventral copulation (Fig. 6) and, occasionally, copulation while the pair was suspended from the bars at the top of the cage. This description, therefore, although more detailed, is, in general, quite similar to that given originally by Fox (1929).

An unusual aspect of the copulatory behavior of the orang utans was the absence of a reliable behavioral indication of ejaculation. That ejaculation, in fact, occurred was indicated frequently by the presence of semen or a semen plug on the genitals of the female or on the cage floor following separation of a copulating pair. One published report on copulatory behavior of captive orang utans stated that a 'quiver' occurred in the male's body at the termination of copulation (Coffey 1975). This was observed occasionally in the males of the laboratory study, but it was not reliably related to the observation of semen.

Data on the frequency and distribution of mating in the cycle are summarized in Table 1 for the two different series of tests. On the initial series to the 30-min criterion, three of the four orang utan pairs copulated on every day and the fourth pair also copulated quite frequently. Thus, there was no clear indication of cyclicity in mating when the animals were tested in this manner. The second series of tests was conducted to determine whether multiple copulations per day would occur if the animals were permitted to remain together for a longer period of time, i.e., 5–6 h. Under these conditions, multiple copulations did

Table 1. Percentage of days of copulation by orang utans during three phases of the menstrual cycle (adapted from Nadler 1977, Table II).

Subject pairs	Test duration	Cycle phase		
		Early	Midcycle	Late
1 (B)	30 min[a]	100	100	100
	5–6 h[b]	0	57	25
2 (S)	30 min[a]	43	63	14
	5–6 h[b]	0	50	0
3 (S)	30 min[a]	100	100	100
4 (S)	30 min[a]	100	100	100
	5–6 h[b]	50	60	0

B – Bornean orang utans
S – Sumatran orang utans
[a] – Criterion of 30 min without copulation (see text)
[b] – Only percentage of multiple copulations is indicated

occur on single days and for the three pairs tested, they occurred primarily during the midcycle phase. These data suggest that despite daily mating by orang utans in the laboratory, there is also some evidence of cyclicity in mating, characterized by an increased number of copulations per day during the midcycle phase.

Other evidence of increased sexual responsiveness by the orang utans was found in the behavior of the females. All of the females spent a certain amount of time during the tests suspended from the bars at the top of the cage, apparently as a way to avoid the males (Table 2). This negative response, in terms of sociality, occurred least frequently, however, during the midcycle phase for all the females. Conversely, several examples of positive social response occurred most frequently during the midcycle phase (Table 3). Grooming and social contact by the female, such as pulling or slapping the male, was observed with three of the females and these forms of behavior occurred most frequently at midcycle. Four examples of female sexual initiative or sexual proceptivity (Beach 1976) were observed, twice each by two females, and these all also

Table 2. Median (and range) number of one-minute intervals (out of 30) female orang utans remained suspended above floor during three phases of the menstrual cycle.

Subject pairs	Number of days tested	Cycle phase		
		Early	Midcycle	Late
1 (B)	29	6.0 (n = 9) (0–10)	3.0 (n = 10) (0–14)	6.0 (n = 10) (0–21)
2 (S)	22	29.0 (n = 7) (21–30)	16.5 (n = 8) (3–30)	29.7 (n = 7) (9–30)
3 (S)	26	8.0 (n = 8) (0–22)	6.5 (n = 10) (0–18)	6.5 (n = 8) (2–12)
4 (S)	22	11.0 (n = 7) (4–15)	4.5 (n = 8) (0–16)	14.0 (n = 7) (3–17)

B – Bornean orang utans
S – Sumatran orang utans

Table 3. Percentage of days female orang utans groomed (and socially contacted) males during three phases of the menstrual cycle.

Subject pairs	Number of days tested	Cycle phase		
		Early	Midcycle	Late
1 (B)	29	66 (89)	70 (100)	40 (100)
2 (S)	22	14 (29)	50 (50)	0 (0)
3 (S)	26	0 (13)	20 (40)	0 (0)
4 (S)	22	0 (0)	0 (0)	0 (0)

B – Bornean orang utans
S – Sumatran orang utans

occurred during midcycle. One of the females pulled the male around to face her and then rubbed her genitals against his, following which he initiated thrusting. The female repeated this behavior following copulation, but was then ignored by the male. The same female rubbed her genitals on the male's back and head on the following day, but was completely ignored on that occasion. A second female initiated thrusting during copulation on two consecutive midcycle days when the male interrupted his thrusting. These examples of sexual proceptivity are relatively few, but their occurrence only during the midcycle phase provides some indication of cyclicity in sexual behavior of orang utans, regulated by the female. The final examples of heightened sexual responsiveness by the females during the cycle were found in their masturbatory activity. Two females were involved in masturbation following copulation with the males. One repeatedly inserted her fingers and toes into her vulva on one occasion, whereas the other female on two occasions stimulated her genitals with a piece of wire. These instances of masturbation, moreover, all occurred during midcycle.

Although sexual proceptivity occurred relatively infrequently in this laboratory study, this is not the only evidence of this behavior by female orang utans. In a recently published report, cyclicity in proceptive behavior was described for a captive female orang utan living continuously with a male over a 90-day period of time (Maple et al. 1979). The male initiated sexual interactions frequently during the entire period, but the female was proceptive during only three intervals of 4–6 days each, separated by 26 and 30 days. That such behavior is not restricted to captive environments is indicated below.

Discussion

As stated at the beginning of this article, the research on the orang utans initiated investigation of three related issues: cyclicity, periovulatory restriction and hormonal regulation of sexual behavior. The issues are related in that cyclicity in sexual behavior (during the cycle) frequently reflects some degree of periovulatory restriction (or enhancement) of mating, which is regulated by hormones in many species that have been investigated (Beach 1948; Young 1961). These issues are of interest in the present context, in part because of their relationship to male and female sexual initiative as proposed by Yerkes (1939). Yerkes (1939) reported that among chimpanzees, female initiation and control of mating resulted in periovulatory restriction of mating, whereas male initiation accounted for mating dissociated from the periovulatory period. He proposed that the periovulatory restriction of mating reflected the influence of physiological factors, such as hormone levels of the female, which he believed had some degree of influence on behavior of all mammals, including the human. He proposed that the performance of mating at times in the cycle other than the periovulatory phase reflected some degree of emancipation of behavior from normonal regulation, which occurred as a result of the chimpanzee's (i.e., the ape's) advanced taxonomic status and degree of encephalization.

The data on sexual behavior of the gorilla (Nadler 1975b, 1976) accord with Yerkes' (1939) hypothesis in that the periovulatory restriction of mating in this species was associated with female sexual initiative. The data on the orang utan are similarly supportive in that the absence of periovulatory restriction of mating was associated with male sexual initiative. These data suggest, therefore, that advanced taxonomic status or encephalization does not assure emancipation of mating from hormonal regulation. Rather, it is the degree to which the male initiates mating that determines whether or not taxonomic status is a factor in the regulation of mating. The data available on menstrual cycle hormone patterns of female orang utans (Blakely 1969; Collins et al. 1975) are consistent with those of the other apes (Graham et al. 1972; Reyes et al. 1975; Nadler et al. 1979), but are based on very few subjects. Interpretation of the hormonal regulation of mating in orang utans at this point, therefore, is indirect and quite tentative.

The laboratory data on sexual behavior of the orang utan, in conjunction with related data on the chimpanzee and gorilla, can be described as reflecting a continuum of sexual responsiveness during the cycle (Table 4; also see Nadler 1977b). The gorilla represents one end of the continuum, with pronounced female proceptivity and the clearest restriction of mating to the periovulatory period. The orang utan represents the other end of the continuum, with primarily male sexual initiative and the least restriction of mating in the cycle. The chimpanzee is intermediate to the other apes with respect to both of these considerations. The inverse relationship between male sexual initiative and female sexual proceptivity in these species, and their inverse and direct relationships, respectively, to the degree of periovulatory restriction of mating in the cycle, as proposed by Yerkes (1939), are clearly evident.

The laboratory research on sexual behavior of orang utans, therefore, has relevance for the theoretical issues discussed above, and further research on menstrual cycle hormone patterns of female orang utans and patterns of sexual behavior studied in relation to the periovulatory phase of the cycle will contribute to their further elucidation in this species. The relevance of the laboratory data to reproductive function, however, is not clear. The four pairs of orang utans that participated in the study were tested during a total of nine cycles. All the males and females were proven breeders, the females were cycling (as indicated by periodic menstruation) and, yet, despite frequent copulations, only a single pregnancy resulted. It is not known why more conceptions failed to occur and certainly, many possibilities exist. One such possibility is that a reduction in potency of the male occurs as a result of repeated copulation prior to the periovulatory phase of the female's cycle, sufficient to reduce the

Table 4. Ranking of the great apes with respect to the regulation of mating during the menstrual cycle[a]. Number 1 indicates the highest rank per category (adapted from Nadler 1977b).

	Gorilla	Chimpanzee	Orang utan
Periovulatory restriction of mating	1	2	3
Male sexual initiative	3	2	1
Female sexual initiative	1	2	3

[a] Mature individuals tested under particular laboratory conditions.

probability of conception. No direct evidence is available that the potency of male orang utans changes (decreases) as a function of repeated ejaculations. If, however, the reproductive systems of the apes are differentially adapted to function in accordance with the species-typical patterns of sexual interaction (Harcourt and Stewart 1977; Nadler 1977b; Short 1977), then marked deviations from the species-typical pattern could well account for reduced conceptions. Considered in this context, it becomes important to determine whether the patterns of sexual behavior recorded in the laboratory are essentially similar to the pattern of sexual relations that result in fertile matings under natural conditions, i.e., the species-typical pattern.

Although there have been by now a considerable number of field studies conducted on orang utans, they have not provided sufficient data regarding the sexual relations of adult males and females to permit resolution of the issue. There is no longer any doubt that forcible copulation or rape is a form of behavior engaged in by male orang utans living in their natural habitat (MacKinnon 1971, 1974; Rodman 1973; Horr 1975; Rijksen 1978; Galdikas 1979). This indicates that rape *per se* is not merely an artifact of the laboratory. Whether or not conception typically occurs following rape and, more specifically, following repeated rape, such as that observed in the laboratory, are additional questions. The answers to both these questions seem to be negative. Rape typically occurs opportunistically by highly mobile males, and the female's response is to avoid or escape. This reduces the probability that the same male will rape a particular female repeatedly over the course of several days. The temporal relationship between incidences of rape and subsequent parturition by the females, moreover, suggests that rape does not generally result in conception (Galdikas 1979). What, then, is the function of rape among orang utans and what is their species-typical pattern of mating?

Two points of view have been expressed regarding these questions, although not necessarily as mutually exclusive alternatives. Far and away, the most common interpretation is that rape is engaged in by sub-adult males and does not represent the species-typical pattern engaged in by fully mature males and females (Rodman 1973; Horr 1975; Rijksen 1978; Galdikas 1979). According to this position, sexual activity is initiated by the female who seeks out a fully mature male when she is sexually responsive and locates him on the basis of his long-call (MacKinnon 1971, 1974). Rijksen (1978) proposed that rape by male orang utans served a dominance function rather than a reproductive function and represented attempts by the sub-adult males to acquire consorts with which to mate when the females eventually became sexually responsive. He reported that the most common conditions under which rape occurred in the wild were 'that the female is newly met or met again after a period of separation' (Rijksen 1978, p. 274). The conditions of testing in the laboratory, therefore, replicated certain aspects of the conditions under which rape typically occurs in the natural habitat. On each day of testing, the female was taken to the male's cage and 'met the male again after a period of separation'. The fact that rape was engaged in by fully mature males in the laboratory, in contrast to its predominance among sub-adult males in the wild, suggests that the condition under which the male and female come together, rather than age *per se*, is the critical factor for evoking rape by male orang utans. Fully mature males were reported to engage in rape with 'some female they encounter' (MacKinnon 1979, p. 265), but in the wild, this occurs much less frequently than among the sub-adult males.

In contrast to the interpretation presented immediately above is the proposal by MacKinnon (1979) that rape by young adult (sub-adult?) males does have reproductive significance and is, in fact, an appropriate reproductive strategy during this stage of development for maximizing their genetic fitness. He proposed, further, that fully mature males are not reproductively active because they become infertile and impotent at a relatively young age, i.e., less than 25 years. Although the evidence to date seems to favor the position of the majority of field workers, MacKinnon's position cannot be discounted. This issue is further complicated by the fact that there are two subspecies of orang utan that may differ between themselves with regard to their patterns of mating (MacKinnon 1979). No subspecies differences were apparent in the laboratory study of orang utans, but probably too few animals were investigated to permit such a distinction. Similarly, our experience in breeding orang utans at the Yerkes Regional Primate Research Center does not support the contention by MacKinnon (1979) that fully mature males develop infertility. Since all our males are estimated to be only in their early 20s, however, they may not have reached the stage to which MacKinnon referred.

Despite the controversy that exists regarding the male orang utan's reproductive strategy, there is increasing evidence that the female plays a more important role in initiating sexual interactions than suggested by the earlier studies (Rijksen 1978; Galdikas 1979; Schürmann 1982). It was reported recently that sexual proceptivity was displayed by female orang utans in the wild, but that it occurred only among sub-adult females in courtship with fully adult males (Galdikas 1979; Schürmann 1982). It is unclear, therefore, whether adult females under natural conditions also seek out mature, resident males for copulation, or whether the adult males seek out adult females periodically to determine their sexual status and to form consortships.

In the absence of clear evidence regarding the species-typical pattern of mating by orang utans, it is impossible to interpret completely the data from the laboratory study. In other words, if the pattern of mating recorded for the orang utan deviated substantially from the species-typical pattern, then interpretation of the regulation of mating would be confounded. On the other hand, certain conclusions can be supported and others are susceptible to experimental testing in the laboratory. There is evidence that under the conditions of laboratory testing, the influence of the male on the pattern and distribution of mating in the cycle was inordinate in comparison to the natural environment (Nadler 1981). This appears to result from 1) male dominance over females, 2) confinement of the pair under conditions that prevent the female from avoiding or escaping from the male, and 3) relatively increased frequency of male sexual initiative in comparison to mating in the wild. This inordinate influence of the male, moreover, was found in laboratory studies on the chimpanzee and gorilla, as well as the orang utan.

In order to investigate the role of the female orang utan in sexual initiation, examine its potential for sexual proceptivity and investigate the periovulatory restriction and hormonal regulation of such behavior in the laboratory, it would be desirable to reduce the male's influence. Yerkes (1939), in fact, originally made this recommendation regarding research on sexual behavior of chimpanzees. Laboratory research on orang utans is currently underway in which the male's influence over the female is reduced and in which the female has the option of whether or not to establish contact with the male. The modified conditions of testing are such that the male and female are in separate compartments of a cage at the start of the test, and the door connecting the two compartments is too small to permit passage by the male, but sufficiently large to permit passage by the female.

Although only a relatively few number of animals have been tested so far under these conditions, it is apparent that the minor physical modification of the environment produced a dramatic effect on the behavior of the orang utans. The females enter the compartment of the male primarily during the midcycle period, touch, pull and slap (playfully) the male, rub their genitals on him and position him for copulation on his back. The males exhibit a 'penile display' in which the erect penis is directed toward the female, either from a reclining position or while the male slowly climbs toward the female. The females mount the erect penis and thrust upon it, dismount, inspect the genitals of the male, remount and reinitiate thrusting (Fig. 7). The male, during all this, so highly aggressive in the earlier test conditions, remains essentially passive under these conditions, permitting the female complete freedom of activity. This extraordinary reversal of behavior by both the males and the females supports the view that female orang utans are capable of considerable sexual proceptivity under certain conditions. Of considerable significance to this research is that

Fig. 7. Female orang utan 'mounted' on male for ventro-ventral copulation.

remarkably similar behavior to that described above is also reported for the first time in this volume for Sumatran orang utans living in the wild (Schürmann 1982)!

It is always encouraging to behavioral researchers when the observations they make in the laboratory are supported by similar observations made in the natural environment. This is especially true in the present case because female sexual proceptivity has been so rarely observed among captive or wild orang utans, and the male penile display has never before been reported for this species. Confirmation that such behavior is part of the species-typical pattern of mating by orang utans is clearly relevant to the continued research in the laboratory on periovulatory restriction and hormonal regulation of sexual behavior. Such confirmation suggests that the modified conditions for testing sexual behavior of orang utans which permit the female the option of whether or not to mate with the male improve our ability to describe the relationship between reproductive behavior and endocrinology of orang utans.

Acknowledgments

Preparation of this article and the work of the author described herein were supported by NSF GB-30757 and BNS 75-06287 and PHS Grant RR-00165 from the National Institutes of Health.

References

Asano, M. (1967). A note on the birth and rearing of an orang-utan at Tama Zoo, Tokyo. Int. Zoo Yearbook, 7: 95–96.

Beach, F.A. (1948). Hormones and behavior. P.B. Hoeber, Harper & Bros., New York.

Beach, F.A. (1976). Sexual attractivity, proceptivity, and receptivity in female mammals. Horm. Behav., 7: 105–138.

Blakely, G.A. (1969). Vaginal cytology and luteinizing hormone levels in Pongo pygmaeus (orang-utan). In: H.O. Hofer (ed.), Recent advances in primatology, Vol. 2. Karger, Basel.

Chaffee, P.S. (1967). A note on the breeding of orang-utans at Fresno Zoo. Int. Zoo Yearbook, 7: 94–95.

Clark, G. and H.G. Birch (1948). Observations on the sex skin and sex cycle in the chimpanzee. Endocrinology, 43: 218–231.

Coffey, P.F. (1972). Breeding Sumatran orang-utan Pongo pygmaeus abelii Lesson 1827. Ann. Rep. Jersey Wildl. Preserv. Trust, 1972: 15–17.

Coffey, P.F. (1975). Sexual cyclicity in captive orang-utans Pongo pygmaeus with some notes on sexual behavior. Ann. Rep. Jersey Wildl. Preserv. Trust, 1975: 54–55.

Collins, D.C., C.E. Graham and J.R.K. Preedy (1975). Identification and measurement of urinary estrone, estradiol-17β, estriol, pregnanediol and androsterone during the menstrual cycle of the orangutan. Endocrinology, 96: 93–101.

Eaton, G.G. and J.A. Resko (1974). Ovarian hormones and sexual behavior in Macaca nemestrina. J. Comp. Physiol. Psychol., 86: 919–925.

Elder, J.H. (1938). The time of ovulation in chimpanzees. Yale J. Biol. Med., 10: 347–364.

Fox, H. (1929). The birth of two anthropoid apes. J. Mammal., 10: 37–51.

Galdikas, B.M.F. (1979). Orangutan adaptation at Tanjung Puting Reserve: mating and ecology. In: D.A. Hamburg and E.R. McCown (eds.), The great apes. Benjamin/Cummings, Menlo Park.

Graham, C.E., D.C. Collins, H. Robinson and J.R.K. Preedy (1972). Urinary levels of estrogens and pregnanediol and plasma levels of progesterone during the menstrual cycle of the chimpanzee: relationship to the sexual swelling. Endocrinology, 91: 13–24.

Harcourt, A.H. and K.J. Stewart (1977). Apes, sex and societies. New Scientist, 76: 160–162.

246

Harcourt, A.H. and K.J. Stewart (1978). Sexual behaviour of wild mountain gorillas. In: D.J. Chivers and J. Herbert (eds.), Recent advances in primatology, Vol. 1, Behaviour. Academic Press, London.

Heinrichs, W.L. and L.L. Dillingham (1970). Bornean orang-utan twins born in captivity. Folia Primat., 13: 150–154.

Herbert, J. and M.R. Trimble (1967). Effect of oestradiol and testosterone on the sexual receptivity and attractiveness of the female rhesus monkey. Nature, 216: 165–166.

Hess, J.P. (1973). Some observations on the sexual behaviour of captive lowland gorillas, Gorilla g. gorilla (Savage and Wyman). In: R.P. Michael and J.H. Crook (eds.), Comparative ecology and behaviour of primates. Academic Press, London.

Horr, D.A. (1975). The Borneo orang-utan: population structure and dynamics in relation to ecology and reproductive strategy. In: L.A. Rosenblum, (ed.), Primate behavior: developments in field and laboratory research, Vol. 4. Academic Press, New York.

Johnson, D.F. and C.H. Phoenix (1976). Hormonal control of female sexual attractiveness, proceptivity and receptivity in rhesus monkeys. J. comp. physiol. Psychol., 90: 473–483.

Keverne, E.B. (1976). Sexual receptivity and attractiveness in the female rhesus monkey. In: D.S. Lehrman, R.A. Hinde and E. Shaw (eds.), Advances in the study of behavior, Vol. 7. Academic Press, New York.

Lippert, W. (1974). Beobachtungen zum Schwangerschafts- und Geburtsverhalten beim Orang-utan (Pongo pygmaeus) im Tierpark Berlin. Folia Primat., 21: 108–134.

MacKinnon, J.R. (1971). The orang-utan in Sabah today. Oryx, 11: 141–191.

MacKinnon, J.R. (1974). The behaviour and ecology of wild orang-utans (Pongo pygmaeus). Anim. Behav., 22: 3–74.

MacKinnon, J.R. (1979). Reproductive behavior in wild orangutan populations. In: D.A. Hamburg and E.R. McCown (eds.), The great apes. Benjamin/Cummings, Menlo Park.

Maple, T.L., E.L. Zucker and M.B. Dennon (1979). Cyclic proceptivity in a captive female orang-utan (Pongo pygmaeus abelii). Behav. Processes, 4: 53–59.

Michael, R.P., J. Herbert and J. Welegalla (1967). Ovarian hormones and the sexual behaviour of the male rhesus monkey (Macaca mulatta) under laboratory conditions. J. Endocrinol., 39: 81–98.

Nadler, R.D. (1975a). Cyclicity in tumescence of the perineal labia of female lowland gorillas. Anat. Rec., 181: 791–798.

Nadler, R.D. (1975b). Sexual cyclicity in captive lowland gorillas. Science, 189: 813–814.

Nadler, R.D. (1976). Sexual behavior of captive lowland gorillas. Arch. Sex. Behav., 5: 487–502.

Nadler, R.D. (1977a). Sexual behavior of captive orang-utans. Arch. Sex. Behav., 6: 457–475.

Nadler, R.D. (1977b). Sexual behavior of the chimpanzee in relation to the gorilla and orang-utan. In: G.H. Bourne (ed.), Progress in ape research. Academic Press, New York.

Nadler, R.D. (1981). Laboratory research on sexual behavior of the great apes. In: C.E. Graham (ed.), Reproductive biology of the great apes: comparative and biomedical perspectives. Academic Press, New York.

Nadler, R.D., C.E. Graham, D.C. Collins and K.G. Gould (1979). Plasma gonadotropins, prolactin, gonadal steroids and genital swelling during the menstrual cycle of lowland gorillas. Endocrinology, 105: 290–296.

Napier, J.R. and P.H. Napier (1967). A handbook of living primates. Academic Press, London.

Reyes, F.I., J.S.D. Winter, C. Faiman and W.C. Hobson (1975). Serial serum levels of gonadotropins, prolactin, and sex steroids in the nonpregnant and pregnant chimpanzee. Endocrinology, 96: 1447–1455,

Rijksen, H.D. (1978). A field study on Sumatran orang utans (Pongo pygmaeus abelii Lesson 1827). Ecology, behavior and conservation. Meded. Landbouwhogeschool Wageningen, 78–2: 1–420.

Rodman, P.S. (1973). Population composition and adaptive organisation among orang-utans of the Kutai Reserve. In: R.P. Michael and J.H. Crook (eds.), Comparative ecology and behaviour of primates. Academic Press, London.

Schultz, A.H. (1938). Genital swelling in the female orang-utan. J. Mammal., 19: 363–366.

Schürmann, C. (1982). Mating behavior of wild orang utans. In: L.E.M. de Boer (ed.), The orang utan. Its biology and conservation. Junk, The Hague.

Seitz, A. (1969). Einige Feststellungen zur Pflege und Aufzucht von Orang-utans, Pongo pygmaeus Hoppius 1763. Zool. Garten (N.F.), 36: 225–245.

Short, R.V. (1977). Sexual selection and the descent of man. In: J.H. Calaby and C.H. Tyndale-Biscoe (eds.), Reproduction and evolution. Griffin Press, Netley, South Australia.

Wallen, K. and R.W. Goy (1977). Effects of estradiol benzoate, estrone, and propionates of testosterone or dihydrotestosterone on sexual and related behaviors of ovariectomized rhesus monkeys. Horm. Behav., 9: 228–248.

Yerkes, R.M. (1939). Sexual behavior in the chimpanzee. Hum. Biol., 11: 78–111.

Yerkes, R.M. and J.H. Elder (1936). Oestrus, receptivity and mating in chimpanzee. Comp. Psychol. Monogr., 13: 1–39.

Young, W.C. (1961). The hormones and mating behavior. In: W.C. Young (ed.), Sex and internal secretions, Vol. II (3rd Ed.). Williams and Wilkins, Baltimore.

Young, W.C. and W.D. Orbison (1944). Changes in selected features of behavior in pairs of oppositely sexed chimpanzees during the sexual cycle and after ovariectomy. J. comp. Psychol., 37: 107–143.

Zuckerman, S. and J.F. Fulton (1934). The menstrual cycle of the primates. Part VII: The sexual skin of the chimpanzee. J. Anat., 69: 38–46.

Author's address:
R.D. Nadler
Yerkes Regional Primate Research Center
Emory University
Atlanta, GA 30322
U.S.A.

13. Social potential expressed in captive, group-living orang utans

Sara D. Edwards

From the literature the general implication is that orang utans lead rather solitary lives that limit the incidence of social interaction. Presumably, this lack of sociality is a consequence of survival strategies which meet environmental demands. Features of the habitat that are socially relevant appear to be influencing the extent to which the social behavior of wild orang utans may be expressed.

This study primarily investigates two questions. What type of social system do orang utans incorporate, and to what extent does this species behave socially?

Background

If orang utans are placed in a situation void of their present environmental influences, e.g. one that provides them with an environment that promotes the maximum expression of the species' potential to interact socially, two alternative outcomes can be predicted. First, if the orang utan is essentially a solitary species, then placing them in such a situation should indicate a lack of social adjustment. Secondly, if orang utans adjust to such a situation, the type of social behavior this species will incorporate and the extent to which orang utans will behave socially can be explored.

Two groups of orang utans at the Henry Vilas Park Zoo, Madison, Wisconsin, USA, were observed to investigate this question. These groups were housed indoors in glass fronted enclosures that measured $5.9 \times 3.0 \times 2.0$ m with an alcove ($3.0 \times 1.5 \times 1.8$ m) for the adult group and $5.9 \times 9.0 \times 2.0$ m for the juvenile group. Ceilings were metal-barred permitting suspension and brachiation; ground platforms provided floor elevation. Chains and tires provided climbing elements and additional suspension. The adult group consisted of one adult male and three adult females, all of which were wild-born. Two of the females had infants which were temporary members of the group during the study. The juvenile group consisted of three captive born juveniles, one male and two females. All animals, adult and juvenile, had been living as separate groups for at least three years when the study was initiated.

It could be argued that grouping a solitary species like the orang utan may

The orang utan. Its biology and conservation, edited by L.E.M. de Boer
© *1982, Dr W. Junk Publishers, The Hague, ISBN 90 6193 702 7*

produce abnormal behavior. However, such groupings can provide information about the social potential of the species that may not be expressible in the current wild habitat of orang utans.

Social or solitary behavior

The first question asked was whether or not the quality of the behavior of these orang utan groups was of a basically solitary or social nature. Focus is on the adult group results. However, equivalent measures were taken for both groups. After 500 min of recorded observation were collected on each individual from both groups, it appeared that the animals had adjusted to the group-living situation since levels of aggression were extremely low and reproduction was successful. The repetoire for both groups included 180 different types of behavioral events, all of which were condensed into 14 behavioral categories (see Fig. 1). A further distinction of interactive versus solitary behavior was made to assess the social versus solitary nature of the groups. This was considered to be a conservative distinction since in order for an interaction to be scored either an animal made contact, showed distinct coordination of movement, or showed patterns of response to another

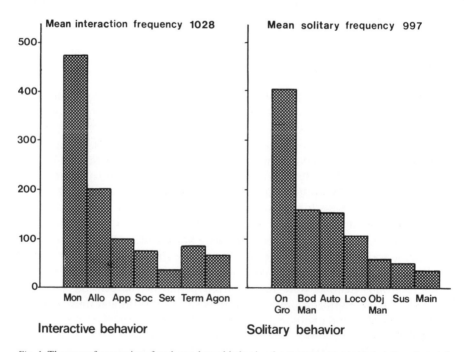

Interactive behavior   Solitary behavior

Fig. 1. The mean frequencies of each condensed behavioral category separated into interactive and solitary behavior for the adult group of orang utans. Overall means for interactive and solitary (non-interactive) behavior are also shown. (Mon = monitoring, Allo = allogrooming, App = approach, Soc = non-grooming social, Sex = sexual, Term = termination, Agon = agonistic, On Gro = on-ground behavior, Bod Man = body manipulation, Auto = autogrooming, Loco = locomotion, Obj Man = object manipulation, Sus = suspended, Main = self-maintenance.)

individual's actions. As indicated in Fig. 1, interactive behaviors included those found for other social primates: monitoring, allogrooming, approach, non-grooming social, sexual, termination, and agonistic behavior. Monitoring was a high frequency social behavior incorporated by adult group animals. Monitoring was defined as an individual directing eye contact to another individual. Monitoring was not distributed randomly across individuals, but in fact followed closely the distribution of other social interactions, a distribution which indicated that relationships were attended to through monitoring.

Solitary behavior, on the other hand, included any on-ground behavior, body manipulation, autogrooming, locomotion, object manipulation, suspended behavior and self-maintenance.

The results of the data suggest strongly that these animals incorporate a high degree of social interaction in their behavioral repetoire, and that indeed the quality of their behavior was social. Half of the total behavioral events scored included animal interactions for both groups (50.8% adults, 50.6% juveniles), whereas solitary activities comprised less than half (49.2% adults, 49.4% juveniles). Proximity measures also showed that orang utans maintained high levels of contact with individuals of their group. Perhaps the absolute quantity of social interaction found in this study is a consequence of the living conditions of captivity (Rowell 1972), but the results do demonstrate the types of social behavior orang utans can display and indicate that, at least when confined to groups, orang utans show behavioral dispositions that are social.

Allocation of space

Another question asked when comparing the orang utan's social system under these different environmental conditions was, are there any similar tendencies of social behavior being expressed by these group-living orang-utans that are comparable to findings of their behavior patterns in the wild? The present results produced findings that correspond to the spacing behavior of wild orang utans. Figure 2 shows a diagram of the adult group compartment. Individual location within compartments was scored and from the total number of locations scored for each animal, that animal's percentage in each area was calculated. For example, S was found in area four 64.4% of all locations scored for S. Over 50.0% of the locations scored for each female were in one specific area, whereas the male's highest frequency in any specific area comprised only 38.4% of the locations scored for him (see Fig. 2). Maintenance of differential use of space within group compartments was more clearly defined for adult females than for juvenile females and more clearly defined for females than for males. Females spent a greater percentage of time in a specific location of the compartment, in contrast to the male who did not show preference for any specific compartment location. This does not mean that males interacted with females exclusively in their most frequently located area, but that when not interacting with another animal, females preferred a specific area of the compartment to other areas, and that females maintained allocation of space much more strongly than did the male.

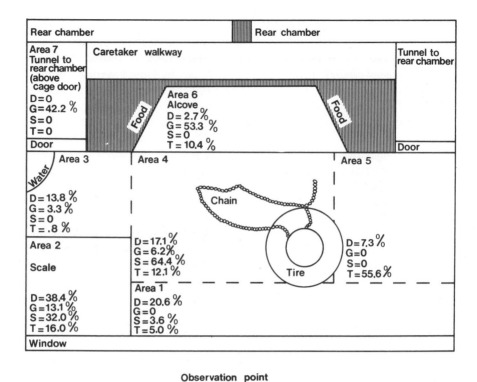

Fig. 2. Diagram of adult compartment and use of space by adults. Notched lines symbolize the division of areas. Animal D is an adult male, animals G, S and T are adult females. Total number of locations scored for adults is 2,209 (from Edwards and Snowdon 1980, redrawn with permission of Plenum Publishing Co., New York).

These findings suggest a spacing tendency that appears to be similar to the ranging behavior observed by Rodman (1973) and Horr (1977) in the field, who found that females established smaller ranges while a male will establish a larger home range covering the ranges of several females. Although the captive group social setting did not allow for ranging and other necessary requirements this behavior may fulfill in the wild, it is interesting that these orang utans were somehow influenced to display a similar behavioral spacing in a captive setting.

Social development

When examining the orang utan social system and the possible determinants of the system, the development of sociality with age can express social tendencies. Field reports (Rijksen 1975) indicate that juveniles behave relatively socially whereas adults minimize social contacts. When comparing adult and juvenile group data of the present study, the results do not suggest that social behavior decreases with adulthood, but that the type of social behavior incorporated does change with age (see Fig. 3). While both groups showed equivalent

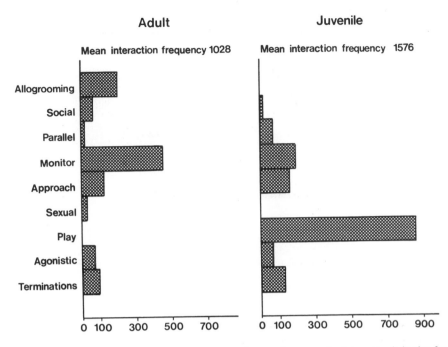

Fig. 3. Mean frequencies for components and overall mean frequency for interactive behavior for both adult and juvenile groups of orang utans.

proportions of social behavior, the juveniles showed a greater rate of behavioral event occurrence (both interactive and solitary). Therefore the mean interaction frequency is greater for the juvenile group. Major differences were found, however, between the types of social behavior incorporated by the two groups. As shown in Fig. 3, allogrooming, sex behavior and play behavior contributed to major differences found. Although allogrooming was never observed among the juveniles, it was the second most prevalent adult social interaction. The most prevalent adult group interaction was monitoring. Sexual behavior was never observed among juvenile animals, which could be due to sexual immaturity of the male or perhaps the influence of captivity on factors of socialization in the juvenile male and females. Play behavior was not observed among adult group animals outside of a sexual context, but play behavior comprised the highest frequency of all behavioral events scored for the juvenile group.

The social behavior observed among the adult group animals was less conspicuous than the active play behavior observed among juveniles. Adults incorporated different types of social behavior, but maintained equally high levels of social interaction. The comparison of groups shows that differences in sociality evolve with development. In examining the social potential of these group-living orang utans the complexity of their social development is indicative of their social capacity. In view of these groups sharing limited space, it is interesting that levels of aggression were low, and that relationships became well defined with age. A sociogram was constructed (see Fig. 4) to display patterns of relationships among individuals of a group. Adult animals showed

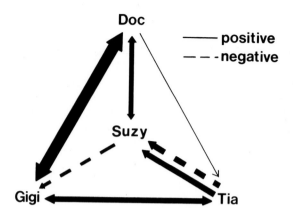

Fig. 4. Posited attention structure for the adult group of orang utans. Thickness of lines is proportional to degree of relationship. Arrows indicate directionality of relationships. Doc is the adult male, Gigi, Suzy and Tia are adult females (these animals correspond to D, G, S, and T in Fig. 2). (From Edwards and Snowdon 1980, redrawn with permission of Plenum Publishing Co., New York.)

social sophistication by displaying an intricate transfer of behavior. Interaction frequencies between the male and each female more than doubled interaction frequencies between females. Monitoring was found to occur more frequently among females than between males and females. All relationships between the male and the females of the adult group were basically positive. On the other hand, relationships between females were more diverse.

Rather than a simple linear hierarchy, the frequency and direction of social behavior defined a complex triadic interactive process which involved discrete relationships and spacing behavior.

Conclusion

The above findings imply that the social system of orang utans can demonstrate a plasticity in social behavior. The complexity of social behavior found in this study indicates that orang utans may be more social than has previously been described. Although social interaction may not always be obvious in the wild because of limitations presented while conducting field research, wild orang utans could possibly interact socially, although these interactions may be dispersed over time and distance.

The social system of orang utans may be undergoing a continuous process of change due to environmental perturbation. Environmental influences alone may determine the extent to which the social potential of this species is expressed. It is important, therefore, to investigate the grouping patterns of this species in light of environmental factors which may tend to modify and determine the minimum requirements for an intact social system, imperative for the orang utan's survival.

Acknowledgements

This report is based on part of a senior thesis submitted by the author. The research was supported by USPHS Grant MH29, 775 and NIMH Research Scientist Development Award to Charles T. Snowdon, and by funds donated from Mr and Mrs Herbert Malzacher. Dr Charles T. Snowdon is thanked for his continual interest and expert guidance, Mr and Mrs Herbert Malzacher for their enthusiasm and support, T. Smith for his contributions, and D. Goldfoot for his information and valuable comments.

References

Edwards, S. and C. Snowdon (1980). Social behavior of captive, group-living orang-utans. Int. J. Primat., 1: 39–62.

Horr, D.A. (1975). Orang-utan maturation: growing up in a female world. In: S. Chevallier-Skolnikoff and F.E. Poiriet (eds.), Primate bio-social development: biological, social and ecological determinants. Garland, New York.

Rijksen, H.D. (1975). Social structure in a wild orang utan population in Sumatra. In: S. Kondo, M. Kawai and A. Ehara (eds.), Contemporary primatology. Karger, Basel.

Rodman, P.S. (1973). Population composition and adaptive organization among orang-utans of the Kutai Reserve. In: J.H. Crook and R.P. Michael (eds.), Comparative ecology and behavior of primates. Academic Press, London.

Rowell, T.E. (1972). Social behavior of monkeys. Penguin Books, Harmondsworth.

Author's address:
S.D. Edwards
Wisconsin Regional Primate Research Center
Madison, WI 53706
U.S.A.

14. Orang utan behavior and its management in captivity

Terry L. Maple

Introduction

Sumatran orang utans have been living in groups at the Atlanta Zoological Park since 1968, when Davenport and Rogers arranged their transfer from the Yerkes Regional Primate Research Center. Since 1975, my associates and I have conducted research at the Zoo and the Primate Center, during which time we have observed orang utans of all age and sex classes. These efforts have led us to conclude that orang utans are considerably more sociable in captivity than had been previously reported (cf. Maple 1980). Our studies indicate that, even in adulthood, there is considerable potential for social interaction in captive orang utans.

The orang utan's locomotor habits and bodily adaptations are the most narrowly specialized of all the great apes. Moreover, the western world has had a lengthy acquaintance with this creature dating back to at least the 18th century (cf. Cuvier 1811; Abel 1818). In light of these details, it is curious indeed that orang utans have been so rarely exhibited in a manner which is consistent with their needs and specializations. Elsewhere (Maple 1979, 1980) I have more completely reviewed the problems encountered in exhibiting orang utans in captivity. In this chapter, my aim is to offer some further generalizations regarding the behavior of orang utans in captivity, and to suggest ways to achieve adequate, if not optimal, living standards for this anthropoid under captive conditions.

The behavioral repertoire

Great apes do not make adjustments to captivity equally well. The extroverted chimpanzee responds to social restriction by exhibiting a wide array of abnormal behavior patterns reminiscent of human psychopathology (Rogers and Davenport 1969). With respect to gorillas, captivity seems to especially affect their reproductive success (Nadler 1974). Ironically, orang utans, though traditionally exhibited under the least appropriate conditions, have adapted best to captivity. Relatively more solitary and stoic than *Pan* or *Gorilla*, when suffering from extreme social deprivation, the visible effects are considerably

The orang utan. Its biology and conservation, edited by L.E.M. de Boer
© *1982, Dr W. Junk Publishers, The Hague, ISBN 90 6193 702 7*

less conspicuous (cf. Puleo et al. 1982). As Robert and Ada Yerkes (Yerkes and Yerkes 1929) suggested: 'The content of the orang-utan literature would suggest ... its relative independence of the factors of social environment and its ability to live contentedly and adapt itself with the usual measure of success in nature or in captivity as an isolated individual.' (p. 140)

Comparative studies of the frequency of social interaction in *Pan*, *Pongo* and *Gorilla* have not yet been conducted. However, it is generally agreed that orang utans are less frequently in contact than their close anthropoid relatives. Representative of the problem in making such comparisons is the inherent difficulty of measuring social behavior. Orang utans appear to communicate with each other in very subtle ways (Schürmann 1982) but we have no way of balancing the comparison to account for such a characteristic. Thus, our impression of orang utan behavior will always contain an element of subjectivity.

Be that as it may, we have observed that orang utans are relatively more peripheral and slow moving than are representatives of *Pan* (cf. Maple 1980; Nadler and Braggio 1974). These characteristics appear to reliably generalize from natural to captive settings. However, the physical environment can greatly influence the human perception of both. For example, peripheralization is potentiated by enclosure fencing and the lack of suitable internal climbing structures. In their characteristically arboreal fashion, captive orang utans cling to any available network of vertical apparatus. In many enclosures, the only suitable substrate is the fencing itself and, by necessity, the animal is therefore peripheral. Similarly, orang utans are most awkward when on the ground, a self-fulfilling prophecy when they are forced to inhabit a barren enclosure.

In these and other ways, the nature of the habitat can influence our perception of the animals that inhabit them. These findings aside, however, we have found that many, if not all, of the behaviors which have been attributed to feral orang utans also occur in captive specimens. To be sure, captivity elevates some behaviors and depresses the expression of others. For example, play is probably more frequent in captivity than in the wild. Captive orang utans also repeatedly regurgitate their food, a behavior which is absent in animals that daily must locate their scattered but more abundant food sources. By contrast, it appears that male long-calling is much less common in captivity.

In our studies of captive orang utans, we have recorded the behaviors which are listed in Table 1. We have found that these behaviors are adequate for any long-term study of orang utans. However, for specific projects, mother–infant interactions for example, a more specialized list was developed (cf. Maple et al. 1978).

Playful behavior

In our studies of play, we have always been surpised at the breadth of the repertoire and its overall similarity to play in *Pan* and *Gorilla* (cf. Maple and Zucker 1978; Zucker et al. 1978). Young orang utans vigorously wrestle with each other, and invite play by the metacommunicative signals of extending their

Table 1. Types of behavior recorded in studies of captive orang utans.

Locomotor behaviors
WK – walk quadrupedally
BI – walk bipedally
BR – brachiate
CL – climb
SW – 'spider walk'; movement with all fours on ceiling
SG – swings; walking with support of bars
SD – slide across the floor with arms on floor in front of animal
CR – crawl
LM – locomote with contact or support of another animal
DR – drop; from ceiling or bars
VR – ventral ride; on another animal's ventrum
DR – dorsal ride; on another animal's back (or on head)
AR – arm ride; clinging to a moving animal's arm
LR – leg ride; clinging to another animal's leg while locomoting
PV – ventral push; pushing animal along floor with belly down
PD – dorsal push; pushing animal along floor on its back
Postural behaviors
SI – sit
SQ – squat
LA – lay
ST – stand
ZZ – sleep
SS – stand with support
HN – hang; by arms or legs and arms combined (see DN – dangle)
VH – 'V' hang; supported by legs but with body upright
NH – net hang; hanging by arms and legs horizontally, arms and legs extended
BH – 'box' hang; sloth hang; suspended by arms and legs, arms and legs parallel
CS – cling; no support provided by other animal
CA – cradle another animal in arms
CF – cradle another animal on floor with arm/hand support
HA – hold another animal in the air
HF – hold another animal down on the floor or ground
DN – dangle; suspended by 2 legs, any one appendage, or 1 arm and 1 leg
PB – place another animal on bars or chain
PC – place another animal on ceiling bars
PF – place another animal on floor
SN – stand with contact with or support by another animal
Self-care behaviors
GS – groom self
SC – scratch self
PT – pat self on head or back
ET – eat
DK – drink; with lips off wall, with hair on arm, with hand cupped, or from spout
SR – stretch
RG – regurgitate food
UR – urinate
DF – defecate
Facial expressions
FF – funnel face; full extension of pursed lips
BT – bare teeth; top teeth exposed
GM – grimace; corners of mouth pulled back, mouth slightly open; 'grin'
OM – open mouth; teeth showing but lips not pulled back
YN – yawn
PY – play face; mouth open, lower teeth exposed
Vocalizations
LC – long call
SK – squeak
KS – kiss-squeak
KW – kiss-squeak with wrist or back of hand
TC – tongue gulp or click

hands, and emitting a play-face. However, we have questioned the predictive value of the latter. We do not know whether orang utans have as much difficulty as we do in reading the play-face. In comparing young orang utans with young chimpanzees (2–4 years of age) we learned that while chimps tend to chase, beat, and hit in play, orang utans prefer to firmly grasp the partner and bite. Typically, the objects of these bites are the appendages of the play adversary. We have interpreted this mode of play in light of the arboreal propensities of *Pongo*. In a more recent study (Zucker, Puleo and Maple, unpublished; reviewed in Maple 1980), we detected evidence of early emerging sex differences in playful behaviors. In heterosexual play interactions, females tended to withdraw from play, engaging in more escapes and less grasping than the males. In every way, male and female play, respectively, resembled the behaviors characteristic of copulating adults, e.g., aggressive, pursuing males, and reluctant, escaping females.

We also learned that adults play. In documenting, for the first time, proceptive female behavior (Maple et al. 1979), we observed that proceptivity is initially comprised of many playful elements. As we pointed out (p. 55): 'Typically, a few days prior to and on the first day of [female] mounting, the female's activity level began to increase, especially solitary play activity. She was observed to belly-slide on the floor, turn forward somersaults, swing vigorously by her feet while slapping the ground, and repeatedly stomp in the water trough. ... These behaviors were succeeded by following the male, slapping at the male while hanging from the bars ... and a vigorous rough-and-tumble play bout.'

More surprising, perhaps, was our observation that adult males occasionally engage in vigorous play bouts with their offspring (Zucker et al. 1978). In these interactions, the male generally reciprocates the play invitation of the younger animal, but Lyn de Fiebre (pers. comm.) informed me about a young adult male that regularly initiated play from his offspring at the San Francisco Zoo. The age of the father may therefore determine the direction of paternal play.

Parental behavior

We have observed the parental behavior of six different females, and two different males, and in several instances can compare their responses to a succession of offspring. Our research on this topic represents a long-term developmental effort, now approaching six years in duration. Our first published report concerned the early development of the male *Merah* (cf. Maple et al. 1978) and the parental behavior of the male *Lipis* and the female *Sungei*.

One of the outstanding features of the maternal behavior of *Sungei* was her oral manipulation of *Merah*'s appendages, but particularly his external genitalia. We first observed this during his fourth week of life, but subsequent observations have revealed that it occurs earlier, even at birth. This behavior frequently preceded maternal *mounting* of the infant, whereby *Sungei* typically placed her offspring supine onto the cage floor and held in place his arms and legs with her hands and feet. In this position, she repeatedly made pelvic thrusts onto his genitals. As we determined, this behavior was typically short in duration, but exceptionally vigorous. *Merah* was first mounted in the presence of observers during his second week of life, but the female *Hati* was subsequently mounted by *Sungei* within five hours after her birth. Despite the fact that *Hati* was mounted so early in her life, later observations (Maple and Steele in preparation) have determined that she was mounted much less frequently than was *Merah*. Thus we must agree with Hess (1973) that a mother's interest in its offspring's genitalia is differentially manifest according to its gender.

Our observation of the female *Sibu* with the juvenile male *Lunak* has revealed that a captive female may still be mounting her offspring in his fourth year of life. Moreover, *Lunak* was observed to attempt coitus with his mother, although awkwardly. In the same fashion, the male *Molek* attempted to penetrate his mother at the age of 14 months. Hand-raised male orang utans typically thrust against their human caretakers at ten months of age and even younger.

We have argued (Maple et al. 1978) that these mothers derive some degree of pleasurable stimulation from genital contact with their offspring. Moreover, we have suggested that the offspring may also benefit from such contact, since it may contribute to sexual adequacy later in life. Such benefits can be defended on the grounds that, as Horr (1973) has pointed out, young orang utans grow up in a predominantly female world. Unlike chimpanzees and gorillas, they lack the primary contact of proximate peers, and the lingering presence of familiar fathers. I am suggesting here that what we have seen in captivity also occurs in the wild. To validate this assertion, an intensive study of feral mother–infant interaction must be eventually carried out.

From our observations of males occupying the same enclosure with an adult female and their offspring, we have learned that males often respond to the maternal mounting of offspring. We have films of the male *Lipis* rushing to investigate the genitals of *Sungei* after she repeatedly thrust against *Merah*. Previously, the 20-year-old *Lipis* had seriously injured *Sibu* in the presence of her then almost 5-year-old male offspring *Lunak*. Although we did not actually witness the dispute, we believe that young males, after they have reached the age of four, are sufficiently strong to effectively inhibit the forceful copulations attempted by their fathers. In this instance, therefore, we hypothesized that *Lunak* interfered with *Lipis'* advances toward his mother, resulting in her subsequent injury. From a management perspective, 4-year-old males cannot long be housed in a small enclosure with their fathers.

As we have seen, adult males can be both playful and aggressive with their offspring. For this reason, zoos often elect to isolate males from their mates when the females become pregnant. As we have argued, and here we are in agreement with Brambell (1975), enforced isolation of the mother is very stressful and can lead to birth complications, most notably abandonment of offspring. Nadler (1974) has also made this point with respect to gorillas. Here again, a well-designed habitat could permit the female to separate herself into a familiar area inaccessible to the male. This can be engineered by regulating the size of the entryway. When the mother can regulate her social contact, a more natural and healthful birth is the likely outcome. Moreover, because such separations are briefer in duration, they result in less frenzied and, hence, less dangerous reunions. I hasten to add here that we have observed nothing but tolerance on the part of our adult males which have witnessed births. Moreover, the male *Bukit* has on many occasions been the puzzled but gentle recipient of the infant *Penari* when *Sungei* has placed it on his back, or in close proximity to him. On a few occasions, the 5-month-old infant has remained with the male while *Sungei* moved off to inspect the cage for food (Maple and Steele in preparation).

In concluding my comments on this topic, it is necessary to emphasize that our males are members of the Sumatran subspecies. Although I have previously played down the alleged subspecies differences in social behavior (cf. Maple 1980), Markham (pers. comm.) has now acquired some compelling data to support John MacKinnon's (1975) earlier generalization that Borneans and Sumatrans differed in their sociability. In Markham's analysis, Bornean orang utans are characterized as a greater risk to injure infants. Thus, where Borneans are concerned, more caution must be exercised in permitting them exclusive access to mother and infant.

Natural hand-rearing

One of the troublesome problems recognized by Perry and Horseman (1972) was the apparent inadequacy of primiparous females, particularly those born in captivity. We have asserted that the relevant variable which influences later maternal behavior is mother-rearing (cf. Maple et al. 1978; Maple 1980; see also Bond 1979). Furthermore, we have argued that, unless one is familiar with the idiosyncrasies of orang utan motherhood, these behaviors may be mis-interpreted as abuse. For example, if a mother 'sits' upon her offspring, attempts to put it onto the fence, or puts it on her head, is this abnormal behavior? Is it abandonment if she leaves it on the floor? We have observed all of these behaviors in otherwise proficient and experienced mothers. Obviously, a mother should not leave an infant too long, nor should she leave it on her head or sit on it the entire day, but it is not abnormal for her to manipulate the infant in ways which seem to us inhuman. After all, orang utans are by definition inhuman.

Recently, scientists have enjoyed some success in modifying the habits of inexperienced gorilla mothers (cf. Joines 1977; Keiter and Pichette 1977). These investigators provided the respective females with dolls and food reinforcement for engaging in the proper handling of infants. With time and prompting, the gorillas exhibited some improvement in their care. Presumably, experience with living juvenile or infant offspring could contribute even more to their successful acquisition of appropriate maternal behavior. A recent example of successful conditioning has been described by Fontaine (1979) who trained a female orang utan to permit supplemental feeding of her infant. Initial steps in the training sequence were described as follows (p. 169): 'Once the trainers had become familiar to the animal and their presence had become reinforcing, they began to increase the amount of physical contact occurring in these sessions. Thus, by gradual accommodation over a one-month period, Suzie allowed hand holding; arm holding; and, eventually, stroking of her midlateral torso in proximity to the infant.'

Eventually, the trainers were able to feed the infant while it remained in contact with its mother. This promising procedure illustrates, among other things, the malleability of the orang utan parent. Where supplemental feeding is required, and Fontaine's procedure is followed, premature separation can be avoided. Needless to say, some habitats are more amenable than others to necessary interactions between caretaker and orang utan.

Of similar utility is the experience of Cole et al. (1979) who successfully reintroduced a hand-reared infant to its mother. The reintroduction was accomplished at the Metro Toronto Zoo by tranquilizing the mother and gently rubbing the infant with the mother's fecal material. Subsequently, the mother retrieved the infant and cared for it in the appropriate fashion. A similar reintroduction, using food and verbal rewards, was successfully accomplished at the Phoenix Zoo by Dr Howell Hood (Schmidt pers. comm.).

During the short hand-rearing phase, Cole and her associates maintained close contact with the infant, even permitting the infant to sleep with them. As the caretakers carried out their work, they frequently carried the infant orang utan in contact with them, much as a mother would do. Elsewhere (Maple 1979, 1980), we have recommended that, when hand-rearing is necessary, the caretakers should mimic as closely as possible the normal care-giving behaviors of orang utan mothers. This includes, first and foremost, and as much as possible, continuous contact with a living caretaker. A fur vest can be worn

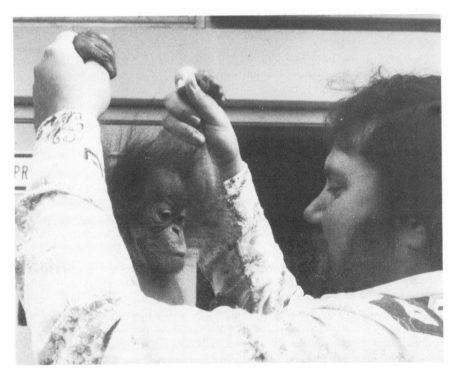

Fig. 1. Animal caretakers can facilitate locomotion and independence by mimicking the behavior of orang utan mothers.

(preferably red in color) to facilitate clinging. As orang utan mothers do, the infant should be encouraged also to cling onto fences and hang onto poles, chains, or tree branches (Fig. 1). By providing the infant with vestibular, locomotor, and social stimulation, it will be prepared for its eventual reintroduction to conspecifics. Hand-rearing should be as minimal in duration as the circumstances will permit, and peer interactions should be encouraged where there are sufficient animals to permit this.

We are currently engaged in developing hand-rearing exercises which permit the caretaker to duplicate the mother's normal repertoire of interaction (Maple and Bradfield in preparation). A description of these exercises will soon be available. They are designed to inhibit the development of stereotyped rocking and self-clinging, behaviors which interfere with reintroduction (cf. Puleo et al. 1981). When apes develop abnormal behavior in the nursery, it is the direct result of neglect.

The physical environment

As we have seen, the social environment is an important determinant of adequate behavior in adulthood. In a recently completed study, Susan Wilson reported that not only was the social environment important, but so also was the physical environment. By sampling orang utan behavior in a number of European zoos, she found that the availability of cage furniture facilitated activity. This is the first empirical demonstration of a generalization which has been asserted by others.

In a previously published chapter (Maple 1979), I suggested a number of ways to improve the existing housing conditions of apes. An effective source of manipulatable material is browse — either branches, straw, sawdust, or other materials. In the presence of these materials, orang utans will make nests, and spend time fooling with it, time which might otherwise be spent in boredom, regurgitation, or coprophagy.

By attending to the vertical component of an enclosure, arboreal locomotion can be effectively simulated. Enclosures should be designed for this, but even barren facilities can be modified by the use of interesting rope, poles, or chains. Even animals which for years have been deprived of vertical pathways are responsive to these new opportunities. Examples of complex vertical habitats can be seen at Atlanta Zoological Park (Fig. 2), Chicago's Lincoln Park Zoo (Fig. 3), the Phoenix (Arizona) Zoo (Fig. 4), and the San Diego Zoo.

Privacy and cover should also be provided in captivity so that animals can separate themselves from their cagemates. This may be an especially important consideration where orang utans are concerned.

As Lethmate (1977) has convincingly demonstrated, orang utans are highly manipulative creatures, and it may be necessary to provide them with more sophisticated opportunities to interact with objects. Markowitz (1979) has devised complex apparatuses for inducing activity and introducing complexity in captive animals. I propose that where space is adequate, an elevated series of automatic feeders could be designed to stimulate sequential feeding. Such an

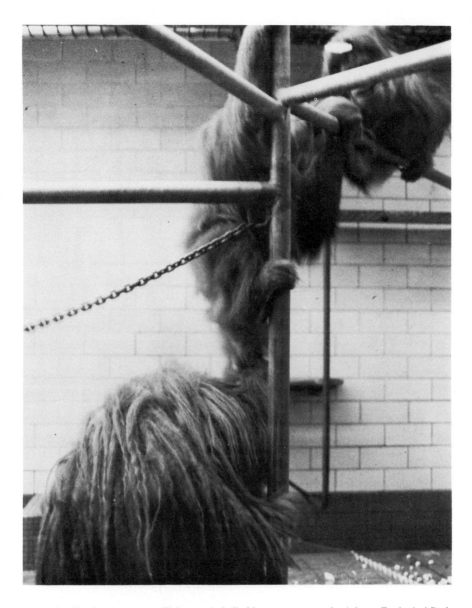

Fig. 2. Family of orang utans utilizing vertical climbing structures at the Atlanta Zoological Park.

apparatus would be especially likely to stimulate movement, and simulate foraging activity.

Finally, it should be stated that we are morally obligated to exhibit the orang utan in humane and stimulating surroundings. If we cannot house it properly, then we should not house it at all. We can no longer plead ignorance either about its habits in the wild, or the effects of captive environments on husbandry. We know what we should do and we know what we can do. It now only remains

Fig. 3. High vertical components at Chicago's Lincoln Park Zoo encourage locomotion in captive orang utans.

for concerned institutions to mobilize their resources and begin to improve those facilities which are outmoded remnants of an unenlightened era.

There is much that the zoo world can do to save this endangered ape. By properly exhibiting orang utans in naturalistic, responsive, and complex surroundings, we can educate the public to contribute to their welfare in the wild state. Moreover, the reproductive efforts of captive specimens are greatly enhanced by increasing available space, and enriching the habitat with stimulat-

Fig. 4. Unique wooden climbing furniture in use at the Phoenix (Arizona) Zoological Park.

ing structures and social possibilities. I trust that, in some small way, this chapter contributes to the development and further proliferation of habitats of this kind. In this way, the aims of conservation agencies and zoological parks are properly synchronized. We are partners in the same enterprise, the preservation of the orang utan for all to see and appreciate.

Acknowledgements

The author gratefully acknowledges the support and assistance of the following sources and individuals: N.I.H. Grant HD00165 to the Yerkes Regional Primate Research Center, Atlanta Zoological Society, E. Bradfield, J.S. Dobbs, R. Jackson, S.P. Puleo, S. Steele, D. Stokes, B. Swenson, S.F. Wilson, M.E. Wilson, and E.L. Zucker.

References

Abel, C. (1818). Narrative of a journey in the interior of China, and of a voyage to and from that country, in the years 1816 and 1817. Longham and Co., London.
Bond, M.R. (1979). Second-generation captive birth of an orang-utan. Intern. Zoo Yearb., 19: 165–167.

Brambell, M.R. (1975). Breeding orang-utans. In: R.D. Martin (ed.), Breeding endangered species in captivity. Academic Press, London.

Cole, M., D. Devison, P.J. Eldridge, K. Mehren and W.A. Rapley (1979). Notes on the early hand-rearing of an orang-utan and its subsequent reintroduction to the mother. Int. Zoo Yearb., 19: 263–264.

Cuvier, F. (1811). Descriptions of an orang outang with observations on its intellectual faculties. (Translation of the 1810 publication). Philosoph. Mag., London, 38: 188–199.

Erwin, J., T. Maple and G. Mitchell (eds.) (1979). Captivity and behavior: Primates in breeding colonies, laboratories and zoos. Van Nostrand-Reinhold, New York.

Fontaine, R. (1979). Training an unrestrained orang-utan mother to permit supplemental feeding of her infant. Int. Zoo Yearb., 19: 168–170.

Horr, D.A. (1977). Orang-utan maturation: growing up in a female world. In: S. Chevalier-Skolnikoff and F.E. Poirier (eds.), Primate biosocial development and ecological determinants. Garland, New York, 259–322.

Joines, S. (1977). A training program designed to induce maternal behavior in a multiparous female lowland gorilla at the San Diego Wild Animal Park. Int. Zoo Yearb., 17: 185–188.

Keiter, M. and P. Pichette (1977). Surrogate infant prepares a lowland gorilla for motherhood. Int. Zoo Yearb., 17: 188–189.

Lethmate, J. (1977). Werkzeugherstellung eines jungen Orang-utans. Behaviour, 62 (1–2): 174–189.

MacKinnon, J. (1975). Distinguishing characters of the insular forms of orang-utan. Int. Zoo Yearb., 15: 195–197.

Maple, T. (1979). Great apes in captivity: the good, the bad and the ugly. In: J. Erwin, T. Maple and G. Mitchell (eds.), Captivity and behavior. Van Nostrand-Reinhold, New York.

Maple, T. (1980). Orang-utan behavior. Van Nostrand-Reinhold, New York.

Maple, T. and E.L. Zucker (1978). Ethological studies of play behavior in captive great apes. In: E.O. Smith (ed.), Social play of primates. Academic Press, New York, 113–142.

Maple, T., M.E. Wilson, E.L. Zucker and S.F. Wilson (1978). Notes on the development of a young orang-utan: The first six months. Primates, 19: 593–602.

Maple, T., E.L. Zucker and M.B. Dennon (1979). Cyclic proceptivity in a captive female orang-utan. Behavioural Processes, 4: 53–59.

Markham, R. (unpublished). An investigation into the differences in social behaviour between the Sumatran and Bornean subspecies of orang-utan in captivity. School of Environmental Sciences, University of East Anglia.

Markowitz, H. (1979). Environmental enrichment and behavioral engineering for captive primates. In: J. Erwin, T. Maple and G. Mitchell (eds.), Captivity and behavior. Van Nostrand-Reinhold, New York.

Nadler, R.D. (1974). Determinants of variability in maternal behaviour of captive female gorillas. Symp. 5th Congr. Int. Primat. Soc.: 207–215.

Nadler, R.D. and J.T. Braggio (1974). Sex and species differences in captive-reared juvenile chimpanzees and orang-utans. J. Hum. Evol., 3: 541–550.

Perry, J. and D.L. Horseman (1972). Captive breeding of orangutans. Zoologica, 00: 105–108.

Puleo, S.G., E.L. Zucker and T. Maple (1982). Social rehabilitation and foster mothering in captive orang-utans. Zool. Garten (in press).

Rogers, C.M. and R.K. Davenport (1969). Effects of restricted rearing on sexual behavior of chimpanzees. Develop. Psychol., 1: 200–204.

Schürmann, C. (1982). Mating behaviour in wild orang utans. In: L.E.M. de Boer (ed.), The orang utan. Its biology and conservation. Junk, The Hague.

Verstraete, A.P. (1977). Orang utan breeding in the United Kingdom. Int. Zoo News, 24: 20–24.

Wilson, S.F. (unpublished). Environmental influences on the activity of captive great apes.

Zucker, E.L., S. Puelo and T. Maple (unpublished). The development of sexual behavior through play in captive young orang utans.

Zucker, E.L., G. Mitchell and T. Maple (1978). Adult male-offspring play interactions within a captive group of orang-utans. Primates, 19 (2): 99–103.

Author's address:
T.L. Maple
School of Psychology
Georgia Institute of Technology
and
Yerkes Regional Primate Research Center
Atlanta, GA 30322
U.S.A.

15. Mating behaviour of wild orang utans

Chris Schürmann

Introduction

Despite the fact that the first observations on the behaviour of the orang utan were made two hundred years ago, the orang utan was, until very recently, considered to be the least known of the great apes. Field studies over the past 20 years have produced a wealth of data, but in spite of this not much is known of the social behaviour of wild orang utans. Since the orang utan is an animal with a predominantly solitary life-style and a generally low frequency of social interaction, it is difficult to obtain quantifiable data on its social behaviour in the wild. Our present knowledge of the social life of the species is still based mainly on the descriptive and qualitative studies of field investigators.

Between June 1975 and April 1979, the author studied a population of wild orang utans in Sumatra, focussing particularly on social behaviour. In this paper the observations on reproductive behaviour made during this study are presented.

History of behavioural studies on the orang utan

The first description of the behaviour of captive orang utans dates from the 18th century. The first field observations on the behaviour of wild orang utans were made in the 18th and 19th century by hunters and travellers. However, these observations were usually made from behind the barrel of a shotgun. Attention focussed on the anatomical description of dead animals. In the first half of the 20th century many living specimens were caught, placed in zoos and observed in captivity.

The first, brief, field study was undertaken by Carpenter in 1937 (Carpenter 1938). By 1960 it became known that the orang utan had become an endangered species and subsequently interest in the animal increased. Barbara Harrisson set up the first rehabilitation centre for orang utans, where she was able to observe them under semi-wild conditions (Harrisson 1960). During the 1960s, several field studies on feral animals were initiated. All our current knowledge of wild orang utans has been collected since 1959. We can divide the period 1959–1979 into roughly three phases as illustrated in Fig. 1.

The orang utan. Its biology and conservation, edited by L.E.M. de Boer
© *1982, Dr W. Junk Publishers, The Hague, ISBN 90 6193 702 7*

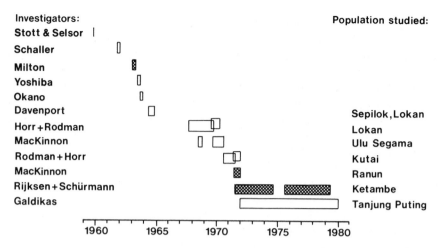

Investigators: Population studied:
Stott & Selsor
Schaller
Milton
Yoshiba
Okano
Davenport   Sepilok, Lokan
Horr + Rodman   Lokan
MacKinnon   Ulu Segama
Rodman + Horr   Kutai
MacKinnon   Ranun
Rijksen + Schürmann   Ketambe
Galdikas   Tanjung Puting

1960   1965   1970   1975   1980

Fig. 1. Duration of field studies on orang utans. White rectangles represent studies on Bornean orang utans, hatched rectangles on Sumatran orang utans.

The first phase, from 1959 to 1963, was the phase of preliminary studies. Surveys were carried out in which the status of the orang utan, the distribution and density of populations and the possibilities for behavioural studies were investigated. Stott and Selsor (1961), Schaller (1961), Milton (1964), Yoshiba (1964) and Okano (1965) all carried out surveys of this kind. No field work during this period lasted longer than two months, but this was long enough for the investigators to realise how difficult it was to locate and observe un-habituated wild orang utans. Very few direct observations were made. Only Schaller mentioned in his report the length of time during which he actually had orang utans under observation, five hours and forty-five minutes.

In the second phase, between 1964 and 1971, more comprehensive field studies were carried out by Davenport (1967), Horr (1972), MacKinnon (1974) and Rodman (1973). They conducted more intensive work for longer periods on certain selected populations and obtained a quantity of basic data on the ethology and ecology of the species. These data led to the first tentative models of the animal's social organization. Their work already contained some quantification of data, but was still mainly of a qualitative and descriptive nature. Their studies varied in duration from 7 to 27 months, and their observations took between 100 and 1,640 h, making a total of 4,700 h.

The third phase, from 1971 onwards, could be called the period of long-term studies. In my opinion, a long-term study is a study which sets out to follow the life history of the individuals in a certain population and which therefore lasts long enough to cover the major phases in that life history and the cycles in the ecology of the species. In common with other apes, the orang utan has a slow ontogeny and a long life-span. A long-term study of orang utans therefore must necessarily extend over a period of many years. Long-term studies began in 1971 on two orang utan populations. H. Rijksen and I have undertaken two successive parts of a long-term study on the Ketambe population in Sumatra. Rijksen worked for three years on this project (Rijksen 1978) and I for four.

270

B. Galdikas (1978) has been carrying out a long-term study for the last eight years on the Tanjung Puting population in Kalimantan. These long-term studies together include over 17,000 h of observation, which amounts to 78% of all observations on wild orang utans.

Since 1960, orang utans have been observed for 22,000 h, 75% of this time having been given to the Bornean and 25% to the Sumatran subspecies. In view of the amount of observation done in over 20 man-years of field work, one would expect a reasonable amount of knowledge and understanding to have been obtained on the basis of firm data. Washburn and Hamburg (1965) estimated that it would take three observers two years and 2,000 hours of observation to obtain a useful understanding of a species. Although orang utan field work carried out so far accounts for ten times this amount, many aspects of orang utan social behaviour are still poorly understood. This may come as a disappointment to those who do not realise how long it takes to accumulate data on this comparatively solitary species. One of the main reasons for this is the low frequency of social interactions. Nevertheless, one might expect the recent long-term studies to have made a more substantial contribution, not only because of the amount of work which has been done, but also because investigations become more fruitful the more familiar the investigators become with the members of the population over the years. Part of the long-term field work has been published in the form of theses (Rijksen 1978; Galdikas 1978), but more than half of the work, including the present study, so far has not yet been published.

This study

Between June 1975 and April 1979 the author studied a wild orang utan population in Ketambe, in the Gunung Leuser Reserve in Sumatra. The orang utan population in the research area which covered 5 km^2 consisted of at least 34 animals, 17 of which were studied intensively by following each individual for several days in succession and recording its behaviour throughout the day. In the four years of field work a total of 4,200 h of direct observation were completed; Indonesian students participating in this project made additional observations for a further 5,500 h.

The purpose of the study was primarily to investigate the social behaviour and the structure and dynamics of social relations in an orang utan population. This paper deals with one of the most important aspects of social behaviour, namely reproductive behaviour, the understanding of which is also of importance to the conservation of the species in its natural habitat and to zoo management.

During the field work, I had good opportunity to conduct a detailed study of the animals' courtship and mating behaviour and was able to trace how this behaviour developed in a few individuals over a period of four years. The present paper merely gives a qualitative description of some of the findings. Papers dealing with quantification of the field data are in preparation.

For those accustomed to observe captive orang utans it has to be noted that

Fig. 3. The young female *Yet.*

Fig. 2. The adult male *Jon.*

the observed animals in Sumatra were almost completely arboreal. The wild orang utan rarely comes on the ground. All behaviours described took place up in the trees.

Courtship and mating

In all field studies the orang utan is described as an animal with a predominantly solitary life-style and consequently a low frequency of social interaction. In the course of my field work I found that orang utans can be unexpectedly gregarious under certain conditions, particularly in connection with reproductive behaviour. A few individuals in the population played a key role in our understanding of social relations in this species. On a number of occasions their behaviour deviated from the pattern expected on the basis of previous reports and my own initial observations. Consequently, observations on these animals were intensified and repeated at specific intervals. The animals concerned were the adult male *Jon* (Fig. 2), which can be considered to be the highest ranking resident adult male in the study area, and three females. One of these females was the young female *Yet* (Fig. 3), which was an adolescent female during Rijksen's study period and developed into a young primiparous female during the subsequent period of my observations.* On the basis of the data collected by Rijksen and myself, a picture has been developed of the socio-sexual development of this female over the past seven years. This development was a slow and very gradual process during which a growing interest of the males in the maturing young female *Yet* was observed.

The young sub-adult males were the first to show interest in the young female *Yet*. Very often when a sub-adult male encountered *Yet*, he tried to form a consort with her. A consort has been defined as a relationship in which a male and a female travel together for several days in succession and show varying degrees of coordination and cooperation in their behaviour (cf. MacKinnon 1974; Rijksen 1978). Early consorts in which *Yet* was involved were initiated by sub-adult males. A sub-adult male prolonged his encounter with *Yet* simply by following her and adapting his daily schedule to hers, travelling when *Yet* travelled, resting when *Yet* rested and feeding when *Yet* fed, generally consuming the same kind of food (Fig. 4). In these early consorts *Yet* did not always seem to be very interested in the sub-adult males. In later consorts, particularly in those with older sub-adult males, she showed greater interest and a greater willingness to coordinate her behaviour with that of the male.

The purpose of consorting sub-adult males was generally obvious. The male tried to make sexual contact, he touched *Yet*'s genitals with his hand and sniffed his fingers or sniffed her genitals directly. In response to this touch and smell behaviour *Yet* sometimes presented to the male. The male then usually mounted her immediately and copulated with her. Some consorts, however, were surprisingly devoid of sexual behaviour for days on end. Such apparently 'non-sexual' consorts gave the impression that mating was not the primary goal.

* *Jon* and *Yet* are the animals Rijksen (1978) called *O.J.* (*Old Jon*) and *Jet*.

Fig. 4. Consort of *Yet* (left) and sub-adult male *Boris*, feeding together.

Fig. 5. Yet approaches *Jon.*

However, observation of a single consort of a dyadic relation could be misleading. It was observed that sexual behaviour and cooperative mating did in fact occur in later consorts of the same dyads. The 'non-sexual' consort probably represents an early phase in the establishment of socio-sexual relationships. On other occasions, however, copulations were more forcefully

274

initiated by the sub-adult male. Such copulations were termed 'rapes' by MacKinnon (1974). Rapes typically occurred during short encounters, not within an existing consort. Cooperative copulations typically occurred during a consort. These observations are in accordance with the findings of Rijksen (1978) and Galdikas (1978).

The female *Yet* was involved in a considerable number of consorts. Her partners included at least four sub-adult males. The consorts formed a discontinuous series. We were almost certain that the early consorts were mainly with younger sub-adult males and that later on it was also the older, the bigger and the higher ranking sub-adult males which formed consorts with *Yet*. When there was a change of consorting males, it was always a higher ranking male which took over from a lower one. *Yet* herself was also more eager to consort with high ranking males.

Yet showed a marked preference for the most dominant male orang utan in the area, the adult male *Jon*. Rijksen (1978) mentioned the likelihood that a female would move to higher ranking partners in successive consorts. He also suggested that this might result in consort formation and mating of the female with the adult male. At the time he did not have conclusive data on the subject. This consort formation and mating of the adult male has now been confirmed by these observations on the Sumatran orang utan. Consorts and copulations involving adult males have also been observed in Bornean orang utans by Galdikas (1978).

Yet was already keenly interested in the adult male *Jon* during her adolescent period, but it took a long time — about five years — for her to build up a relationship which resulted in courtship and mating. As an adolescent *Yet* often tried to attract the attention of *Jon* by using various kinds of soliciting behaviour. She moved about in the vicinity of *Jon*, looked at him from a distance, approached and looked at him more closely, particularly while he was feeding (Figs. 5 and 6). *Yet* would also touch *Jon*'s arm or pull his hair slightly, and sometimes, often as if it were accidentally, she would present to *Jon*. At first *Jon* avoided all these advances, but gradually began to show a growing tolerance towards *Yet*. In view of the solitary attitude and usual intolerance, particularly of the adult male towards conspecifics, this tolerance was unusual and even conspicious. *Yet* would often feed very close to *Jon* and sometimes she was allowed to take pieces of fruit directly out of his hand or even from his mouth (Fig. 6). This kind of tolerance is seen commonly in the mother–infant relationship, but is otherwise rare in the orang utan world and occurs only in very close social relationships. Even when consorted by a sub-adult male, *Yet* still seemed to seek company of the adult male *Jon*. She often fed in the same fruit tree as *Jon*, and followed him around for part of the day, obviously trying to engage him in a consort relationship. The sub-adult male, which was originally in consort with *Yet*, often remained in the vicinity. Over time the assertiveness of *Yet* increased and her soliciting behaviour turned into more overt proceptive behaviour. From this time onward, *Yet* presented more often to *Jon*, sometimes exposing her genital area only 30 cm away from his face. *Jon* usually just looked at her and occasionally sniffed at her.

Yet had already been seen masturbating occasionally during adolescence.

Fig. 6. A: *Yet* looks at *Jon*, while he is feeding; B: she takes a piece of fruit from his mouth.

Now, she seemed to use masturbation as a form of invitation display. *Yet* masturbated, sitting on a branch in front of *Jon*. She also touched *Jon* more and more, even touching his penis with her hand and later with her genital area, but still obtaining almost no reaction. *Jon* usually tolerated the actions of *Yet*, sometimes he turned away or faced away and on rare occasions gave her a kind of rebuff, a soft push with his arm. *Yet* became more obtrusive in her soliciting behaviour. She seemed to become quickly irritated and slightly aggressive when *Jon* avoided or rebuffed her advances. As a reaction she might tug his hair or even hit him but as a rule this was always accompanied by submissive 'bared-teeth' display. Finally, after months of soliciting behaviour, *Yet* succeeded in forming a consort with *Jon*, which meant that *Jon* followed her for several days. During that consort *Jon* clearly and frequently reacted to genital touch by *Yet* with 'male presenting', which can be defined as follows: from a sitting position on a branch, the male bends backwards until he almost lies on his back, often extending his arms holding onto a branch for support, and often extending his legs. Logically, one would say: thus revealing his penis; but *Jon*'s penis, even when erect, was generally not visible among the long hair on his belly. In any case he made his penis accessible by male presenting and thus gave a clear invitation to the female. Male presenting was the usual way in which *Jon* reacted when *Yet* touched his genital area but he also reacted in this way to other proceptive behaviour. Sometimes the mere approach of *Yet* was sufficient to elicit male presenting. Male presenting was only observed in the adult male *Jon*; it was never seen in sub-adult males.

As a reaction to male presenting, *Yet* looked at *Jon*, touched him again and positioned herself in a hanging posture above *Jon* and lowered herself onto his penis (Fig. 7a). In most copulations there was no immediate intromission but a

Fig. 7. Yet mating with *Jon*. A: *Jon* is male presenting while *Yet* mounts ventro-ventrally; B: *Yet* aids intromission manually; C: latero-ventral; D: dorso-ventral copulations.

period during which *Yet* rubbed her genital area against *Jon*'s, before she inserted his penis, often aiding intromission with her fingers (Fig. 7b). Not only did *Yet* initiate the copulation, she was also the active partner, assisting intromission and performing all the pelvic thrusts. whereas *Jon* remained completely motionless (cf. Fig. 7a, b, c and d). On a few occasions it was noticed that he shifted his arm or leg slightly and very occasionally that he made some pelvic movements, which were too vague to be called thrusts. Although *Jon* was inactive in these first copulations, it was clear that he had become more interested in *Yet*. He showed more vigour in chasing away approaching sub-adult males when he was in consort, and he made a greater contribution to the coordination of behaviour, which is characteristic for a close consort.

Several months after the first copulations between *Jon* and *Yet*, their relationship climaxed in two consort periods with a very high level of mutually coordinated behaviour. The first consort lasted 15–17 days and the second one 20–22 days. The consorts were separated by a solitary period of 8–10 days. (The observations of these consorts are incomplete, so there is an inaccuracy of two

277

days in the length of the consorts.) During the first consort, copulations occurred on nine consecutive days and during the second consort on ten days, with one exception. The copulations occurred in the middle of the consort, never near its beginning or end. In the middle of these copulation periods there were a few days in which numerous and lengthy copulations took place. There were 30–31 days between these mid-consort multiple copulation days in the two consorts.

An orang utan female in oestrus does not show any visible oestrus signs. There may be a slight swelling of the vulva, but this is hardly visible and is subject to wide individual variation. At least for the observer the female cycle can be determined only by behavioural criteria, such as the willingness of the female to consort and mate, and the amount of soliciting and proceptive behaviour. Thus, a cyclicity was found in consort behaviour and mating behaviour, which most probably followed *Yet*'s sexual cycle. Thirty days is the average length of the sexual cycle of captive orang utans (Blakley 1969). If the days with high mating activity did indeed coincide with ovulation and the solitary days with menstruation, then the two consorts of *Jon* and *Yet* were related to *Yet*'s sexual cycle. Presuming a 30-day cycle, the two consorts are represented schematically in Fig. 8.

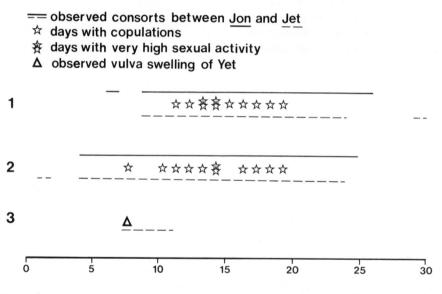

Fig. 8. Consorts between *Jon* and *Yet* during two successive 30-day cycles (1 and 2) and the subsequent first days of pregnancy of *Yet* (3).

The sexual behaviour of the female as well as that of the male showed some variation over the sexual cycle. The impression was obtained that the genital odour of the female was a determining factor in the amount of interest of the male.

Almost all copulations between *Jon* and *Yet* were initiated by *Yet*, only a very few by *Jon*. Most copulations performed by *Yet* were in a ventro-ventral

278

position (Figs. 7a and b). During the copulations, however, *Yet* often changed position, so that latero-ventral (Fig. 7c) and dorso-ventral (Fig. 7d) positions also occurred. *Yet* often interrupted the copulation for a short while, manipulating *Jon*'s penis with her hand, licking it or putting it in her mouth before mounting again. On the few occasions that *Jon* initiated and performed the copulation, both animals were hanging from branches, *Jon* with his body in a vertical position and *Yet* usually in a latero-horizontal position, which generally resulted in a latero-ventral copulation position.

All the copulations involving *Yet* which were observed during these two months were with the adult male *Jon*; no copulations with other males were observed. During her consort with *Jon*, copulations with other males were virtually impossible because every approaching male was chased away by *Jon*. One sub-adult male did try to mate with *Yet* while *Jon* was chasing away another sub-adult male, but he was very nervous and before he could mount he had to flee from *Jon*. In the inter-consort period matings may have occurred with other males, but this does not seem very likely, since *Yet* avoided contact with males, and the males did not seem interested in her.

After termination of the second consort, *Yet* tended to become rather more solitary. When she was observed two weeks later her nipples and vulva were obviously swollen (Fig. 9). This indicated pregnancy, which was proven eight months later when she gave birth to an infant (Fig. 10). Thus, pregnancy was detectable some three weeks after conception. Calculated from the estimated last menstruation, the gestation period was 265 days.

Sexual strategy of the orang utan

The above description is a case-history of two animals only, but from it some facts emerge which are of importance for our general picture of the orang utan's sexual strategy. *Jon* is the father of *Yet*'s first infant, which means that the adult male *Jon* had more reproductive success in mating with *Yet* than four sub-adult males. This is all the more remarkable because we know that *Yet* was already menstruating five years before she conceived and had mated on numerous occasions with sub-adult males before she mated with *Jon*. Obviously, however, none of the copulations with sub-adult males, whether they were forceful or cooperative, resulted in conception, at least in as far as can be ascertained from our field observations. A possible miscarriage could occur unnoticed in the wild, unless it was accompanied by obvious behavioural changes.

What causes the reproductive success of the adult male, or rather the more remarkable lack of success of the four sub-adult males? There appears to be a difference in the chance of fertilization by the adult male and the sub-adult males. The adult male *Jon* was the most dominant orang utan in the study area. He could easily oust all known sub-adult males, so he was able to maintain a consort with a potentially receptive female for a long period without being disturbed by others. The adult male was even able to reserve the most receptive period of the female exclusively for himself. Once the receptive female is in consort with the adult male, the sub-adult males do not have any chance of

Fig. 10. Yet and her infant *Yop* at the age of two months.

Fig. 9. Yet with swollen vulva in the fifth month of pregnancy.

inseminating her. Sub-adult males were not able to maintain a consort for such a long period, their consorts were much shorter, three days on the average. The consort of a sub-adult male was easily ended because he was ousted by a higher ranking male, or because he easily lost the female when she showed insufficient willingness to consort.

Another difference between the adult male and the sub-adult males is in the intensity of their sexual contacts, and this might influence conception. Since sub-adult males are reported to be sexually more active than adult males (MacKinnon 1974), field workers have been puzzled about the role of the adult male. From my observations it is seen that sub-adult males are indeed sexually more active than adult males, whose sexual life is generally more passive. The sexual activity of the sub-adult males is spread out over the year, whereas the copulations of the adult male are concentrated in a short period, which is probably the most receptive period of the female. Although the numbers of interactions are still awaiting statistical analysis, it seems that the sub-adult males, in spite of their higher sexual activity, do not show a higher degree of sexual interaction over the year than the adult male. In one consort, *Jon* and *Yet* mated as often as 25 times in the course of ten days. This is more than the total number of copulations observed between *Yet* and the sub-adult males over the course of four years.

Sub-adult males were always the initiators and active partners in the copulations, which generally lasted from 5 to 15 min. Between *Jon* and *Yet*, when *Yet* was the active partner, copulations often lasted longer than 15 min, once even an hour. On the few occasions that *Jon* initiated the copulation with *Yet*, his actions were very fast, sometimes faster than those of sub-adult males. It is striking that the latter copulations were always preceded by part of a 'long-call' (MacKinnon 1974). The consorting *Yet* reacted to the long-call by approaching *Jon* and presenting to him. *Jon* approached *Yet*, sniffed her genitals, mounted her and copulated with her. Once I saw this sequence of long-call, approach, sniff, mount, copulation and separation completed within 2 min.

Some observations suggest that female orang utans probably know how to induce this sequence. A long-call is often produced as a reaction to the sound of a falling tree or other loud sounds. Once I noticed that *Yet* was soliciting and presenting to *Jon* without getting any reaction. She then went away and pushed a dead tree until it crashed to the ground. Subsequently *Jon* produced a long-call, *Yet* approached him and he mounted and copulated with her.

The long-call is the most impressive sound made by male orang utans. A long-calling male is in a very excited state, which is apparent because the vocalization is often accompanied by impressive movements, tree shaking and pilo-erection. During a long-call, the male is apparently also sexually aroused, as penis erection was observed several times and on a few occasions even ejaculation before intromission. I have also observed penile erection and ejaculation after a long-call in a solitary situation. These observations indicate that the long-call is probably much more closely connected with sexual behaviour than we thought. In addition to being a long-distance call to attract receptive females, as supposed by Rodman (1973), and to space the adult males, as supposed by MacKinnon (1974), the long-call appears to function also as a short-distance communication preceding copulation, probably attracting the consorting female to sexual interaction.

The reproductive success of the adult male is probably mainly the result of the female's behaviour. The female selects the adult male as her sexual partner and she chooses when to consort and to mate with him. In spite of the fact that the female has this great influence, it is amazing that *Yet* had not conceived by a sub-adult male in any of the approximately 60 oestrus periods that she must have had before conception.

The above description is based on one relationship. I also observed part of the development of a relationship between *Jon* and another young nulliparous female, *Sina*; the formation of this relationship followed almost the same pattern. The described pattern may not be anomalous as far as the behaviour of the young nulliparous females is concerned, but it is probably not representative of the sexual relationships of all orang utan females, as is indicated by my observations of an older multiparous female with offspring. I observed the consort formation of *Jon* with *Bel*, a female with a juvenile son. *Bel* showed much less soliciting and proceptive behaviour in courtship than *Yet*. Consorts between *Jon* and *Bel* lasted for a long time, about two weeks or longer. *Jon* was clearly much more interested in *Bel* than he was in *Yet*. He took a very active part in the consorts with *Bel* and maintained them. Only a few copulations were observed; these were of a short duration and always initiated by *Jon*. Preceding the copulations *Bel* engaged in some proceptive behaviour, but this was much less obvious than was seen in the case of *Yet*. Apparently, the young nulliparous female has to put a lot of energy into attracting the attention of the adult male and leading him to consort and mate. It appears that the adult male and the sub-adult males were generally much more interested in older females with offspring than in young nulliparous or adolescent females. It seems as if the fact that a female has produced offspring makes her more attractive to the males.

Conclusion and Summary

Despite 20 years of field work little knowledge has been gained about the reproductive behaviour of wild orang utans. Early data from the field suggested that copulations were mostly forcefully initiated by the male and were called 'rapes' by MacKinnon (1974). Later it was found that cooperative copulations also occur (Rijksen 1978) and that these generally occur in consort relationships. I found that forceful copulations do in fact occur, but that the majority of sexual interactions are cooperative copulations which generally occur in consort relationships. An established social relationship, through consorts, seems to be of importance for cooperative and also, probably, for successful copulation. These findings for Sumatran orang utans are in accordance with those of Galdikas (1978) for Bornean orang utans.

In the field data there is a remarkable lack of observations of copulations of adult males (MacKinnon 1974; Rijksen 1978). Only Galdikas (1978) reported copulations of adult male Bornean orang utans. Consort formation and mating of Sumatran orang utans have now been observed. Females were found to prefer to consort and mate with the highest ranking male. Consort formation between a female and an adult male can be a slow process and it may be a long time until it leads to cooperative mating. Of two consorts, including many matings between the adult male and a female which were observed, there was an obvious cyclicity in consort behaviour and mating activity. This is in contrast with data from captivity, where very little evidence has been found of cyclicity in behaviour (Nadler 1977). The proceptive behaviour of the female and the resulting mating activity were almost certainly related to the sexual cycle of the female. In their relationships with the adult male, female orang utans may display marked proceptive behaviour and take an active role in sexual behaviour under certain conditions. The amount of proceptive behaviour observed in young nulliparous females in the wild is very high and this differs from the captive situation (cf. Nadler 1977). Only very recently, under special conditions, proceptive behaviour in captive females has been observed by Nadler (1982). Since the orang utan female does not show any visible oestrus signals, this proceptive behaviour seems necessary, particularly on the part of the young nulliparous female, to induce the adult male to mate.

In view of the frequent sexual contacts of the female with sub-adult males, the reproductive success of the adult male is remarkable and cannot yet be fully explained. The adult male, which is the highest ranking and biggest male, obviously does have a reproductive function and is selected as a sexual partner by the receptive females. The sexual dimorphism in this respect therefore is no longer a problem. Although other explanations may still remain valid, sexual selection on the part of the female can explain the evolution of the marked sexual dimorphism in orang utans.

Acknowledgements

This study was supported by a grant from the Netherlands Foundation for the Advancement of Tropical Research (WOTRO) to Dr J.A.R.A.M. van Hooff, to whom I am greatly indebted for his support and encouragement, and for reading and criticising earlier drafts of this paper. I would like to thank the Indonesian Nature Conservation Service (P.P.A.) and the Indonesian Institute of Science (L.I.P.I.) for giving us permission to do orang utan research in Sumatra. I am grateful to the late Dr J. Westermann and the Netherlands Gunung Leuser Committee for their support and interest. I owe special thanks to Mrs N. Sulaiman of Universitas Nasional and the biology students who assisted in the field work.

References

Blackley, G.A. (1968). Vaginal cytology and luteinizing hormone levels in *Pongo pygmaeus* (orang utan). Proc. 2nd Int. Congr. Primat. Soc., vol. 2: 22–29.
Carpenter, C.R. (1938). A survey of wild life conditions in Atjeh, North Sumatra. The Netherlands Committee for International Nature Protection, Amsterdam, Communication no. 12: 1–34.
Davenport, R.K. (1967). The orang-utan in Sabah. Folia primat., 5: 247–263.

Galdikas, B.M.F. (1978). Orangutan adaptation at Tanjung Puting Reserve, Central Borneo. Ph.D. thesis, University of California, Los Angeles.

Harrisson, B. (1960). A study of orang-utan behaviour in semi-wild state, 1956–1960. Sarawak Mus. J., 9: 422–447.

Horr, D.A. (1972). The Borneo orang-utan: population structure and dynamics in relationship to ecology and reproductive strategy. Paper presented at the Meeting of the American Association of Physical Anthropologists, April 1972.

Horr, D.A. (1975). The Borneo orang-utan: population structure and dynamics in relationship to ecology and reproductive strategy. In: L.A Rosenblum (ed.), Primate behaviour, 4. Academic Press, New York.

Leyhausen, P. (1965). The communal organization of solitary mammals. Sym. Zool. Soc. London, 14: 249–263.

Lippert, W. (1974). Beobachtungen zum Schwangerschafts- und Geburtsverhalten beim Orang-utan (*Pongo pygmaeus*) im Tierpark Berlin. Folia primat., 21: 108–134.

MacKinnon, J.R. (1974). The behaviour and ecology of wild orang-utans (*Pongo pygmaeus*). Anim. Behav. 22: 3–74.

Milton, O. (1964). The orang utan and rhinoceros in North Sumatra. Oryx, 7: 177–184.

Nadler, R.D. (1977). Sexual behavior of captive orang utans. Arch. Sex. Behav., 6: 457–475.

Nadler, R.D. (1982). Reproductive behaviour and endocrinology of orang utans. In: L.E.M. de Boer (ed.), The orang utan. Its biology and conservation. Junk, The Hague.

Okano, T. (1965). Preliminary survey of the orang-utan in North Borneo. Primates, 6: 123–128.

Rijksen, H.D. (1978). A fieldstudy on Sumatran orang utans (*Pongo pygmaeus abelii* Lesson 1827); ecology, behaviour and conservation. Mededel. Landbouwhogeschool Wageningen, 78–2: 1–420.

Rodman, P.S. (1973). Population composition and adaptive organization among orang-utans of the Kutai Reserve. In: R.P. Michael and J.H. Crock (eds.), Comparative ecology and behaviour of primates. Academic Press, London.

Schaller, G.B. (1961). The orang-utan in Sarawak. Zoologica, 46: 73–82.

Stott, K. and C.J. Selsor (1961). The orang-utan in North Borneo. Oryx, 6: 39–42.

Washburn, S.L. and D.A. Hamburg (1965). The implications of primate research. In: I. Devore (ed.), Primate behavior. Academic Press, New York.

Yoshiba, K. (1964). Report of the preliminary survey on the orang-utan in North Borneo. Primates, 5: 11–26.

Author's address:
C. Schürmann
Laboratory of Comparative Physiology
University of Utrecht
Jan van Galenstraat 40
3572 LA Utrecht
The Netherlands

284

16. Orang utans as seed dispersers at Tanjung Puting, Central Kalimantan: implications for conservation

Biruté M.F. Galdikas

Introduction

The name 'orang utan' which originates from the Malay words for 'person of the forest' reflects two realities: the relative closeness of this Asian great ape to humankind and the fact that its range is essentially restricted to primary tropical rain forests on Kalimantan (Borneo) and northern Sumatra. Orang utans have long been considered endangered (Harrisson 1961; Milton 1964) and major efforts have been expended to ensure their survival. For instance, national parks have been established in Malaysia and Indonesia to help safeguard remnant wild orang utan populations and rehabilitation programs initiated for ex-captive individuals. Yet the question still lingers: why should precious resources be channelled to save a species of no demonstrable economic importance when thousands, perhaps millions, of humans remain destitute throughout South East Asia as a whole.

Many would simply argue that each species has a right to exist. Since human encroachment on and destruction of orang utan habitat as well as outright killing of orang utans has brought the species to its present imperiled condition, some would agree that the moral responsibility for orang utan survival lies in human hands. However, altruism is not always the most convincing argument; it is better to demonstrate that in the long run the presence of wild orang utans serves a useful function for humankind.

My argument is simple. Undisturbed tropical rain forest functions as a vast storehouse of plant species, only a tiny fraction of which is currently used by humans. Most plant species remain unstudied; thousands remain uncatalogued. As the properties of various rain forest plants are investigated and examined it is certain that products and derivatives from tropical rain forests will play an increasingly crucial role in advancing modern technology and in enhancing the quality of human life (Whitmore 1975). Undoubtedly, more tree species will become marketable for lumber. The forest will probably become a major source of biochemical materials (such as drug plants and pesticides) not easily obtained elsewhere. Future attempts to improve indigenous tropical fruit trees, most of which differ little from ancestral wild strains, will depend on utilizing the variation present in the gene pools of the tropical rain forests (Whitmore 1975).

The orang utan. Its biology and conservation, edited by L.E.M. de Boer
© *1982, Dr W. Junk Publishers, The Hague, ISBN 90 6193 702 7*

The following paper demonstrates that orang utans, a primarily frugivorous species (MacKinnon 1971, 1974; Horr 1972, 1975; Rodman 1973, 1977; Galdikas 1975, 1978; Rijksen 1978), function as important seed dispersal agents for a variety of tropical rain forest plants, some of which are presently, and others which someday may be proved to be, valuable for civilization. Annihilation of orang utans from the great seed storehouse of the tropical rain forest could conceivably lead to the decline or perhaps even extinction of plant species which depend on orang utans for dispersal. While the consequences of orang utan extinction, or the extinction of any species, for that matter, are bound to be profound, such consequences cannot yet be fully understood, since the intricacies of multiple interactions between co-existing plants and animals in tropical rain forests have yet to be unravelled (Whitmore 1975). This paper suggests that orang utans inadvertently play an important role in shaping and even preserving the tropical rain forests which constitute their habitat.

Orang utan effects on forest

The basic units of orang utan populations consist of 1) solitary adult males, 2) lone adult females accompanied by one or two dependent offspring, and 3) independent immatures occasionally in association with their natal unit (MacKinnon 1971, 1974; Horr 1972, 1975; Rodman 1973; Galdikas 1975, 1978; Rijksen 1978). Immatures tend to be more social than mature individuals (Rijksen 1978; Galdikas 1978). However, while the nature of orang utan adaptation to the forest has been the subject of several intensive studies, there has been little cognizance of the fact that orang utans, as the world's largest frugivorous and arboreal species, affect and shape the forests they inhabit. Rijksen (1978) has a brief discussion of orang utan influence on the habitat as does Rodman (1977).

Orang utan effect on the forests probably takes many forms. Simply by moving through the canopy, orang utans may bend and snap numerous branches. During agonistic displays orang utans may break so many branches that once at Tanjung Puting an entire tree was denuded of its limbs. In addition, adult males (occasionally, individuals of other age/sex classes as well) topple snags, which, as they crash, bring down other vegetation and create small openings on the forest floor. (This maintains sites suitable for the growth of some plant species. For instance, some dipterocarp saplings need gaps in the canopy for growth spurts to occur.) The total effect may not be negligible as one habituated adult male at Tanjung Puting pushed over snags at the rate of once every 18 h of observation. Further orang utans daily bend, break and arrange branches to construct nests in a complicated procedure which may require over 20 min.

While feeding, orang utans frequently consume vast quantities of foliage, buds and flowers, while they bend and break branches and twigs to bring food within reach (MacKinnon 1971; Rodman 1977; Rijksen 1978). Although the effects of such orang utan actions have not been systematically studied, they must be considerable. For instance, a Barro Colorado Island site occupied by white-faced monkeys (*Cebus capucinus*) had trees with significantly more branches than a comparable site on the Canal Zone mainland without monkeys (Oppenheimer and Lang 1969), probably due to the removal of terminal buds during feeding, which in turn released lateral buds from apical dominance and resulted in increased branching. Since orang utans are much larger and probably consume more than *Cebus*, orang utans would probably have an even greater impact in this regard. At Ketambe, Rijksen (1978) notes that some climbing Araceae have a remarkably bushy shape; he attributes this to removal of terminal shoots by orang utans.

Further, at Tanjung Puting there was some dramatic evidence of orang utan influence on the forest in the form of two or three dead trees which had been killed by orang utans stripping the bark around the girth of the trunk. This was also noted at Ketambe (Rijksen 1978). However, since tropical rain forest trees seem incredibly resistant to damage of this sort, usually damaged trees exhibited new growth below the bare spots stripped of bark.

While such effects merit further study, probably the most important and far-reaching consequence of orang utan presence in the tropical rain forest consists of their role as seed dispersers. Botanists have argued that angiosperm dominance over much of the earth's land surface resulted from the diversification of birds and mammals, the major seed vectors for flowering plants, during the Cretaceous (Regal 1971). While previously only reptiles, wind and water were available as vectors, consumption of seeds and deposition in feces by birds and later, bats and other mammals must have increased the efficiency of seed dispersal manyfold so that seeds were transported beyond immediate concentrations of seed predators and widely scattered (Regal 1977). Thus, many tropical rain forest fruits have evolved to be eaten upon ripening (Whitmore 1975). Although detailed comparative data with other frugivorous species are lacking, orang utans may well be the most important nonflying mammalian seed dispersers at Tanjung Puting due to their arboreality, large size, long-distance male ranging patterns and ability to span virtually all habitat types within the general region of the Reserve.

Method

The following paper is primarily based on data gathered over a six-month period (January–June 1978) at the Tanjung Puting Reserve, Kalimantan Tengah (Central Indonesian Borneo) during the course of an uninterrupted long-term orang utan study now in its ninth year (1971–1980). In total over 15,000 h of direct observation on target wild orang utans have been collected by a research team headed by the author in a 35 km^2 study area which encompasses lowland dipterocarp forest (63%), peatswamp forest (27%), tropical heath forest (5%) as well as secondary forest, shallow lakes and grass fields (5%).

A large study area was initially chosen so that 1) the home ranges of at least several members of each sex would be partially or totally encompassed, and 2) data on a representative sample of the orang utan population could be gathered enabling interactions between individuals of different age/sex classes to be observed and ecological and other comparisons among classes facilitated. The

287

entire study area is covered with a grid of small transects initially placed 500 m apart running north-south and east-west. The combined length of all transects and trails exceeds 125 km. The transect system is staked every 25 m allowing for the accurate mapping of orang utan daily movements and home ranges.

The principal difficulty in studying orang utans is finding them. Once located, an animal is followed, if at all possible, until it nests for the night. The next morning, an observer(s) arrives at the nest before dawn. The bulk of the data consists of whole days (when the animals are followed continuously from leaving their night nest in the morning to the time they make a nest for the following night). Since only one orang utan (or an adult and her dependent offspring) is usually present, sampling of behavior is automatically limited to the 'focal' individual or target (Altmann 1974). Target individuals have been observed for periods ranging from 1 min to 65 consecutive days. However, usually at least two or three whole days of consecutive observation are obtained before an individual is released from scrutiny.

Since orang utan behavior is slow, it is usually possible to accurately monitor a target individual throughout the day, to log the occurrence of each major activity (resting, feeding, travelling, nesting, displaying, fighting, chasing, allo-grooming and mating) and to record times for bout initiation and termination. With the exception of very infrequent events such as allogrooming, the smallest observation interval utilized is one minute. In addition, the food type eaten (if known), vocalizations uttered or heard, drastic changes in the target's height from the ground (in 5 m intervals), the target's descent to the ground, duration on the ground and the presence of other orang utans, primates or large animals in the vicinity are always noted. Descriptive notes of behaviors involved in activity bouts as well as on postures, locomotory and feeding techniques, self-grooming and playing are made but are not consistent in quality throughout the day due to observer fatigue.

During January–June 1978 general observations as described above were collected by the author assisted by Rod M.C. Brindamour, Universitas Nasional biology students Mohamad Boang and Djoharly Debok and local assistants. In addition, supplementary data were collected in an attempt to assess orang utan impact on the habitat in terms of seed dissemination. First, detailed observations were made of processing modes while orang utans foraged on fruit.

Second, an attempt was made to collect fecal material from the forest floor whenever orang utans under observation (target individuals and others) defecated. Once feces were recovered, they were brought back to camp, weighed, washed through a 1 mm mosquito net wire sieve and dried. Intact seeds were sorted out from the dried material, identified (whenever possible), counted and weighed. It proved extremely difficult to collect complete samples of orang utan defecations since feces frequently scattered over wide areas or fell into swamp puddles. Sometimes no samples could be gathered while in many other cases, only partial samples were collected. Although all observers collaborated in collecting orang utan feces in the field, washing and processing of fecal material was initially conducted by the author and then subsequently by Djoharly Debok.

288

Results

From January through June 1978 various observers accumulated approximately 644 h of direct observation on target orang utans. Orang utans were feeding predominantly on fruit during this period and were observed eating 79 known food types consisting of 50 different fruits, one flower, 13 species of young leaves, seven kinds of barks, termites, four different kinds of other insects, two ground plants and one small vine. Of the 50 fruits, some represented fruits eaten only once during the six months. Detailed observations on 48 fruit species indicated seeds were usually either spit out and/or swallowed whole for 34 fruits (71%) and usually chewed and destroyed in order to extract nutrients for 14 others (29%). However, orang utan modes of consumption were not always consistent. Frequently during a feeding bout seeds of the same species were destroyed as well as spit out and/or swallowed. On some occasions orang utans swallowed seeds which they normally spit out and vice versa. Indeed, on the rare occasion orang utans swallowed intact seeds of fruits whose seeds were normally destroyed.

At Tanjung Puting orang utan dissemination of seeds takes at least four forms: three direct and one indirect. The first form involves consumption of fruit skins and/or flesh with intact seeds swallowed and passed through the digestive tract to be deposited as fecal matter. Mean orang utan day range is 800 m (Galdikas 1978) with a maximum of 4 km observed for adult males. Since orang utans sometimes do not defecate for long periods exceeding several days, an adult male orang utan could conceivably be 8–10 km from a parent tree before expelling its seeds. Further, the frequent scattering of orang utan feces while dropping from the canopy facilitates very local seed dispersal. While the feces of terrestrial mammals were sometimes observed on the forest floor with a clump of seedlings sprouting in a miniature thicket, concentrated orang utan droppings of this sort were rarely encountered.

Second, orang utans frequently chew a mass of fruit flesh and fibre for as long as 15–20 min before spitting out a wad of intact seeds and (sometimes) chewed fibres. Often individuals leave a parent tree while chewing and travel up to 75 m before finally expelling a wad of seeds from their mouths.

Third, orang utans may carry intact fruits some distance in their mouths or, less frequently, hands or feet before consuming the fruits. Fruits observed carried in this manner are usually large or hard, requiring concentrated effort to open. In one case *Mezzettia leptopoda* fruits (whose seeds are consumed) were carried 200 m lodged inside an orang utan's mouth before being consumed in his night nest several hours later. Orang utans also occasionally carried branches with adhering fruits from the parent tree to an adjacent tree where the fruits were consumed and seeds spit out.

Finally, there is at least one indirect method of seed dissemination. While feeding, orang utans frequently drop defective and/or unripe fruits, and an occasional ripe fruit, to the ground. Such litter is frequently eaten by terrestrial mammals such as mouse deer (*Tragulus napu* and *Tragulus javanicus*), barking deer (*Muntiacus muntjak*), sambar deer (*Cervus unicolor*) and pigs (*Sus barbatus*), who otherwise would not have immediate access to the fruit, as well as

Malayan honey bears (*Helarctos malayanus*). Seeds may pass intact through the digestive tracts of these animals. All of the above-mentioned species were observed eating fruit litter under trees where orang utans had been foraging. Casual inspection indicated intact seeds were frequent components of deer and bear feces.

During January–June 1978 one hundred and one defecation bouts were observed. Orang utans defecated up to four times per day although sometimes they did not defecate for several days. Samples from 64 defecations (63% of total) were collected. Due to difficulties in collection it seemed that most samples were partial, not whole. These 64 samples came from 18 individuals (5 adult males, 8 adult females, 3 sub-adult males and 2 adolescent females) and were collected over 34 different calendar days. The largest fecal sample came from an adult male and weighed 353 g. The smallest sample which probably represented only part of the fecal material from one defecation bout was 12 g. The mean weight per sample was 141 g per defecation. Mean feces weights differed for age/sex classes being largest for adult males (190 g) and smallest for adolescent females (72 g).

Virtually all samples (94%) contained at least one intact seed. Only four samples (6%) did not. The maximum number of seeds found in one sample was 2,028 (primarily seeds from *Baccaurea pendula*) with a median of 111 intact seeds per defecation. Most samples (78%) contained intact seeds of two or more different species with a mean of 2.5 seed types per fecal sample.

Intact seeds from the vast majority (70%) of fruit species observed consumed by orang utans during the six months were either recovered from feces (46%) or observed spit out during feeding. Examination of fecal material revealed intact seeds of at least 23 species (Table 1) as well as numerous broken or destroyed seeds of the same and other species. Only one intact seed was recovered from a fruit species (*Xylopia malayana*) whose seeds are normally destroyed during consumption.

Discussion

Over the past decade emphasis on conserving single key species has been replaced with an understanding that examples of whole ecosystems must be preserved intact. This attitude comes, at least partially, from a renewed cognizance of the intricate network of plant and animal relationships which exist in nature. Elimination of one species from the habitat may have myriad effects, sometimes leading to marked decreases in, and even extinction of, associated species. The tree *Calvaria major* which is endemic to Mauritius provides a dramatic example. It is almost extinct because its seeds apparently required passage through the digestive tract of the now extinct dodo, *Raphus cucullatus*, for germination to occur (Temple 1977). Although *Calvaria major* is the only documented case in which germination does not occur without passage of seeds through an animal's gut, in many other cases germination rates are extremely low without such treatment. It is likely that germination of some fruit species at Tanjung Puting may be facilitated by passage through orang utan digestive tracts although direct evidence is lacking. However, Rodman (1977) notes that *Koordersiodendron pinnatum* seeds deposited in orang utan feces at Kutai seemed to germinate more readily than seeds of the same species dropped from the mouths of gibbons or squirrels.

Dispersion of seeds away from the parent tree has at least two benefits. If fruits merely fall under the parent tree they may be subject to concentrations of insect or other predators; seedlings that germinate may be subject to crowding (Whitmore 1975). Although many other animals at Tanjung Puting function as dispersal agents, orang utan impact overall may be very important due to the wide variety of fruit species orang utans consume (approximately 200), the large number of fruits consumed daily (for instance hundreds of *Ryparosa kostermans* or thousands of *Baccaurea pendula* fruits in one feeding bout), the long distances travelled and their ability to traverse virtually all habitat types within the general region of the Reserve.

A comparison can be made with the sole other frugivorous primate, the gibbon (*Hylobates agilis*), sharing inland primary tropical rain forests with orang utans at Tanjung Puting. Orang utans are considerably larger and presumably can consume considerably more. Further, while gibbon units utilize very small territories, orang utan female ranges span 5–6 km^2 while males may range over 40 km^2 or more (Galdikas 1978). Also, while orang utans are primary forest dwellers, they also, unlike gibbons, occasionally venture out into shallow lakes and grassy clearings several hundred meters from the nearest patch of forest as well as cross small rivers through overhanging vegetation or on fallen trees. This is not to negate the importance of gibbons to the maintenance of tropical rain forests. On the contrary, their role as seed dispersers is probably crucial for certain plant species. The remains of one gibbon (which apparently died in a fall) found on the forest floor contained over one hundred seeds from at least six different species, most of which are only rarely eaten by orang utans.

While other investigators have also noted that orang utans disperse seeds, patterns of seed dispersal probably vary from region to region and even from season to season. Over a three year period at Ketambe, Rijksen (1978) examined 96 fecal samples. Only 44% contained intact seeds as opposed to Tanjung Puting where 94% of samples collected over six months did so. A number of factors could be involved. For instance, Rijksen's study involved observations of both ex-captive rehabilitant and wild orang utans. Unfortunately, it is not specified what proportion of fecal matter, if any, at Ketambe originated from re-habilitant orang utans. Since only a portion of rehabilitant orang utan diet consists of wild foods, feces at the Tanjung Puting rehabilitant feeding station (located outside the wild orang utan study area) contain few seeds.

However, probably even more important are differential ecologies at the two sites. While figs account for approximately one third of orang utan diet at Ketambe, figs are a negligible source of food at Tanjung Puting where orang utans utilize a much wider spectrum of foods with none assuming the importance that figs do at Ketambe. Also, the fact that Rijksen's collection spanned three years may be important. At Tanjung Puting there are few seeds in feces during periods when orang utans eat predominantly leaves, bark or flowers. Thus, a collection from Tanjung Puting which spans many seasons probably would contain a lower proportion of feces with seeds than reported here.

There is one other striking difference between Ketambe and Tanjung Puting. Rijksen (1978) reports that orang utans destroy the seeds of *Durio oxleyanus* during processing. He concludes that humans, Malayan honey bears and, possibly, tigers are the main dispersal agents for this wild durian. Observations during nine years indicate this is generally not the case at Tanjung Puting. While orang utans may consume the crisp seeds of very unripe fruits, wild orang utans generally eat the flesh of unripe fruits, discarding mouthfuls of intact stones. Wild orang utans function as dispersal agents for this species because they discard stones up to 50 m away from a mothertree. However, rehabilitant orang utans frequently destroy the stones, especially if fruits are few.

Orang utans in other areas have been depicted as destructive feeders (MacKinnon 1971; Rijksen 1978). At Tanjung Puting orang utans routinely destroy the seeds of approximately 30% of fruit species eaten. In some cases, these seeds are so hard (e.g. *Mezzettia leptopoda* or *Liptocarpus sericobalanus*) that it is unlikely other primates eat them in the mature state. Since many of the seeds destroyed by orang utans represent relatively common tree species, probably other animals function as dispersal agents for these species. A tantalizing hint is provided by J.A.R. Anderson (pers. comm. 1975). He identified two intact *Mezzettia leptopoda* seeds recovered from the stomach contents of a rhinoceros* which had died after years of captivity in a British zoo.

The list of species disseminated by orang utans at Tanjung Puting during the six months reported here includes a hefty number of plants which play a role in the local economy. While not one of these plants is crucial locally, cumulatively their effect is not inconsiderable. Orang utans spread the seeds of at least 13 species with edible fruit, six species of which are occasionally sold in the local market.

Other tree species which orang utans disseminate are useful for lumber, frequently being made into boards, beams, etc. for local construction and occasionally for export. Another five species are utilized for other purposes. For example, *Chrysophyllum lanceolatum* is the wood of choice for axe handles while *Santiria apiculata* is used for machete handles. Other species are used for firewood. 'Getah merah' (*Palaquium gutta*) is the source of an expensive wild rubber so extensively collected for export in previous years that large trunks of this species (felled in the process) seem rare anywhere. Local uses of various plant species which orang utans disseminate are summarized in Table 1.

---

* Rhinoceroses may have been present in the general region of the Tanjung Puting Reserve forty or more years ago but are now extinct.

Fig. 1. Adolescent orang utan female is chewing on *Garcinia* sp. fruit after having moved to adjacent *Macaranga* sp. tree on edge of primary tropical rain forest at Tanjung Puting Reserve. (Photograph Rod Brindamour, © National Geographic Society.)

Table 1. Fruit species disseminated by orang utans at Tanjung Puting, January–June 1978.

Plant family	Scientific name	Local name (Kumai dialect)	Seeds recovered from feces	Seeds discarded from mouth	Local uses of plants
Anacardiaceae	Melanochyla sp.	Rangas		x	Lumber
Annonaceae	Xylopia malayana	Penyeluangan laki	x		Highly regarded as firewood
Burseraceae	Dacryodes sp.	Kembayau	x		Edible fruit
	Santiria apiculata	Semongga	x		Wood used for machete handles; firewood
	Santiria sp.	Poga		x	
	Santiria sp.	Bekacang		x	Lumber
Euphorbiaceae	Baccaurea pendula	Jijantik	x		Edible fruit
	Baccaurea sp.	Merinjihan burung	x		
	Chaetocarpus castanocarpus	Lurangan	x		Made into poles for planting rice
	Chrysophyllum lanceolatum	Pulut	x		Wood used for ax handles
Flacourtiaceae	Ryparosa kostermans	Subang		x	
Gnetaceae (gymnosperm)	Gnetum latifolium[a]	–		x	
Guttiferae	Garcinia sp.	Kemanjing	x?		Edible fruit; sap used to poison knife blades
	Garcinia sp.	Madu-maduan	x?		Edible fruit sold in market
Lauraceae	_[a]	Medang	x		Lumber
Leguminoseae		Kedarah[b]	x		Sap used as medicine for dog's eyes
Meliaceae	Aglaia sp.	Sampa bulan		x	Wood used for light constructions

294

Family	Scientific name	Local name			Uses
	Aphanamixis humilis	Duku hutan	x		Edible fruit; firewood
Moraceae	*Artocarpus* sp.	Kubing		x	Lumber
	Artocarpus sp.	Tewadak air		x	Lumber
Myrtaceae	*Eugenia lineata*	Ubar salim		x	Made into charcoal
Sapindaceae	*Nephelium* sp.	Rambutan hutan	x		Edible fruit sold in market; firewood
Sapotaceae	*Palaquium gutta*	Getah merah		x	Sap collected for export; lumber; edible fruit sold in market
	Palaquium rostratrum	Nyetuh	x		Edible fruit sold in market; lumber
		Meritam	x		Edible fruit sold in market; lumber
Tetrameristaceae	*Tetramerista glabra*	Merang	x		Edible fruit
Tiliaceae	*Microcos* sp.	Belimbing demak	x		Edible fruit
Not known	—[a]	Belalai	x		Edible fruit sold in market
		Ketapi hutan	x		Edible fruit
	—[a]	Rangga mahalang[b]		x	
		Keluing		x	Edible fruit; lumber
	Unidentifiable, only from seeds in feces	A	x		
		B	x		
		C	x		
		D	x		
Totals			23	12	

[a] Liana (rest of species are trees).
[b] Name is in the Pasir Panjang (Dayak) dialect.

Orang utans also affect the two most commercially important peatswamp forest trees when these species fruit. 'Ramin' (*Gonystylus bancanus*) is probably the most extensively taken export wood from peatswamp forests in Kalimantan as a whole. Orang utans occasionally feed on the fruits, discarding seeds from their mouths, thus contributing to the dispersal of this economically important species. 'Jelutong' (*Dyera polyphylla*), which is tapped for the sweet white latex used in chewing gum, also has value as an export timber. Since orang utans destroy the seeds while feeding on 'jelutong' fruits, they probably contribute little to the propagation of this tree species.

Conclusions

Tropical rain forests represent a reservoir of genetic material unequalled elsewhere on this planet (Whitmore 1975). Yet the maintenance of these forests depends almost entirely on wildlife. Birds, mammals and insects are essential for the continued reproduction of the incredible variety of flora contained within tropical rain forests. Most tropical rain forest plants depend on animals both for pollination and for seed dissemination (Dasmann et al. 1973). As the *Calvaria major – Raphus cucullatus* case illustrates, interdependence can be very strong. A single species of insect, bird or mammal can be the sole pollinator or the sole seed disperser for a particular plant. Extinction of one may lead to the extinction of the other.

Unfortunately, we humans have no idea as to what resources we will need in the future. Thirty years ago, Malayan foresters were poison-girdling 'ramin' as a weed tree (Whitmore 1975). Subsequently, the wood became marketable. Now it is probably the most important tree in the peatswamps of Indonesia with thousands depending on its exploitation for their livelihood.

Orang utans are the world's largest frugivores. At Tanjung Puting they have been observed foraging on approximately 200 different fruit species. During one six month period, orang utans acted as dispersal agents, in one way or another, for approximately 70% of the fruit species consumed. Some of the plant species orang utans disperse are important in the local economy. The extinction of orang utans from the tropical rain forests they inhabit probably would have profound consequences on those forests, consequences which as yet cannot be fully evaluated. The dodo bird became extinct 300 years ago but the effect of its extinction did not become apparent until this century with the virtual loss of a majestic and important tree from Mauritius. It probably would take humans centuries to inventory and assess the full consequences of orang utan extinction. But by then it would be centuries too late. The time to act is now.

Acknowledgements

The wild orang utan research on which this paper is based was funded by the Wilkie Brothers Foundation, the L.S.B. Leakey Foundation, the National Geographic Society, the New York Zoological Society, the Van Tienhoven Foundation of Holland, the Herz Foundation, the Jane and Justin Dart

Foundation, the World Wildlife Fund and P.T. Georgia-Pacific, Indonesia. I am very grateful to these organizations and to the people associated with them for making this work possible. Dr L.S.B. Leakey's initial support was instrumental in enabling me to begin this research. Lembaga Ilmu Pengetahuan Indonesia (Indonesian Institute of Sciences) and Perlindungan dan Pelestarian Alam branch of the Forestry Department served as our sponsors in Indonesia. I wish to express my deep gratitude to Mr Soedjarwo (Director-General of the Forestry), Mr Siswojo Sarodja, Mr Prijono Hardjosento, Mr Walman Sinaga, Mr Lukito Daryadi, Mr Goenari and Mr Soegito Tirtomihardjo as well as Mr Widajat Eddypranoto, Mr Rombe, Mr Bahrun Harun and Mr Beringan. We are also very grateful to Dr Bachtiar Rifai (Head, L.I.P.I.), Ms Sjamsiah Achmad, Ms Moertini Atmowidjojo, Mr Napitapulu, Mr J. Bima and Mr Rio Rachwartono. The provincial government of Kalimantan Tengah encouraged and supported this work. We are particularly grateful to Mr R. Sylvanus (former governor), Mr Gara (governor) and to Mr G.T. Binti as well as numerous other provincial officials. I also wish to express my deep gratitude to Mrs Joan Travis, Mrs Nina Sulaiman, Mr Sulaiman Sumitakusuma, Mrs Allene Masters, Mrs Mary Pechanec, Mrs Sandy Johnson, Mrs Olive Kemp, Dr Bernard Campbell, Dr Joseph Birdsell, Dr Rainer Berger, Dr Y.P. Chen, General Hoegeng, General Rachman Masjhur, Dr M. Payne, Dr E. Snider, Mrs Mary Griswold Smith, Ms J. Hess, Dr J. Westermann, Dr Wayne King, Mrs Neva Collins, Mr R. Gilka, Ms Mare Tiido, Mrs Jean Brindamour, my parents Antanas and Filomena Galdikas, my sister Aldona Galdikas-Franz, my brother Al Galdikas, Ms Loretta Brickus Moiel, Dr Moiel, Dr Jane Goodall and Dr Barbara Harrisson for their help and support. I thank my students from the Universitas Nasional: Mr Suharto Djojosudarmo, Mr Jaumat Dulhajah, Mr Sugardjito, Mr Endang Soekara, Mr Barita Manulang, Mr Yatna Supriatna, Mr Dwi Sutanto, Mr Benny Djaya, Mr Natasudradjat Amban, Mr Mahfudz Markaya, Mr Dadang Kusmana, Mr Richard Pattan, Mr Pepen Abdullah, Mr Benny H. Ismunadji, Mr Toto Susilarto, Mr Cok Gede Parthasuniya, Mr M. Mudjiono and Mr Mohamad Boang. In particular I thank Mr Djoharly Debok for helping with the processing of seeds. I also thank Gadjah Mada University Department of Forestry lecturers, Mr Kuswanto and Mr Djuwantoko, and students, Mr Darsono and Mr Harijaka, for their help in the field. Botanical specimens were identified at the Herbarium Bogoriense through the courtesy of Dr Mien Rifai to whom I am very grateful. I am also very indebted to Dr J.A.R. (Robb) Anderson who provided much help and expert advice, and who first suggested that a study of seed dispersal by orang utans would be helpful. I am very grateful to Gary Shapiro for his enthusiastic support and all his help. Finally, I wish to express my deepest gratitude to Rod M.C. Brindamour for his seven and a half years of support and devotion to the orang utan research.

References

Altmann, J. (1974). Observational study of behavior: sampling methods. Behaviour, 49: 227–267.
Dasmann, R.F., J.P. Milton and P.H. Freeman (1973). Ecological principles for economic development. John Wiley and Sons, London.

Galdikas, B.M.F. (1978). Orangutan adaptation at Tanjung Puting Reserve, Central Borneo. Ph.D. thesis, University of California, Los Angeles.

Galdikas-Brindamour, B. (1975). Orangutans, Indonesia's 'People of the Forest'. National Geographic, 148: 444–473.

Harrisson, B. (1961). Orang-utan: What chances of survival? Sarawak Mus. J., 10: 238–261.

Horr, D.A. (1972). The Borneo orang-utan. Borneo Research Bull., 4: 46–50.

Horr, D.A. (1975). The Borneo orang-utan. Population structure and dynamics in relationship to ecology and reproductive strategy. Primate Behavior, 4: 307–323.

MacKinnon, J.R. (1971). The orang-utan in Sabah today. Oryx, 11: 141–191.

MacKinnon, J.R. (1974). The behavior and ecology of wild orangutans (*Pongo pygmaeus*). Animal Behaviour 22: 3–74.

Milton, O. (1964). The orang utan and rhinoceros in North Sumatra. Oryx, 7: 177–184.

Oppenheimer, J.R. and G.E. Lang (1969). *Cebus* monkeys: Effects on branching of gustavia trees. Science, 165: 187–188.

Regal, P.J. (1977). Ecology and evolution of flowering plant dominance. Science, 196: 622–229.

Rijksen, H.D. (1978). A fieldstudy on Sumatran orang utans (*Pongo pygmaeus abelii* Lesson 1827): ecology, behavior and conservation. Meded. Landbouwhegeschool Wageningen, 78–2: 1–420.

Rodman, P.S. (1973). Population composition and adaptive organization among orang-utans. In: R.P. Michael and J.H. Crook (eds.), Comparative ecology and behaviour of primates. Academic Press, London.

Rodman, P.S. (1977). Feeding behaviour of orang-utans of the Kutai Nature Reserve, East Kalimantan. In: T.H. Clutton-Brock (ed.), Primate ecology: studies of feeding and ranging behaviour in lemurs, monkeys and apes. Academic Press, London.

Temple, S.A. (1977). Plant-animal mutualism: coevolution with dodo leads to near extinction of plant. Science, 197: 885–886.

Whitmore, T.C. (1975). Tropical rain forests of the Far East. Clarendon Press, Oxford.

Author's address:
B.M.F. Galdikas
Department of Anthropology
University of New Mexico
Alberquerque, NM
U.S.A.

17. Orang utan conservation in Sumatra, by habitat protection and conservation education

Rosalind J. Aveling

Introduction

For many years the populations of both Bornean (*Pongo pygmaeus pygmaeus*) and Sumatran (*Pongo pygmaeus abelii*) orang utans have been declining in the wild due to a variety of human pressures. In this paper the author will concentrate on the status of the Sumatran sub-species, and the conservation measures being undertaken in Sumatra.

Current estimates of the remaining population of *Pongo pygmaeus abelii* range from 4,500 individuals (Borner 1976), to between 5,000 and 15,000 (Rijksen 1978). Estimates are based on known population densities in extensively studied areas, and the calculated amount of suitable remaining habitat. Marked local variation in the orang utan carrying capacity of different forest areas, and the inaccessibility of much of the terrain, make more precise figures impossible to obtain.

Threats to the survival of this population can be summarised as hunting/trade and habitat destruction. In the early decades of the century, hunting and trade were significant threats. Despite the Indonesian Fauna Protection Ordinance of 1925 (which prohibited killing orang utans), and additional regulations in 1931 and 1932 (which prohibited all forms of trade in, or possession of, orang utans), both hunting and trade flourished. Orang utans (particularly infants obtained by killing nursing females) were openly sold within Indonesia; figures are difficult to obtain but, according to Rijksen and Rijksen-Graatsma (1975), 136 orang utans were sold in 1964, in the Alas Valley alone. However, within the last 20 years, destruction of orang utan habitat — primary tropical rain forest — has reached critical proportions and has become the major threat to the species' survival. A rapidly expanding human population has put more pressure on the remaining forest, and 'slash and burn agriculture' has replaced the environmentally less destructive 'swidden agriculture', resulting in unproductive 'alang-alang' grass wasteland. According to Forestry Department figures (in Borner 1979) this barren land now covers approximately 25% of Sumatra island. In addition, expansion of timber extraction, the use of more modern methods by timber companies, extensive oil exploration, transmigration projects, forest clearance for rubber and oil palm plantations and other aspects of 'development' have contributed to the decline of tropical rain

The orang utan. Its biology and conservation, edited by L.E.M. de Boer
© *1982, Dr W. Junk Publishers, The Hague, ISBN 90 6193 702 7*

forest. Almost the only remaining undisturbed forest left in Sumatra lies within the reserves and even they are not inviolate.

When orang utan rehabilitation centres were established in Sumatra (Ketambe, funded by the Netherlands National Appeal of the World Wildlife Fund, in 1971, and Bohorok, funded by the Frankfurt Zoological Society, in 1973), reliable information on the status of the wild population and the extent of the factors threatening it, was limited. The centres hoped to decrease the thriving trade in orang utans and possibly increase the wild population with successful rehabilitants. Government Nature Conservation Department (PPA, 'Perlindungan dan Pengawetan Alam') personnel were encouraged and helped to confiscate captive animals, and the publicity the centres attracted focused both local and international attention on the conservation problems. However, potential negative aspects of rehabilitation were soon realised — mainly the possibility that disease could be passed via rehabilitants to the wild population. Also, as the extent of habitat destruction became known, the possibility of actually stressing the wild population by increasing numbers within a limited area, became evident. Early in 1979 Ketambe stopped rehabilitation and extended its function as a research station; Bohorok has continued its system of rehabilitation which minimises the risk of disease transference (see Appendix), and extended its function as a centre for conservation education.

Current conservation measures concerning the Sumatran orang utan

Law enforcement

A direct result of the establishment of the rehabilitation centres, and the activities of the scientists running them, has been a decrease in the open trade in captured animals. The protected status of orang utans is now more generally known, and PPA makes efforts to confiscate captive animals. However, the recent influx of new arrivals at Bohorok (Table 1 in the Appendix), which indicates increased efficiency of detection and confiscation by PPA, also illustrates that young orang utans are still being poached. Most originate from isolated timber and oil camps in Aceh Province, and while forest destruction continues the capture of displaced animals will also continue. With international agreements restricting trade in endangered species, the lucrative export market for orang utans, that thrived in recent decades, has dwindled and is probably reduced to a few cases of successful smuggling. The internal market, although reduced with increasing awareness, still exists. While the only 'punishment' for keeping an orang utan is confiscation of the animal if it is detected (often very difficult), there is not much of a deterrent for internal trade. Two animals we discovered during September 1979 had been bought for approximately US$ 125 each; the owners were well-to-do people who could easily absorb the loss confiscation represented. One persistent offender is openly flouting the law, and we are currently trying to instigate legal proceedings through PPA. There is a precedent, as in Kalimantan a Chinese shopkeeper was successfully prosecuted for owning an orang utan and although the fine was nominal, the shame of legal proceedings was considerable.

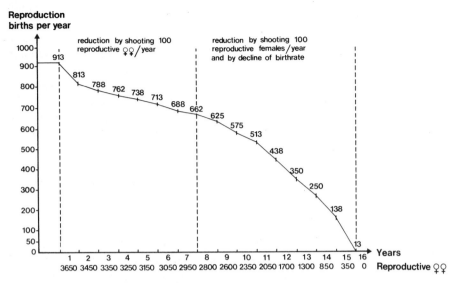

Fig. 1. Calculated decline of reproduction caused by shooting 100 reproductive females per year from a population of 10,000 orang utans (from Frey 1978).

It is important to continue to depress the market, especially for captured infants. Frey (1978) gives a rough model (Fig. 1) to show the calculated effect of killing 100 reproductive females per year from a population of 10,000 orang utans. Making various reasonable assumptions, she shows how the reproductive rate could decline to zero in as little as 16 years. While this is only theoretical, and various factors in the natural situation could influence the curve upwards, it illustrates what a drastic effect even a limited amount of hunting can have on even a comparatively large population. Prosecution of owners, with stiff fines or prison sentences, should be standard practice if the law is to become respected and an effective deterrent. Unfortunately, given the social and administrative structure of Indonesia, this is unlikely to occur, but at least continued vigilance in detecting and confiscating captive animals is essential.

Habitat protection

Almost one third of the distribution range of the Sumatran orang utan falls within the Gunung Leuser group of reserves (Rijksen 1978). The reserve complex (Fig. 2) covers approximately 8,000 km², but is very mountainous with large tracts over 2,000 m elevation. Unfortunately the best orang utan habitat is the lowland forest around the reserve edges, and that is the land most coveted by cultivators and timber companies. Slash and burn agriculture and local timber extraction for sawmills are major problems in the densely-populated Alas Valley. Elsewhere the reserve complex is surrounded by timber concessions, and there is even a timber company operating within the reserve. This latter project is controversial, with political ramifications; a 'Pilot Project for Habitat Improvement' has been set up, under which a commercial timber

301

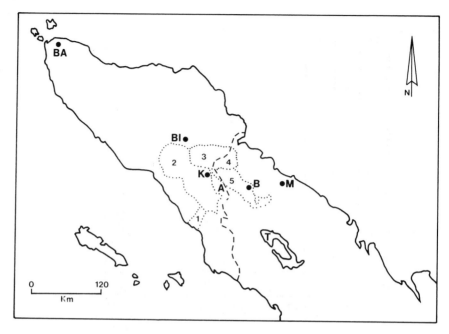

Fig. 2. North part of Sumatra island, showing the Gunung Leuser complex of reserves, and the position of Bohorok and Ketambe orang utan rehabilitation centres. Key:

1   = Kluet Reserve
2   = Gunung Leuser Reserve
3   = Kappi Reserve
4   = Sikundur Reserve
5   = Langkat Reserve
... = reserve boundaries
--- = boundary between Aceh Province
        (to the north-west) and North
        Sumatra Province.

BA = Banda Aceh
Bl   = Blangkejeren
M   = Medan
T   = Lake Toba
A   = Alas Valley
K   = Ketambe Centre
B   = Bohorok Centre

company is being allowed to log selectively 10,000 ha of prime lowland forest and, under the supervision of PPA, schemes for replanting with both hardwoods and fruit trees have been started. The 'Pilot Project' is in the Sikundur Reserve, and covers the only really extensive area of continuous lowland forest remaining in the complex. The justification for the experiment was that the logging could improve the habitat for such animals as elephant and deer, although no detailed surveys had been made.

To obtain data on how logging was affecting different tropical rain forest fauna, we conducted surveys in the Sikundur area (Aveling and Aveling 1979b) comparing the extent and diversity of the fauna in three different habitat types — untouched forest, forest that had been logged up to 10 years previously, and forest that was then being logged. The comparative data for orang utans is represented in Fig. 3; nest counts along survey walks were used to compare orang utan abundance in the different habitat types. Although this method should not be used to derive absolute figures for population densities, it is the only way of assessing orang utan abundance (as direct sightings are rare), and

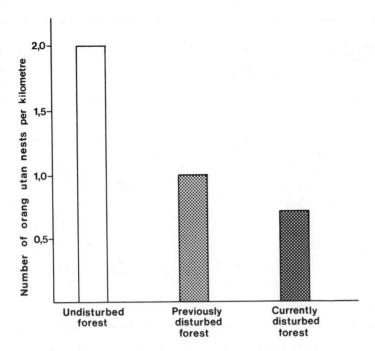

Fig. 3. Surveys in different habitat types, Sikundur Reserve, North Sumatra. Number of orang utan nests seen per km of survey in undisturbed forest (61.0 km), previously disturbed forest (52.5 km) and currently disturbed forest (54.0 km).

can provide useful comparative figures — especially if the same (experienced) observers cover large distances in all compared areas. Despite much better visibility in both disturbed habitats, we still saw twice as many nests per km in undisturbed forest, and the surveys indicated that species such as orang utans, gibbons and hornbills are primary forest species that require untouched habitat to thrive. While it was found that some species, notably artiodactyls, can utilise secondary forest (with its more abundant low-level browse), those species are unlikely to become so rare as that habitat type will always be more abundant. Management measures such as those in Sikundur are therefore directly deleterious to the more endangered species. Despite the political difficulties, both national and international organisations expressed their doubts about the advisability of the scheme and, after a visit to the area, the Minister for Environment and Development Supervision in Indonesia recently cancelled plans for the extension of the Pilot Project to a further 25,000 ha. The situation needs to be monitored, but if the Minister's decision is effected, 25,000 ha of prime orang utan habitat will be reprieved.

At present the Gunung Leuser complex only has 'reserve' status. However, legislation is in progress to found 'national parks', and Gunung Leuser would be one of the first areas to achieve this status. A management plan for the proposed national park has been prepared by the WWF Indonesia Programme (Van Strien 1978) and WWF funds are now being deployed to try and implement the plan, with an expatriate expert cooperating closely with the

Fig. 4. Infant male Sumatran orang utan at Bohorok Rehabilitation Centre.

government departments involved. Both Ketambe and Bohorok centres form important foci in the management of the area — Ketambe as a research station providing the ecological data which should always be a prerequisite of any management measures. Bohorok Centre is a publicly attractive project which directly contributes to orang utan protection via law enforcement, shows the economic potential of an unlogged natural area through local and international tourism, and plays a vital role in conservation education. Centres such as these contribute directly to habitat preservation by protecting the integrity of the reserve in their area, and by the more widespread activity and involvement of the scientists stationed there.

Conservation education

The importance of conservation education for the future survival chances of orang utans and their habitat cannot be overstressed. The need for, and long-term advantages of, conserving natural forest areas must be appreciated by both the rural population and by policy makers if conservation is to succeed. However, conservation education in a developing country like Indonesia is a very difficult field. There is no broad basis of knowledge of the issues involved in conservation, and most people do not appreciate its urgency or importance for the long term. It is difficult to spark the interest of such people, and therein lies the value of a centre like Bohorok. The orang utans and aims of the project attract the interest which can then be developed and channelled.

Rijksen (1978) has stated that visitor programmes and rehabilitation of orang utans are incompatible, as even the occasional attention of visitors was a considerable set-back to the progress of rehabilitants at Ketambe. Certainly, if visitors and rehabilitants are allowed to interact this can occur (and the possibility of transferring disease is increased). However, experience at Bohorok has shown that if the rehabilitation programme is set up and managed in such a way that visitors and orang utans are never in contact — with visitors restricted to certain times, and controlled during their visit to the centre — their presence does not harm the progress of rehabilitants. Under these circumstances it is possible to capitalise on the tremendous interest that the project attracts, and make it a really useful conservation education tool.

For the last three years Bohorok Centre has received about 5,000 visitors annually, of which 80% are Indonesian nationals. A maximum of 50 visitors is allowed across to the centre (a boat crossing over the natural barrier of the Bohorok River facilitates control), and accompanied into the forest to watch the orang utans at each feeding time. Visitors include school and scout groups, villagers, urban dwellers, army and police personnel, and officials from all levels of local to national government. We have established a photographic exhibition (with accompanying leaflets) for visitors to the centre, which outlines the concepts of rain forest conservation and the plight of endangered fauna like the orang utan. Experience has shown that the exhibition is most effective if visitors are accompanied around it, so interest and questions are stimulated. The best way of getting through to people about conservation is by personal discussion;

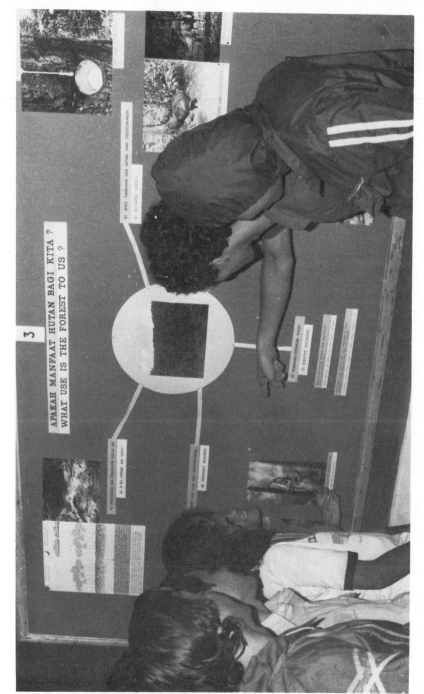

Fig. 5. Indonesian students looking at the photographic exhibition on conservation of tropical rain forest at Bohorok Centre.

Fig. 6. Indonesian students visiting Bohorok Centre cross the river by dug-out canoe.

talking with visitors while they are watching orang utans at the forest feeding site is probably the most effective form of conservation education, judged in terms of interest shown and interchanges stimulated.

To reach a wider audience, audio-visual presentations — preferably with a live commentary — are useful, and Bohorok Centre now has projection facilities and basic slide shows. Regina Frey, while project officer for a WWF Conservation Education Project (No. 1513), set up a mobile audio/visual unit run by two Indonesian students who travelled around towns and villages of North Sumatra and Aceh Province giving slide shows on conservation topics; unfortunately the unit has stopped operating due to lack of funds and guidance.

There is scope to further develop the educational opportunity offered by Bohorok Centre. Although 50 visitors per feeding time is the maximum the centre can absorb (without disrupting the rehabilitation project that the visitors come to see), on many days more visitors have to be turned away. Such people could be catered for by developing facilities in the village of Bukit Lawang near the centre. We have proposed the establishment of a 'Visitors Centre' there (Aveling and Aveling 1979a), with extended educational facilities and permanent personnel. This is only at the planning stage, and international support in terms of funds and expertise would be invaluable; as an initial step, the 'mobile unit' that has had experience in rural educational work could be based at Bukit Lawang, develop the facilities there, and extend contact with schools and other organisations.

At a more general level, the recently established 'Green Indonesia Foundation' ('Yayasan Indonesia Hijau', Tromolpos 3572, Jakarta, Indonesia) is trying to stimulate a wider understanding of conservation at all levels of Indonesian society. It has started to attract interest and support within the country but is in need of financial assistance.

Future prospects

One conflict that could be resolved is that which often occurs between foreign aid programmes for developing countries, and conservation programmes. In Sumatra a road betterment scheme, financed by US AID, will soon bisect the Gunung Leuser complex of reserves; that could have the positive result of making tourism more feasible, providing an income for the proposed national park, but it is more likely to encourage the spread of human activity (logging and agriculture) outwards from the road, further disrupting the reserve. Potential conflicts which harm both aid efforts could be reduced by closer international cooperation and exchange of information at the planning stage of all aid projects. There are glimmerings of ecological responsibility in organisations like the World Bank, but the global situation would obviously improve if all concerned individuals and organisations put pressure on all aid-giving and receiving governments to ensure that cooperation between development aid programmes and conservation aid programmes was standard practice.

In the 'Global Strategy for Primate Conservation' produced by the IUCN/SSC (International Union for Conservation of Nature and Natural

Resources, Survival Service Commission) Primate Specialist Group (Mittermeier 1978), protection of primate *habitat*, and *conservation education* are stressed as the two main priorities for international attention and support. Of course the means of achieving these goals varies with the different situations in different countries; in Sumatra the rehabilitation centres have provided a basis for activity in both fields. Because of doubts about the advisability of rehabilitation *per se*, Rijksen and Rijksen-Graatsma (1979) have suggested that rehabilitation projects within the present range of wild orang utans should be discontinued. It would certainly be ideal if such projects could be moved outside the range of the wild population, but in actuality it is unrealistic. PPA has no intention of stopping rehabilitation at Bohorok, which is a too attractive and potentially lucrative project for the province. (The local government in Aceh Province, concerned with the loss of Ketambe as a tourist attraction, has already expressed the intention of starting another centre.) In that situation it is practical to concentrate on and expand the positive contribution a centre like Bohorok makes to conservation. Early in 1980 the management of Bohorok had become the total responsibility of PPA — in terms of both funds and personnel. Maintenance of the present controls to minimise the risks of disease transfer (see Appendix) is vital and will become the task of the new managers. Scope for international help lies in the vital area of conservation education as outlined above; also in financing the search for, and establishment of, a suitable region outside the present range of the wild orang utan population which could be repopulated with rehabilitants (once they had been prepared for forest life at Bohorok) and translocated wild orang utans (see Appendix). If a suitable area for repopulation can be found, it should not become another rehabilitation centre, as the orang utans released there would either be independent re-habilitants of suitable age or already totally wild. They might initially have difficulty in locating food sources in a new area, and a camp could be established there for scientists to monitor the experiment, but visitors and tourism should not be encouraged.

I would like to stress that despite the positive contribution that rehabilitation centres can make to conservation, there are already enough of them to fulfil the need throughout the orang utan's range, and it would not be advantageous to establish more. Ketambe and Tanjung Puting are undertaking long-term research on wild Sumatran and Bornean orang utans and acting as field training centres for Indonesian student scientists. Bohorok is an accessible centre in Sumatra for conservation education activities — a function now also being developed at Sepilok in Borneo. Establishing more centres for rehabilitation and tourism alone, even in areas where no wild orang utans occur, is not necessary in the present situation, and might divert much needed funds from the urgent priority of protecting large enough areas of untouched habitat.

Sufficient habitat for the continued survival of a healthy and sufficiently large population of Sumatran orang utans would be available if the integrity of the Gunung Leuser complex of reserves could be assured. WWF efforts, chanelled through the WWF Indonesia Programme, are working towards this goal as described above. *If* the Gunung Leuser complex becomes a national park, *if* PPA continues to receive advice and aid to help resolve the conflict of interests that threaten the park, and *if* the government and people can be convinced of the long-term benefit to the country through conservation, compared to the short-term gain from the present pattern of development, then the Sumatran orang utan stands a good chance of survival in the wild. The potential benefit to man of conserving a viable population of any species in the habitat in which it has evolved and continues to evolve, is immeasurably greater than that of continuing the existence of the species only through a genetically limited captive population. With the orang utan, captive breeding for re-release is not a realistic proposition, as it is the remaining *habitat* that is limiting the wild population more than hunting pressure. Thus, while it is necessary to put some funds and effort into captive breeding (in order that the world zoo population remains self-sustaining), most funds and effort should be directed towards *conservation of the wild populations* while there is still a chance of success.

Appendix

Bohorok orang utan rehabilitation centre — methods and results

Between July 1973 when the centre was set up and the time of writing (September, 1979), 98 orang utans have been received at Bohorok (age/sex categories are given in Table 1; the sources [original owners] are given in Table 2). In many cases information concerning captive animals was obtained by the project directors and confiscation achieved through the mediation of PPA. Often animals were owned by high officials (police, army) which made confiscation more difficult for the relatively low status PPA; such people could usually be persuaded to surrender their orang utans for rehabilitation, and would sometimes visit the centre.

All new arrivals at Bohorok are put into quarantine for at least one month and often longer. Blood and stool samples are checked at laboratories in Medan. Routine worm cures are given and specific treatment for any disease that is, or becomes, evident. All orang utans routinely receive B.C.G. vaccination; while this does not eliminate the risk of transferring tuberculosis (Rijksen 1978), we feel it is a worthwhile precaution in a project that incorporates visitors. Proximity to the city of Medan ($2\frac{1}{2}$ h by road), and cooperation from the University Medical Faculty, the General Hospital (X-ray facilities) and various doctors, make invaluable diagnostic help and treatment available for sick orang utans. Quarantined orang utans receive a varied diet with vitamin supplements, and are provided with leafy branches twice a day; almost invariably they start building nests with the branches while in quarantine.

The next stage of rehabilitation is release of the orang utan at a forest feeding site about 15 min walk away from the centre. There all free-living rehabilitants are fed twice a day with a monotonous diet of bananas, and milk for the young ones. A feeding platform about 3 m up in the trees is used, to which the orang utans can gain access by arboreal pathways. The site of the platform is changed every few months so that young trees and saplings broken by the rehabilitants

310

Fig. 7. Sub-adult male Sumatran orang utan at Bohorok Rehabilitation Centre.

can re-generate. If a site is used for too long, the local environment of the platform is destroyed and the orang utans encouraged down to the ground. Provisioning is essential while the orang utans become used to the forest environment, and is necessary anyway while the animals stay in the vicinity as an artificially high orang utan density is created. However, the animals do utilise wild foodstuffs, and are encouraged to seek the variety of natural foods by the deliberately monotonous nature of that provided. The length of this stage varies with the age of the individual, character, duration of prior captivity, and other factors. Typically, adult females seem to be the most independent and often leave the area of their own accord soon after release. Young animals tend not to leave until the time when they would naturally become independent in the wild (4–7 years). Orang utans that arrive as infants go through a slightly different process; initially they are cared for very much as human babies, with bottle feeding and body contact. They then play and start to climb in trees near the centre buildings; the next step is being taken up to the feeding site to mix with the other orang utans at each feeding time. Later they are left out between the

Table 1. Age-sex classes of all orang utans received at Bohorok Orang Utan Rehabilitation Centre between July 1973 and September 1979.

	1973 ♂	1973 ♀	1974 ♂	1974 ♀	1975 ♂	1975 ♀	1976 ♂	1976 ♀	1977 ♂	1977 ♀	1978 ♂	1978 ♀	1979 ♂	1979 ♀	
Infant (0–2½ yr)	2	1	6	0	2	1	1	0	3	0	1	2	4	3	
Juvenile (2½–7 yr)	1	4	9	4	3	6	5	3	5	2	0	1	6	3	
Adolescent (7–10 yr)	0	0	2	3	1	0	0	0	0	0	0	1	0	2	
Adult female (8+ yrs)		0		3		2		1		3		0		1	
Sub-adult male (10–15 yr)	0		1		0		0		0		0		0		
Adult male (15+ yr)	0		0		0		0		0		0		0		
Totals ♂	3		18		6		6		8		1		10		52
Totals ♀		5		10		9		4		5		4		9	46

Table 2. Sources of all orang utans received at Bohorok Orang Utan Rehabilitation Centre between July 1973 and September 1979.

	1973	1974	1975	1976	1977	1978	1979	Totals*
Army/police	0	3	6	1	4	0	1	15
Other government officials	1	3	2	1	1	1	1	10
Plantation officials	0	5	1	0	3	1	0	10
Timber and oil concessions	1	5	2	7	1	0	5	21
Other private owners	4	8	2	0	4	0	2	20
Other (confiscation from dealers, villagers, etc.)	2	4	0	0	0	3	2	11
Ketambe Centre	0	0	0	0	0	0	6	6
Zoos	0	0	0	1	0	0	2	3

* +2 orang utans translocated from concession into reserve in 1975.

312

morning and afternoon feeds, but still caged at night. Only when they start making night nests are they left out at night also. If two young ones of similar age are received at the centre, they are kept together throughout this process; in many cases bonds develop which seem to help both individuals towards independence.

Given the correct circumstances and the opportunity, most orang utans seem to rehabilitate fairly easily in terms of being able to survive on their own in the forest, or 'ecological rehabilitation'. Regarding 'social rehabilitation' — integrating into the wild community and showing appropriate reactions to other orang utans — there is less certainty. The rehabilitants do have experience of other orang utans while they are free-living but still receiving food; the rehabilitant 'community' contains individuals of different ages and sex with whom they can interact. Concerning interactions with wild orang utans we have seen a range of appropriate and inappropriate responses; rehabilitant adult females have been seen mating with wild adult and sub-adult males. While most juvenile or adolescent male rehabilitants retreated from the presence of a wild adult male, two individuals did not and were bitten. The only active 'teaching' given by project personnel is to chase the animals back up into the trees if they come down to the ground; orang utans that have been in captivity have often lost their natural caution towards ground walking and this has to be re-established due to the presence of ground predators in the forest. Of the 98 orang utans that have been, or are going, through this process at Bohorok, the only obvious failures are a few individuals that have become too neurotic due to the conditions of their captivity prior to confiscation.

The final stage of rehabilitation at Bohorok occurs when the individual is felt to be capable of fending for itself. It is then weighed and supplementary food is stopped; many animals then leave the area of their own accord. Others stay around, and if after one month a weight check and examination indicate that they are still healthy, such animals have previously been taken about 20 km further into the reserve by helicopter and released there. Between 1974 and 1977, 25 orang utans were released in that area. Such animals were often followed for a few days, but the nature of the terrain (steep-sided valleys and ridges intersected by rivers and streams) made longer follow-up studies prohibitively difficult: the orang utans could move easily through the canopy and easily evade followers on the ground. Radio-tracking could facilitate a long-term assessment of the rehabilitants' success; however, this would require a powerful enough transmitter to be attached to the orang utan to enable location by air, but which would not itself impede the animal's chances of survival in the arboreal environment of dense tropical rainforest. There is other evidence that the released animals can survive and thrive as at least two have been seen several months after release from the helicopter landing site, in apparently healthy condition. Of the animals that left the centre of their own accord, at least seven have been seen long afterwards, sometimes in the company of wild orang utans.

Since September 1977 supplementary food has been continued for all free-living rehabilitants, and they have not been encouraged to leave the forest in the vicinity of the centre. Meanwhile attempts have been made to find a suitable protected area that is isolated from the present wild population of orang utans, which could be used to reintroduce animals that have been through the stages of rehabilitation at Bohorok. None of the areas we have looked at so far as possible release sites (WWF/IUCN Project 1589 Progress Reports 13–18) meet all requirements regarding size, habitat structure, isolation from agricultural areas, etc. However, if a suitable area could be found and established, it would take the positive step of extending the orang utan's range without endangering the present wild population through the factors mentioned in the introduction. The area could also be used for the transference of wild orang utans from doomed forest patches surrounded by timber concessions; already such 'trans-location' has been tried at Bohorok, where two orang utans were caught from a concession area and transferred into the Langkat Reserve.

From the experience gained at Bohorok, the following are some of the important criteria for successful orang utan rehabilitation:

1. *Physical health.* All orang utans should be thoroughly screened and treated to try to ensure that they are not carrying any of the many diseases they can pick up in captivity, and should only be released when felt to be physically fit.

2. *Mental health.* Orang utans that show neuroses, or are very young, should be given considerable attention and physical contact preferably with one individual 'substitute mother', and then encouraged to transfer the need for companionship to other orang utans as soon as possible.

3. *The site and set-up of the rehabilitation centre* is crucial. Free-living orang utans should be prevented by natural barriers from gaining access to agricultural areas. The feeding platform should be off the ground and accessible through the trees; it must be far enough away from the centre buildings to discourage orang utans from returning there (any that do return to the buildings should never get food there, but be taken back at once, or caged and starved for one or two days, then fed at the forest feeding site). The site of the platform should be changed frequently to discourage local destruction of the forest which encourages ground walking by the orang utans.

4. *Human contact* once the orang utans are released in the forest should be minimal. Project personnel who feed the free-living animals should have regular medical checks, and care should be taken to ensure that food and milk provided for the orang utans is hygienically prepared. Visitors to the centre must be prevented from having contact with the orang utans. This can be achieved by a physical barrier between visitors and the feeding platform, and by always accompanying visitors to prevent interactions. Visiting times must be restricted to the short periods when the semi-wild orang utans come for food, so that they spend most of the day without human disturbance.

Acknowledgements

I would like to thank all those whose support made the Bohorok project possible, in particular the Indonesian Directorate of Nature Conservation (PPA), Dr. R. Faust and the Frankfurt Zoological Society.

References

Aveling, C. and R.J. Aveling (1979a). Bohorok orang utan rehabilitation centre: proposal for development. February 1979. Report to Indonesian Nature Conservation Department, Frankfurt Zoological Society and World Wildlife Fund. Mimeo.

Aveling, C. and R.J. Aveling (1979b). Surveys in Sikundur reserve: the effects of logging on the fauna of lowland tropical rainforest. Internal Report to Indonesian Nature Conservation Department and the World Wildlife Fund Indonesia Programme. Mimeo.

Borner, M. (1976). Sumatra's orang utans. Oryx, 13: 290–293.

Borner, M. (1979). A field study of the Sumatran rhinoceros *Dicerorhinus sumatrensis* Fischer 1814. Ecology, behaviour and conservation situation in Sumatra. Ph.D. Thesis, University of Basel, Switzerland.

Frey, R. (1978). Management of orangutans. In: Wildlife management in Southeast Asia. Biotrop. Spec. Publ. 8.

Rijksen, H.D. (1978). A fieldstudy on Sumatran orang utans (*Pongo pygmaeus abelii* Lesson 1827); Ecology, behaviour and conservation. Mededel. Landbouwhogeschool, Wageningen, 78–2: 1–420.

Rijksen, H.D. and A.G. Rijksen-Graatsma (1975). Orang utan rescue work in North Sumatra. Oryx, 13: 63–73.

Rijksen, H.D. and A.G. Rijksen-Graatsma (1979). Rehabilitation, a new approach is needed. Tigerpaper, 6 (1): 16–18.

Mittermeier, R.A. (1978). A global strategy for primate conservation. Report of the IUCN Survival Service Commission Primate Specialist Group.

Strien, N. van (1979). Management plan for the proposed Gunung Leuser National Park. Report prepared by N. van Strien/WWF Indonesia Programme for the Directorate of Nature Conservation, Bogor. Mimeo.

WWF/IUCN Project 1589 — Progress reports numbers 13–18, submitted by C. Aveling and R.J. Aveling, 1977–1979. Mimeo.

Addendum

Since this paper was submitted in 1979, the 'Gunung Leuser Group of Reserves' has become the 'Gunung Leuser National Park', one of the first National Parks to be gazetted in Indonesia. Early in 1980, PPA took over full responsibility for both management and funding of Bohorok Centre, which has since been under the direction of Suharto Djojosudharmo (Aveling 1980). One of the rangers at Bohorok, James Sipayung, was honoured as a 'Hero of the Environment' by the Republic of Indonesia, on World Environment day in 1981.

Some conservation education work has been undertaken, particularly in the Alas valley, but further funding for this and for educational development at Bohorok is still required (Aveling and Mitchell 1982). Protection of remaining habitat has proved increasingly difficult, although the benefits in human terms of doing so have been dramatically illustrated; forest clearance on steep slopes in the Alas valley led to erosion and floods which caused the deaths of several people in May 1981 (Robertson and Soetrisno 1982).

Aveling, R.J. (1980). Helping orangs and people. IUCN Bulletin New Series II, 9/10: 91.

Aveling, R.J. and Mitchell, A.H. (1982). Is rehabilitating orang utans worth while? Oryx, 16(3): 263–271.

Robertson, J.M.Y. and Soetrisno, B.R., (1982). Logging on slopes kills. Oryx, 16(3): 229–230.

Author's address:
R.J. Aveling
Frankfurt Zoological Society
Alfred-Brehm Platz 16
D-6000 Frankfurt am Main
F.R.G.

18.   How to save the mysterious 'man of the rain forest'?

Herman D. Rijksen

For one reason or another the orang utan (*Simia satyrus*) seems to conjure up mystery: mystery as to his common name, his Latin name, his distribution, his population density and size, his social organisation, his reproductive strategy, and, last but not least, his prospects of survival and the measures to possibly ensure his survival. In this presentation I intend to clear some of this confusion, as it is my belief that the conservation of this species is not served with inadequate information.

To begin with the name, there is no doubt that the Malay translation of 'man of the forest' — *orang hutan* — has never been used locally to denote this ape. Most likely it is a straight concoction by the interpreter of Jacob de Bondt (Bontius), eagerly seeking some acceptable translation of one of the local Bornean names *kehiau, keau, kagiu, kayies, kayas*, or *maias*. Unfortunately Bontius's field notes, posthumously published 27 years after his death in 1631, and in particular the illustration of his '*orang outang*', made the scientific world doubt whether De Bondt had indeed referred to the red-haired ape of Borneo, or to a local woman suffering from some extremely rare, if not hypothetical form of hypertrichosis (Visser 1975; Napier and Napier 1967; Van Bemmel 1968; Jones 1968). Subsequently modern taxonomy easily bypassed the first accurate description of the Sumatran orang utan by Nicolaas Tulp (Tulpius) in 1641, an error possibly caused by ignorance of the fact that the district of Angkola (or Angola) was to be found in North Sumatra — southwest of Lake Toba — and further ignorance of the fact that the Southwest African county of Ndola, later Angola, most likely never harboured chimpanzees (Purchas 1613), the discovery of which ape species was thus wrongly attributed to Tulp.

Whether Linnaeus knew all these facts is uncertain, but he flatly ignored the name his own student, C.E. Hoppius, had given to the red-haired ape (namely *Simia pygmaeus*; Hoppius 1763) in a thesis in 1760, and used the name *Simia satyrus* instead, as a clear homage to Tulp, whose *Homo sylvestris* had been described in a chapter bearing the suggestive title *Satyrus Indicus* (Tulpius 1641). That the man of the forest eventually acquired a corrupted, old Congolese name for the gorilla (Purchas 1613) — *mpungu* — as his generic name seems to ridicule the taxonomy of the anthropoids. For this reason I suggest that we restore the orang utan's old, most correct name *Simia satyrus* Linnaeus 1766 (Linnaeus 1766), with the subspecific addition *S.s.sylvestris* Tulpius 1641

The orang utan. Its biology and conservation, edited by L.E.M. de Boer
© *1982, Dr W. Junk Publishers, The Hague, ISBN 90 6193 702 7*

for the Sumatran, and possibly *S.s.borneensis* Bontius 1658 for the Bornean subspecies; after all the name *satyrus* fits the man of the forest best.

As to the orang utan's distribution range the estimates given in most authoritive references (Groves 1971) generally under-rate the range-size of the Sumatran, and over-rate the range-size of the Bornean subspecies. The former error appears to stem from the hurried and incomplete inquiry results published by Carpenter in 1938, apparently overlooking the very accurate range description by Van Heurn, which was published in the same series by the Netherlands Committee for International Nature Protection only three years earlier (Van Heurn 1935). The over-estimates of the range of the Bornean subspecies may well be due to the general obscurity of the vast, inaccessible Bornean region. Being an animal of the natural non-climax rain forest habitats, the orang utan — like early man — formerly occurred mainly in a wide belt along rivers, streams and swamps of the Sundaland drainage system (Rijksen 1978a), extending into the hill country and lower escarpments deeper inland, but he only sparingly — if at all — inhabited the extensive, flat lowland climax dipterocarp forests, just as the ape is rare at altitudes over 2,000 m. Currently most of the lowland river-valley forest habitats and swamp forests have become occupied by people, and the concomitant widespread forest destruction has pushed the remaining men of the forest deeper inland, and into less favourable habitats.

The present distribution of the Sumatran orang utan covers approximately 30,000 km², of which some two thirds comprise either unsuitable or less suitable habitat in the form of mountain escarpments and summits over 2,000 m altitude. This figure disregards possible relic populations in areas south of Lake Toba (Rudin 1935; Wilson and Wilson 1973; Rijksen 1978a); reliable information suggests that orang utans may still occasionally be encountered as far south as the Wai-hitam forest complex near the source of the river Komering, and the Semangus forest complex near Palembang, both in South Sumatran Province (C. Tantra personal communication 1979). Investigations on the spot are necessary to further elaborate on this information.

Gathering from scant reports and recent landuse maps, the distribution range of the Bornean orang utan covers an area some ten times larger than the range of the Sumatran subspecies. Of this area some two thirds may also be either unsuitable or less suitable habitat due to climax conditions, mountain escarpments over 2,000 m altitude and heavy predation by some indigenous tribal communities (Schlegel and Müller 1844; Hagen 1890; Van der Valk 1941; De Silva 1971; Medway 1976; King 1978). The range (see Fig. 2) excludes the forest complex between the headwaters of the river Barito and river Mahakam, where orang utans have never been recorded (Westermann 1938); most likely they have been eradicated by local hunters in that area before historical times.

Schlegel and Müller in their excellent monograph published in 1844 stated that orang utans are rarely met, a statement assented to by most naturalists and

Fig. 1 (opposite page). The man of the rain forest (*Pongo pygmaeus abelii*, Gunung Leuser Reserve). There is no other place where this ape can develop and exploit his intellectual capacity.

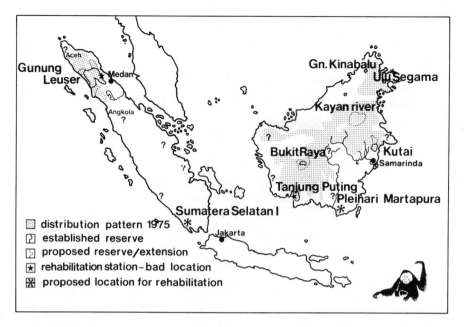

Fig. 2. Distribution range and current conservation situation (1979) of the orang utan (?: distribution unconfirmed).

collectors until the 1960s. Yet Wallace was able to shoot more than nine orang utan specimens in one location in Sarawak during May and June 1855, not counting the apes which appeared too difficult to collect after having been fatally wounded (Wallace 1869).

During the early 1960s surveys by naturalists in Borneo and Sumatra only confirmed the concept of orang utan rareness (Harrisson 1960; Schaller 1961; Milton 1964; Davenport 1967). Obviously too short to be effective, these surveys relied on the inaccurate method of counting nests along transects in order to assess population densities. The density estimates yielded by means of these nest counts were in the order of less than one orang utan per km². These results eventually led to an estimate of 4,000 individuals for the total world population of red haired apes (Harrisson 1961, 1965; Simon 1966). Although totally inaccurate, this figure fortunately elevated the orang utan into the ranks of highly endangered species and set off measures for its protection including a world-wide ban on trade.

Unfortunately it is usually necessary to exaggerate the endangered status of a species or ecosystem in order to raise sufficient awareness and support for urgent conservation measures to become sanctioned. It may therefore perhaps be understandable that the first biologist to mention a density of 1.5 orang utan individuals per km² for the Ulu Segama area in Sabah (MacKinnon 1971), met with aggressive disbelief from conservationists. Nevertheless the subsequent field studies in Borneo and Sumatra revealed that the density of orang utans per km² of rain forest may, in some places, be much higher than previously thought, indeed reaching densities similar to those reported for the more gregarious

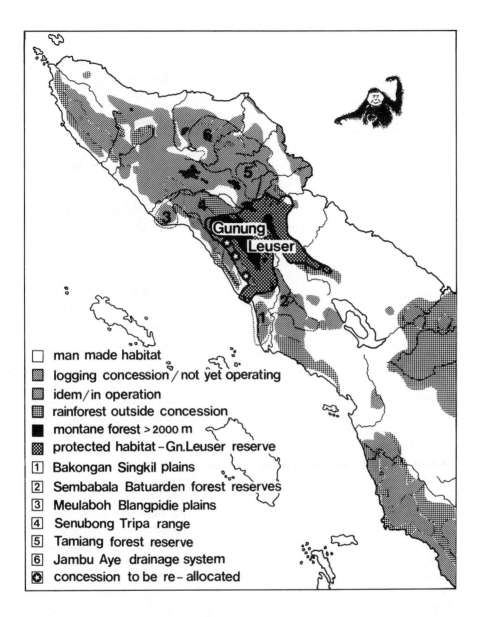

□ man made habitat
▨ logging concession / not yet operating
▧ idem / in operation
▦ rainforest outside concession
■ montane forest > 2000 m
▨ protected habitat – Gn.Leuser reserve
1 Bakongan Singkil plains
2 Sembabala Batuarden forest reserves
3 Meulaboh Blangpidie plains
4 Senubong Tripa range
5 Tamiang forest reserve
6 Jambu Aye drainage system
✪ concession to be re-allocated

Fig. 3. Possible extensions (1–6) for the Gunung Leuser Reserve within the present distribution range of the Sumatran orang utan (currently these extensions are leased as timber concessions):
  – to the south (1,2): the contiguous forest area of the Bakongan-Singkil plain (as was suggested in the original 1928 proposal for the establishment of the Gunung Leuser Reserve), and the Sembabala Barat, Dolok Sembilin, Singkil Utara and Batuardan forest reserves;
  – to the north-west (3,4): the contiguous forest area into the Meulaboh-Blangpidie plain (also originally proposed in 1928), and the Senubong Range, annex drainage system of the River Tripa until the Gunung Abong-abong;
  – to the north-east (5,6): the contiguous forest area into the Segama range, annex Simpang Kanan drainage system (the former Tamiang Forest Reserve), annex Jambu Aye drainage system as far as Gunung Geureudong.
It is noteworthy that most of these forest regions comprise water catchment and drainage systems of *vital importance* for the extensive agricultures in the coastal areas of Aceh.

chimpanzee in East African regions (Wrangham 1977). The Ketambe study area in North Sumatra harboured an average density of 5 orang utans per km^2 during the period 1971–1974, and both the Kutai area in East Kalimantan (Rodman 1973a), and the Tanjung Puting area in southwest Kalimantan (Galdikas 1978) harboured an average density of 3 individuals per km^2. It is significant that the study areas mentioned largely represent mixed forest communities close to large rivers or swamps. Consequently, if we consider the entire distribution range, comprising a variety of forest communities, it is inappropriate to use the high population densities in an assessment of the total population size. Instead we may use a density estimate of 0.5 to 1.5 individuals per km^2 (MacKinnon 1973; Rijksen 1978a). Hence the total Sumatran population may be estimated at between 5,000 and 15,000 individuals, while the Bornean population may comprise six times as many individuals. These figures need not stir up enthusiastic greed in ape-consuming institutions for, as we shall see shortly, the conservation situation for the man of the forest is as grave as ever imagined. Nor do these estimates need to generate defensive concern among conservationists who desperately try to adhere to consistently low population size estimates based on inaccurate data (Borner 1976); nature conservation in Indonesia — where the vast majority of orang utans live — is not served with exaggerations of any kind.

'The orang utan is a solitary animal' is a statement which engenders the idea of a loner or hermit, and indeed it induced biologists to further speculate on the 'nomadic' nature of this ape, whose 'daily and seasonal movements are largely sporadic' (Groves 1971), and whose reproductive success is entirely dependent on 'chance meetings' (MacKinnon 1974). Yet the short life description by Schlegel and Müller (1844) already indicated that the social organisation of the man of the forest is considerably more complex than the term 'solitary' supposes. The social organisation is based on long-term social relationships among individuals who live in large, mutually overlapping home-ranges. In fact, the orang utan's social structure stands out amongst the anthropoid structures only in a relative paucity of daily social contacts. Adult males are generally antagonistic — a trait shared with the lesser apes, and to some degree with the gorilla, the chimpanzee and man; adult females have some friendly relationships, and show mutual indifference in the presence of an adult male — a trait shared with the gorilla and the chimpanzee; and independent pre-adult individuals show a comparatively high degree of sociability, comprising steady friendly relationships. The greatest differences in the social structures of the anthropoids seem to stem from differences in the social attitudes of the males, rather than of the females — as in the cercopithecoids. Orang utan males at the summit of their social status, consider every other male as an 'out-group' conspecific, whereas in the social attitudes of both the gorilla, the chimpanzee and the human male, varying numbers of other males may be included in an 'in-group' concept, to such extent that in chimpanzees and some archaic human forms one may speak of closed male-communities (Bygott 1974; Rijksen 1978b). It is remarkable that these differences in attitude are reflected in the degree of sexual dimorphism of the great apes, relating their evolution to sexual selection.

Independent pre-adult orang utans may regularly engage in social groups for any length of time. In such groups the individuals establish relationships amongst their peers, which comprise both friendship and dominance, and they explore the ranges in which their peers have been brought up, as well as entirely unfamiliar regions. In addition to these relationships with peers, the mature sub-adult individuals attempt to establish relationships with adults. In this process the female has a considerable advantage over the male as she can offer her sex as a political tool. The male however has little to offer; his appearance still lacks the signs of fitness-quality while his social status is unknown among the adults since he is not in the physical condition to compete with the establishment in personal space advertisement ('long calling'). Consequently the sub-adult male advertises himself as a potential breeder to young adult females by means of enforced contact, usually terminating in a copulation.

MacKinnon's first field-reports of these 'rapes' again stirred widespread scepticism and engendered accusations of unscientific anthropocentrism. Whether such accusations are justified is questionable. It is likely that the rape behaviour in orang utans is analogous in almost every socio-ethological respect to the rape behaviour in humans, centring around the motivational theme of social acceptance. Whether rape is a more or less accepted — though as regards the victim apparently undesirable — form of relationship establishment, as it is in orang utans, or, under the human codes of moral, has become a totally unacceptable form of behaviour, is irrelevant in biological terms.

The exceptional development in secondary sexual characteristics of adult male orang utans suggests that female choice plays a predominant role in mate selection, and as the information on fitness in anthropoids is at least partly acquired through personal relationships, the sub-adult male has little option other than advertise his qualities in this way (Rijksen and Rijksen-Graatsma 1982). Later, once he is recognised as an accepted mate, raping behaviour becomes superfluous if not almost impossible due to the energetic constraints during the pursuit phase, imposed by his large size and heavy weight. Once fully adult, the male has to shift the emphasis of his social strategy away from relationship establishment with potential mates, towards improving or maintaining his social status in the deme. In this process his weighty demeanor certainly plays an important, supportive role, although the same features are undoubtedly a severe handicap in the ecological sense. There are indications that orang utan females for their reproduction choose only those adult males with whom they have for a longer time established a good relationship. For a female in heat the long call of her particular male(s) serves as a beacon of love.

It will be difficult to even approximate under any form of captive management the extremely complex patterns of relationships that are built and maintained in wild orang utan populations, while it is quite likely that such a complexity of relationships is indispensible for the psychological well-being of both the captive male and female ape, notably if they are meant to achieve any reproductive success. The extremely high percentage of rapes in captive orang utan pairs (Nadler 1977, 1982) — carried out even by adult males, which is very rare in the wild situation — , the frequent inadequacy of nursing care in captive mothers, and the problems of life-long second generation captives to breed successfully are indications that an artificial environment does not offer the

Fig. 4. The exploratory behaviour of orang utans in the wild also comprises esthetics; a sub-adult male sniffing a wild flower (Gunung Leuser Reserve).

social (and other) stimulation necessary for a normal ontogenetic development in orang utans (Fig. 4).

To complicate matters of management even more, it is possible that apes of Sumatran origin need a higher degree of social contact than those originating from Borneo. MacKinnon was the first to suggest that sociability among Sumatran orang utans is better developed than among their Bornean counterparts. The differential sociability is indeed reflected in a comparison of the encounter frequencies of observer and ape units in the Ketambe area, North Sumatra, and the Tanjung Puting area, southwest Kalimantan (46% lone units against 88.5% lone units, respectively; Galdikas 1978). Of course the value of such a comparison is debatable since the data are strongly dependent on the age/sex compositions of the respective demes. Nevertheless the differences in sociability are equally strongly suggested in the observed behaviour patterns during temporary associations in both areas. Galdikas (op cit.) has reported that 'contacts between (adult) males and all other classes of orang utans … could be characterized (as) avoidance'; adult males as well as adult females, in general, seemed to repel other females as well as pre-adult individuals. Such general avoidance between the age/sex classes was rarely observed in Sumatra where, indeed, the large fruit trees were the meeting places for many members of the deme. During such temporary associations the individuals displayed a scala of interactions which revealed much of their patterns of social relationships. The only general rule was that prime adult male/adult male contacts were invariably antagonistic in both subspecies.

In any case, even the least sociable — Bornean — orang utans spend some 15% of their time in some form of contact with a conspecific, while for the important phase of independent pre-adulthood, when the necessary relationships are being established, the contact period covers more than 40% of their time (Galdikas 1978). Whether this contact should, in case of management, invariably be non-agonistic is questionable; the disadvantage of any form of management will always be that the subject cannot influence the arrangements of his own social contacts, while in the wild situation he may develop a strategy that fits his intellectual capacity and psychologically gives aim to his life.

Since it has been accepted widely that ecological factors are major determinants of the evolution of social organisation, students of orang utan biology have sought for explanations of the 'solitary' nature of their subject, although several overlooked the factor of long-term predation by humans (Rodman 1973b), or, worse, muddled up the concept of behavioural ontogeny with that of evolutionary adaptation (MacKinnon 1974; Galdikas 1978, 1979). Still, although the human occupation of North Sumatra is of a considerably younger date than that of Borneo — suggesting adaptive differences in sociability — it cannot be ruled out that some of the differences can be explained in ontogenetic terms, linking them to a local, or temporary difference in habitats, particularly concerning the distribution of food.

In spite of the many confusing descriptions of the ecological conditions in the Sundaland rain forest, it is obvious that for a frugivorous animal the forest can be described as a patchy habitat as regards the availability of food in space, time, quality and quantity (Whitmore 1975; Rijksen 1978a). That the over-all habitat conditions do not differ much between the Sumatran and Bornean study sites can be extracted from a comparison of the size of the home-ranges and the distance travelled per day by the apes in the respective habitats. In the comparatively poor habitat of the Tanjung Puting area the orang utan individuals occupy home-ranges which rarely exceed 6 km² in extent, while their daily range of travel covers some 790 m on average (Galdikas 1978). These figures correspond closely with those of the home-range size and daily range of orang utans in other, supposedly 'richer' habitats, notably the Ketambe area, where indeed the population density may be little higher.

Though suggestive, this comparison is obviously too coarse to provide any clue to the problem of different sociability of the Sumatran and Bornean apes. Only detailed comparative analyses of the distribution of food in space, time, quality and quantity in both regions can give insight in the development of a feeding strategy related to the social attitude, a comparison which is as yet impossible as the data are not available. Still one important comparison can be made. The Ketambe area is very rich in large fruit trees and groves of fruit trees which regularly provide a substantial, localised amount of high quality food for several individuals over a long period of time, a condition apparently unknown for the Tanjung Puting region (personal observation 1973), where, for example, large strangling fig trees are absent. Under the conditions of the Ketambe area where individuals can associate in large fruit trees while competition — at least for food quantity — is negligible, it is conceivable that the growing orang utan infant is exposed to a much higher degree of social contacts than his Bornean counterpart living under the stress of high competition at any one location. Nevertheless, whether the differences in sociability between Sumatran and Bornean orang utans are indeed ontogenetically determined remains to be demonstrated conclusively. I strongly recommend that rehabilitation pro-grammes and zoos, through their experimental raising conditions study these basic ethological problems in a long-term, cooperative research programme. Better insight in these matters will certainly help us understand the adaptation of limited gregariousness — or the 'solitary' nature — of the orang utan as a species.

The conservation situation of the orang utan is perhaps the most confusing and least understood. I have already mentioned that the endangered status of the man of the forest as laid down in the Red Data Book of IUCN-WWF, is entirely justified, in spite of the fact that the numbers of extant apes may be several times higher than the 4,000 previously estimated.

The threat with the greatest public appeal, namely hunting pressure and illegal trade in orang utan youngsters, is more or less under control in North Sumatra (Aveling and Aveling 1977) and South Kalimantan (Galdikas and Brindamour 1979), a feat accomplished largely by the impact of the re-habilitation programmes. However, since social and economic development of the western type has become a major objective in the policy of Indonesia, the orang utan is facing a far more serious threat, namely the imminent, almost total destruction of his habitat, the Sundaland rain forests of Sumatra and Borneo. If we should fail to understand the present conservation situation and, on the simple ground of numbers, would lift the ape's endangered status, it would be nothing less than disastrous.

On paper, the Sumatran orang utan is comparatively well protected in the almost 10,000 km² large Gunung Leuser Reserve. The reserve covers almost one third of the ape's present distribution range, but unfortunately some two thirds of the reserve comprises unsuitable habitat for orang utans and many other rain forest animals and plants, being situated at an altitude of well over 2,000 m. If the present area of forest in the reserve, in particular the few river valley habitats, the (swampy) lowlands and the medium altitude plateaus can be kept inviolate, the Gunung Leuser Reserve can harbour anything up to 3,000 orang utans. Nevertheless, whether this population size is sufficiently large to support long term visability is uncertain.

Modern island biogeography offers some tantalising suggestions for a conservationist as to the theoretical relationship between area of possible distribution and extinction rate, but unfortunately it cannot yield an accurate way to calculate the minimum size of reserves for a viable population of a certain species to survive, mainly because the variables involved cannot sufficiently be characterised. Consequently the only way to calculate a minimum area of habitat is to assess the minimum size of a viable population, and multiply the number with the known minimum population density of the extant population in the area to be reserved. In order to assess the number of individuals making a viable population it is possible to draw from data of the man of the forest's nearest relative, *Homo sapiens*, in a similar ecological condition. Thus, some ancient ethnic communities of the South East Asian region, namely the Andaman negritos (Ratcliffe-Brown 1922) and the Tasmanian australoids (Diamond 1978), were able to survive in equilibrial isolation since the Late Pleistocene with a population size of some 5,000 individuals (of population densities between 1.5 and 10 individuals per km²), although typically both populations perished only decades after their uninvited contacts with Europeans. Smaller extant aboriginal populations on islands have not been found. If the same genetic criteria of viability apply to the survival of the orang utan, it is clear that, in order to bring the number of protected Sumatran apes into the range of the viable minimum, the Gunung Leuser Reserve should be extended with a contiguous area of rainforest habitat to eventually comprise 15,000 km².

In theory there are still possibilities for such an extension (Fig. 3), but whether theory can be turned into practice depends on the political powers of conservation in Indonesia. Most areas suitable as extensions of the reserve have been leased as timber concessions and consequently will soon be turned into more or less severely depleted, if not entirely unsuitable habitat for orang utans, if the necessary actions to legally re-allocate the land use objectives of these areas will not be taken immediately (Fig. 5).

The conservation situation in Borneo at present is more precarious, but there are some prospects for a significant improvement. Currently two medium-sized reserves are in existence in Borneo, both situated in Kalimantan (Indonesian Borneo). The Tanjung Puting Reserve, covering some 2,500 km² of which almost half comprises suitable habitat for orang utans; and the Kutai Reserve, covering some 2,000 km² of which one third has been destroyed by commercial logging, leaving some 1,300 km² unimpaired (UNDP/FAO 1973). Possible

Fig. 5. The main threat to the orang utan's survival is large-scale habitat destruction, mainly caused by commercial logging, but also by local, human encroachment ('ladang' agriculture within the Gunung Leuser Reserve).

extensions of these reserves meet with the same problems as those concerning the possible extensions of the Gunung Leuser Reserve, namely that nearly all surrounding forest area has been leased as timber concessions.(Fig. 6).

Recently, two new reserves have been proposed for the Bornean region (see Fig. 2). Declared as a nature reserve by the Provincial Government in 1973, the extremely valuable Bukit Raya Reserve — some 1,400 km^2 — awaits ratification including a substantial extension which is meant to bring the total protected area up to some 12,000 km^2. The other area proposed as a reserve is the Kayan catchment system (UNDP/FAO 1977). With its proposed 16,000 km^2 in extent, the Kayan River Reserve may become the largest, and — as regards floral and faunal components — the most valuable Sundaland rain forest reserve. However, for its establishment to be ratified will need time, while encroachment will continue unhampered. Nevertheless, if legal establishment of these two immensely valuable major reserves can be achieved before the ecosystems have been destroyed by logging and encroachment, the Indonesian government has taken a major step in the conservation of the orang utan, and innumerable other components of the heritage of this country. Perhaps the establishment of these reserves will also set an example for Malaysia to follow suit; both Sarawak and Sabah do not have reserves exceeding 1,000 km^2 which may be considered as less than 10% of the preferable minimum size for a rain forest reserve.

Still, having a reserve legalised on paper does not implicitely guarantee its integrity in the field — a considerable proportion of rain forest reserves in Indonesia have been irreparably destroyed by commercial logging activities and illegal encroachment. If, for reasons of space and time, we disregard the problems of conservation management in Indonesia — such as the effectiveness of protection of ecosystems against local encroachment — it is important to realise that the world demand for cheap wood has put such economic pressures on Indonesia (FAO 1977) that virtually every stretch of forest area is meant to be utilised eventually for timber exploitation in the current strategy of the powerful forestry. Recent allocations of timber concessions inside long-established nature reserves (such as the valuable Sikundur area in the Gunung Leuser Reserve [Aveling and Aveling 1979], and the Kutai Reserve) are indicative in this respect. Then, Kalimantan may even present a special case as it is relatively thinly populated and many ethnic communities still live in what is generally considered as 'primitive tribal existence'. Consequently, the land use objectives other than timber exploitation, which might accrue to the indigenous population of ethnic minorities, are grossly neglected in view of development priorities elsewhere. In such cases timber exploitation is seen as especially beneficial since it also opens up the primitive hinterland for development. Added to the supposedly beneficial effects of removing the rain forest are the vastness and poor access to the Bornean region, factors which make that control on the logging activities and their extent are insufficient if not virtually absent.

In view of this situation it is important to denote one frequently made ecological error as regards forest management of ecosystems harbouring orang utans — whether inside, or outside reserves. Time and again it has been stated

Fig. 6. After the commercial logging practice, local people may occupy the area and eventually turn it into derelict land, prone to erosion; the first stage of planting after the area has been abandoned by the logging company (Gunung Leuser Reserve).

that commercial forestry activity does relatively little damage to the habitat of orang utans (Stott and Selsor 1961; Wilson and Wilson 1975; Chivers 1977), and may even be a favourable form of habitat management. Ironically this idea may draw upon conclusions from ecological studies in undisturbed regions, showing that orang utans may attain highest population densities in areas with a non-climax forest community of trees. For two reasons this idea of beneficial forestry activity is a false supposition, even apart from the fact that surveys have conclusively demonstrated that orang utans are either absent or extremely rare in logged-over areas. First, it attempts to cover up a basically destructive activity in supposedly beneficial concepts — not seldom worded as 'habitat improvement' — and essentially disregards the delicate interrelationships comprising the rain forest-ecosystem of which orang utans are an inextricable component. Second, it ridicules the scale and impact of commercial activities. Even the most careful forms of selective logging, extracting some 10% of the large boled trees, results in the destruction of at least 65% of the forest (Burgess 1971), in effect leaving behind a patchwork of isolated micro-island habitats unsuitable for at least 60% of the original flora and fauna (Harrison 1965; Rijksen 1978a; Aveling and Aveling 1979). Thus, whereas the orang utan, and numerous other rain forest animals and plants may benefit from relatively small scale disturbances — such as those caused by elephants, or by the subsistence activities of a human population comprising less than 10 individuals per km^2 — there can be no doubt about the disastrous effects of commercial logging and other modern developments. It is easy to foresee how the current forestry strategy in Indonesia, under the economic and political pressures of the wealthy consumer nations, will cause the total destruction of what is known as the richest terrestrial ecosystem in the Old World, namely the Sundaland rain forests of Sumatra and Borneo.

Fortunately there is a glimmer of hope. The awareness of an imminent ecological disaster is dawning in Indonesia and the Nature Conservation Department (PPA) is rapidly gaining political and effective strength. The increase in status of the conservation movement is largely owing to the active support by the new State Ministry of Development Control and the Environment, while the World Wildlife Fund (WWF) and the Food and Agricultural Organisation of the United Nations (FAO) provide joint technical assistance to the Nature Conservation Service (PPA). Recently also the Netherlands Bilateral Aid Program (DITH) added technical assistance in the important field of training in Environmental Conservation Management for PPA personnel and foresters. Nevertheless the dawning concern for the conservation of the rain forest is to be illustrated best with the public promise of the Director-General of Forestry in August 1979, that all commercial logging will be banned from reserves and proposed reserves.

Every effort should be made to save the man of the rain forest, but the objectives should be sound, and the efforts be concentrated according to priorities. Further mysteries or confusion of thoughts should be avoided if the efforts are really meant to save the ape — notably objectives of pure biomedical or pure biological research should not be muddled up with objectives of applied ecology which conservation in fact is. Among other objectives this also applies to the captive management of species and rehabilitation. It is certainly true that some bird and ungulate species on the verge of extinction have been raised to substantial numbers again by means of captive management (Fitter and Scott 1965). Following these examples one may feel captive management to also be a means of saving apes, notably orang utans; in particular the supposed successes of rehabilitation programmes seem to have strengthened the idea that a combined captive breeding/rehabilitation programme can save the man of the forest from extinction. However, whether such a development is feasible is highly doubtful, while its desirability is debatable.

Initially the rehabilitation of orang utans centred around the objective of restocking the supposedly depleted wild population with breeding potential, in combination with the animal welfare idea of saving individuals from a miserable existence in captivity. Consequently all rehabilitation stations have been established in areas where wild orang utans occur. Ironically, the first project founded by Barbara Harrisson in Bako National Park was considered a bad choice of location since wild orang utans were absent in the Park.

In general the practice of rehabilitation consists of finding the illegally kept ape; confiscate him; transport him to the station; check him medically and cure him for verminosis and other possible ailments; put him in quarantine for at least one month; introduce him to the group of feral peers and; once released, provide him with a feeding regime according to his physical condition and his assessed dependence on provisioned food. Special training, such as demonstrating to the ape what kinds of food are edible appears not to be necessary for orang utans — and would indeed be impossible for a manager lacking the detailed environmental knowledge of, for example, an archaic gatherer-hunter of the rain forest. Newcomers easily learn from the more experienced rehabilitant apes, while many ex-captives may well remember the identities of many foodplants from their youth in wild conditions. In the two Sumatran rehabilitation stations the practice was to transfer those apes which were considered as rehabilitated — i.e. had become largely independent of the provisioning and were able to establish relationships with peers — to areas inside the Gunung Leuser Reserve where they were left devoid of provisioning and would most likely not easily come into contact with humans again. In both the Bornean stations the provisioning is continued, and there it is expected that the rehabilitant individuals eventually become 'wild' on their own account.

If the development of rehabilitation in individuals is closely monitored, the following conclusions can be drawn: (a) ecological adjustment to the habitat in ex-captive apes is possible and comparatively easily achieved by the majority of subjects; (b) social integration of ex-captives into a peer-group of rehabilitants is easily achieved, but social integration of such subjects into the wild, established population is a difficult process, feasible only for a few individuals, notably adolescent females. Nevertheless there are some important complicating factors in the rehabilitation process which should be considered carefully. Firstly, it can be conjectured that even an ecologically fully rehabilitated individual, released into a totally unfamiliar area will experience terrible problems as regards food gathering, since he is familiar neither with the topography nor with the distribution of possible food plants. If we then consider that the ape is particularly liable to stress when encountering hostile, unfamiliar residents, it can be envisaged that only the fittest individuals will survive such an ordeal.

Secondly, there are the much more serious problems as regards the wild population in which such rehabilitants are dumped, notably the probability of disease-introduction and transfer. It has been demonstrated conclusively that the human attendants and visitors of rehabilitation projects are a source of diseases which can easily be passed on through rehabilitant apes to the wild population (Rijksen and Rijksen Graatsma 1975). Also there is the possibility that serious diseases, notably tuberculosis and poliomyelitis, are being introduced into the wild population with rehabilitated apes (Figs. 7 & 8). The stress during the early phases of total independence in an unfamiliar, hostile environment can cause sub-clinical stages of a disease to become highly infectious clinical stages. And last but not least, the introduction of a number of strangers into a wild population in equilibrium with its habitat, will cause serious social and ecological problems for these wild individuals, in the sense of xenophobic actions and increased food competition.

It will be clear that these problems, added to the present knowledge of the distribution and the population size of extant orang utan populations, demand a revision of the original objectives of rehabilitation. After some 15 years of experience in orang utan rehabilitation the values of the programmes for the conservation of the man of the forest have proved to be nothing more than: (a) enforcement of the law concerning illegal hunting and trading animals of endangered species, and (b) conservation propaganda and fund raising potential.

Considering the value of rehabilitation for law-enforcement, and realising that the ongoing destruction of habitat will inevitably cause further availability of illegal captive orang utans in the coming years, it is justified that rehabilitation programmes continue to operate in both Sumatra and Borneo, though preferably not in the present locations.

Unfortunately the value of rehabilitation for conservation propaganda is often interpreted as touristic attraction, and the rehabilitation stations, notably the Bohorok station in North Sumatra and the Sepilok station in Sabah, have attracted considerable numbers of foreign and local visitors. It should be clear, however, that rehabilitation of apes is incompatible with the objective of tourist, or visitor attraction; the attention of visitors sets back the rehabilitation process significantly, and the frequent contact with visitors increases the probability of disease-transfer manyfold (Rijksen 1978a) (Fig. 9).

In view of the serious problems related to the rehabilitation of orang utans in areas where wild apes occur, I strongly recommend the liquidation of the existing programmes and their re-location in rain forest areas where orang utans have been exterminated by local hunting long ago. For the Kalimantan region the Pleihari-Martapura nature reserve, in the area between the river Barito and the river Mahakam, may be a suitable alternative site for the present Tanjung Puting rehabilitation project; for the Sumatran region the Sumatera Selatan I nature reserve may be a suitable alternative site for the present Bohorok rehabilitation project (see Fig. 2). The former Ketambe rehabilitation project was liquidated in April 1979 with the transfer of all rehabilitant apes to the Bohorok station. The Ketambe station has become a permanent research

Fig. 7. The practice of rehabilitation consists mainly of food provisioning. In the foreground a wild sub-adult male gathers the scraps of leftover food (Ketambe Rehabilitation Centre).

Fig. 8. Adult male orang utan climbing the feeding cage of the juvenile rehabilitant orang utans; contacts between the wild and the rehabilitant apes are inevitable (Ketambe Rehabilitation Centre).

Fig. 9. Contacts between visitors and rehabilitant orang utans are inevitable, causing a serious health hazard for both the rehabilitant and the wild population. (Note bipedalism in the rehabilitants, a habit learned during their former captivity.) (Ketambe Rehabilitation Centre)

station for forest ecology and conservation, an example which could easily be followed by the Tanjung Puting station. An alternative function for the Bohorok station could be as a main interpretation centre for the Gunung Leuser Reserve, a function it has already begun to explore with great success during the last few years.

Apart from avoiding the deliberate contamination of the wild orang utan population with rehabilitants, the transfer of rehabilitation programmes to areas where wild orang utans are absent has the advantage that the rehabilitants can socially integrate themselves into a more or less permanent group and ecologically adjust themselves in an area where they can live permanently without the stresses of unfamiliarity and xenophobic residents.

It is noteworthy that the present rehabilitation programmes concern only those ex-captives which have been extracted from the wild population, the majority of which have been in captivity for a relatively short period of time, and many of which, as said, retain some knowledge of their food plants. The apes who had been in captivity for a longer period of time — over 4 years — experienced obvious problems in their rehabilitation, both their ecological adjustment and their social integration, even in the rehabilitant group; an observation which suggests that the rehabilitation of life-long captives, such as excess apes from zoos and institutions, meet with difficulties as yet unimagined. Moreover, I hope to have made it clear that the main function of rehabilitation is law-enforcement in order to save what is left of the wild population, rather than re-introducing apes to living under wild conditions, which should be considered as nothing more than the effect of it.

I hope that every intellectual, educated in biology, will support the concept that the conservation of the orang utan primarily concerns the preservation of the rain forest ecosystem, of which the ape is a biotic component. Losing the original habitat of the man of the forest will be an ethical disaster for the world, apart from the fact that such a loss will be an ecological disaster also for the Indonesian people. It is my belief that an orang utan as a being isolated from his natural selection pressures is little more than a living museum specimen, whose ethical value is educational only as long as his wild counterparts still exist in their natural habitat. Otherwise it represents a shameful example of human incompetence and destructiveness.

Currently there is a growing army of conservationists fighting for the cause of conservation of the rain forest and its inhabitants against the short-term economic interests of people, groups of people, or even nations. In order to be able to win their case they need every support that can possibly be raised — moral, political, physical, and financial.

Obviously, there are few institutions as well equipped as zoos with expertise and material for the task of educating large numbers of people into the marvels of nature and propagandising the cause of nature conservation. It is therefore my opinion that the role of zoos in the conservation of the orang utan should be mainly in the field of education, interpretation and propagandising, rather than in wasting effort and money in attempts to breed apes in captivity for the replenishment of a — fortunately not depleted — wild population, which can still be preserved.

Fig. 10. Much is to be improved in the captive management of orang utans in Indonesian zoos (Medan Zoo); a clear prospect for fruitful technical assistance from Western zoos.

One possibility for zoos in developed countries to actively support conservation of orang utans, still largely unexplored, is to provide technical assistance in the fields of management, education and interpretation to zoos in tropical countries where rain forest ecosystems still exist, and eventually establish a fruitful cooperation with such zoos. Notably in Indonesia the zoos are immensely popular recreation resorts which would provide an excellent field for large-scale public education in biology and environmental conservation.

Moreover, technical aid aimed at improving the management of Indonesian zoos would certainly help curbing the still substantial drain on wild populations of animals, including orang utans, since many zoos need a steady influx of fresh animals for replenishing the considerable mortality of their stock due to mismanagement (Fig. 10), not to mention the illegal trade. Most likely the impact of technical expertise from western zoos in these fields will yield a much greater effect for the conservation of the orang utan, than the implementation of the well-meant, but poorly-founded ideas of captive breeding and rehabilitation of the domesticated end-product can ever achieve. Conversely, the idea of captive breeding will probably divert the attention of the public away again from the real problems of conservation, namely large-scale habitat destruction, as has been one of the propagandistic flaws of rehabilitation. There should be no doubt that the man of the forest can only be saved by preserving his forest.

References

Aveling, C. and R.J. Aveling (1977). Orang Utan Rehabilitation Centre, Bohorok. Report WWF project 1589 (unpublished): 13–14.

Aveling, C. and R.J. Aveling (1979). The effects of logging on the fauna of lowland tropical rainforest. Joint PPA/WWF surveys in Sikundur Reserve, unpublished report.

Bemmel, A.C.V. van (1968). Contribution to the knowledge of the geographical races of *Pongo pygmaeus* (Hoppius). Bijdr. Dierk., 38: 13–15.

Borner, M. (1976). Sumatra's orang-utans. Oryx, 13: 290–293.

Burgess, P.F. (1971). The effect of logging on hill dipterocarp forests. Malay. Nat. J., 24: 231–237.

Bygott, J.D. (1974). Agonistic behaviour and dominance in wild chimpanzees. Ph.D. thesis, Cambridge University.

Carpenter, C.R. (1938). A survey of wildlife conditions in Atjeh, North Sumatra. Med. Ned. Comm. Int. Nat. Besch. Amsterdam, 12: 1–33.

Chivers, D.J. (1977). The lesser apes. In: G. Bourne (ed.), Primate conservation. Academic Press, New York.

Davenport, R.K. (1977). The orang utan in Sabah. Folia Primat., 5: 247–263.

Diamond, J.M. (1975). The island dilemma: lessons of modern biogeographic studies for the design of natural reserves. Biol. Conserv., 7: 129–146.

Diamond, J.M. (1978). The Tasmanians: the longest isolation, the simplest technology. Nature, 273: 185–186.

FAO (1977). Yearbook of forest products 1964–1975. FAO, Rome.

Fitter, R. and P. Scott (1978). The penintent butchers; 75 years of wildlife conservation. Collins, London.

Galdikas, B. (1978). Orang utan adaptation at Tanjung Puting Reserve, Central Borneo. Ph.D. thesis, University of California, Los Angeles.

Galdikas, B. (1979). Orangutan adaptation at Tanjung Puting Reserve: mating and ecology. In: D.A. Hamburg and E. McCowwn (eds.), The behaviour of the great apes. Benjamin/Cummings, Menlo Park.

Galdikas, B. and R. Brindamour (1979). The Tanjung Puting Rehabilitation Center and orangutan conservation in Kalimantan Tengah. Report WWF project 1523 (unpublished).

Groves, C.P. (1971). *Pongo pygmaeus*. Mammalian species, 4: 1–6. (Am. Soc. Mammol.).

Hagen, B. (1890). Die Pflanzen und Tierwelt von Deli auf der Ostküste Sumatras: Naturwissenschaftliche Skitzen und Beiträgen. 1890.

Harrison, J.L. (1965). The effect of forest clearance on small mammals. IUCN Publ. NS., 10: 153–158.

Harrisson, B. (1961). Orang utan: what chances of survival? Sarawak Mus. J., 10: 238–261.

Harrisson, B. (1965). Conservation needs of the orang utan. IUCN Publ. NS., 10: 294–295.

Harrisson, T. (1960). A remarkably remote orang utan. Sarawak Mus. J. NS., 9: 448–451.

Heurn, F.C. van (1935). Biologische aantekeningen over eenige belangrijke diersoorten van Java, Sumatra en Borneo; de anthropoide apen, de rhinocerossen en het baardzwijn. Med. Ned. Comm. Int. Nat. Besch. Amsterdam, 10: 1–28.

Hoppius, C.E. (1763). Anthropomorpha (1760, September 6). Section CV of C. Linnaeus, Amoenitas Academiae, 6: 63–76. Uppsala.

Jones, M.L. (1968). The geographical races of orang-utan. Proc. 2nd Int. Congr. Primat., Atlanta., vol. 2: 217–223.

King, V.T. (ed.) (1978). Essays on Bornean societies. Hull/Oxford, London.

Linnaeus, C. (1766). Systema Naturae. 12th ed. (also 1758, 10th ed.). Uppsala.

MacArthur, R.H. and E.O. Wildon (1967). The theory of island biogeography. Princeton, University Press.

MacKinnon, J.R. (1971). The orang-utan in Sabah today. Oryx, 11: 141–191.

MacKinnon, J.R. (1973). Orang-utans in Sumatra. Oryx 12: 234–242.

MacKinnon, J.R. (1974). The behaviour and ecology of wild orang-utans. Anim. Behav., 22: 3–74.

Medway, Lord (1976). Hunting pressure on orang utans in Sarawak. Oryx, 13: 332–333.

Milton, O. (1964). The orang utan and rhinoceros in North Sumatra. Oryx, 7: 177–184.

Nadler, R.D. (1977). Sexual behavior of captive orangutans. Arch. Sex. Behav. 6: 457–475.

Nadler, R.D. (1982). Reproductive behaviour and endocrinology of orang utans. In: L.E.M. de Boer (ed.), The orang utan. Its biology and conservation. Junk, The Hague.

Napier, J.R. and P.H. Napier (1967). A handbook of living primates. Academic Press, New York.

Purchas, S. (1613). Hakluytus Posthumus, or Purchas his Pilgrims. (Edition 1625). London.

Ratcliffe-Brown, A.R. (1922). The Andaman Islanders. (Edition 1964). Free Press, New York.

Rijksen, H.D. (1978a). A fieldstudy on Sumatran orang utans (*Pongo pygmaeus abelii* Lesson 1827), ecology, behaviour and conservation. Meded. Landbouwhogeschool Wageningen, 78–2: 1–420.

Rijksen, H.D. (1978b). Hunting behaviour in hominids: some ethological aspects. In: D.J. Chivers and K.A. Joysey (eds.), Recent advances in primatology, vol. 3. Karger, Basel.

Rijksen, H.D. and A.G. Rijksen-Graatsma (1975). Orang utan rescue work in North Sumatra. Oryx, 13: 63–73.

Rijksen, H.D. and A.G. Rijksen-Graatsma (1982). Bosmens zonder Toekomst. Centr. Uitg. Maastricht.

Rodman, P.S. (1973a). Synecology of Bornean primates; with special reference to the behavior and ecology of orang-utans. Ph.D. thesis, Harvard University, Massachusetts.

Rodman, P.S. (1973b). Population composition and adaptive organisation among orangutans of the Kutai Reserve. In: J.H. Bourne and R.P. Michael (eds.), Comparative ecology and behaviour of the primates. Academic Press, London.

Rudin, W.F. (1935). Over den wildstand in het Ranau gebied. De Tropische Natuur, 24: 151–157.

Schaller, G.B. (1961). The orang utan in Sarawak. Zoologica, 46: 73–82.

Schlegel, H. and S. Müller (1844). Bijdragen tot de natuurlijke historie van de orang-oetan (*Simia satyrus*). In: C.J. Temminck, Verhandelingen over de natuurlijke geschiedenis der Nederlandsche Overzeesche Bezittingen, door leden der Natuurlijke Commissie in Indië en andere schrijvers. Zoologie, 2: 1–28. Leiden.

Silva, G.S. de (1971). Notes on the orang-utan rehabilitation project in Sabah. Malay. Nat. J., 24: 50–77.

Simon, N. (ed.) (1966). Red data book, vol. 1, Mammalia. IUCN, Morges.

Stott, K. and C.J. Selsor (1961). The orang-utan in North Borneo. Oryx, 6: 39–42.

Tulpius, N. (1641). Observationem Medicarum, Libri tres. Amstelredamensis: 274–279.

UNDP/FAO (1973). Conservation in Indonesia. Unpublished report.

UNDP/FAO (1977). Proposals for the establishment of conservation areas in East Kalimantan. Unpublished report.

Valk, A.C. van der (1941). Vangen en jagen in Sumatra's wildernis. Amsterdam.

Visser, R.P.W. (1975). De ontdekking van de orang oetan, 1641–1840. Spiegel Historiael, 10: 258–265.

Wallace, A.R. (1869). The Malay Archipelago. (10th ed.). Dover, London.

Westermann, J.H. (1938). Natuur in Zuid en Oost Borneo. In: 11de Jaarverslag Ned. Ind. Ver. tot Natuurbesch.: 3 Jaren Indisch Natuurleven, Batavia (1936–1938): 334–411.

Whitmore, T.C. (1975). Tropical rainforests of the Far East. Oxford, London.

Wilson, W.L. and C.C. Wilson (1973). Census of Sumatran primates. Unpublished report, and personal communication.

Wilson, C.C. and W.L. Wilson (1975). The influence of selective logging on primates and some other animals in East Kalimantan. Folia Primat., 23: 245–274.

Wrangham, R.W. (1977). Feeding behaviour of chimpanzees in Gombe National Park. In: T.H. Clutton-Brock (ed.), Primate ecology. Academic Press, London.

Author's address:
H.D. Rijksen
School of Environmental Conservation Management
ATA 190
Kotak Pos 109
Bogor
Indonesia

19. Epilogue

Leobert E.M. de Boer

Conservation

During the course of history of life on earth there has been a continuous process
of extinction of animal and plant species, a process that has more or less held
pace with the development of new species by mechanisms of speciation and
radiation. The present picture of flora and fauna merely represents a tran-
sitional stage of these processes; many of the extant species are on their way to
becoming extinct, while many others are in the course of speciation. As such, in
evolutionary terms, extinction is a fully natural process.

   Although man is one of the currently living biological species, a species that
after its initial development competed with other species for its existence in a
completely natural way, with the increase of his intelligence and technological
attainments he began to threaten an ever-growing number of contemporary
species. He became an unequal competitor to all his fellow species and he
probably exterminated more contemporaries than did any other species before
him. Particularly during the last decades he became supermighty and his
influence on all other life on earth has become disastrous. It seems appropriate
therefore not to speak of a natural process anymore when species are threatened
with extinction by their greatest predator and competitor, man.

   It is this admission which logically leads to the conviction that we should
reduce the threat that we impose on the animal and plant kingdoms and that we
should put every effort in conserving endangered species. It goes without saying
that there is no reason to intervene in natural processes of extinction, but
whenever man is the prime cause of threat to a species an attempt should be
made to prevent its total disappearance from this earth. It may, however, not
always be clear at first sight whether a given species is endangered solely because
of man's activities or whether it is undergoing a natural process of extinction
because of competition or predation by other species or because of changing
environmental factors not influenced by man. It is true that even whole groups
of species, such as the proboscids, the perissodactyles and also the anthropoid
apes, seem to be 'on their way out' when we compare the few extant species with
their limited ranges to the many various species with vast distribution areas
representing these groups in palaeontological times. Such groups may be
deemed to disappear completely, but even then this would be a process of many

The orang utan. Its biology and conservation, edited by L.E.M. de Boer
© *1982, Dr W. Junk Publishers, The Hague, ISBN 90 6193 702 7*

thousands or millions of years. In most such cases human expansion and destruction of ecosystems seem to cause an unnatural acceleration of the process of extinction resulting in fatal decreases in population sizes within a few decades.

Unfortunately, the conviction that we should not destroy our fellow species is not shared by everyone, at least not when the price that has to be paid to conserve them (in terms of personal or national income, luxury or freedom of action) is considered too high. It may even be that those truly convinced of the necessity to conserve wildlife stand against a majority of 'pseudo conservationists', speaking of conservation but not actually willing to pay the price, and of people who do not really care. Nevertheless, since the earth cannot be considered the possession of mankind only, the voice (or rather the silence) of the majority should not necessarily be decisive and the request to put every effort into conservation in the interest of life on earth, even if expressed only by a minority, is valid.

Even in the circles of those who truly want to conserve, there are, however, differences of opinion regarding the question of how to do so. Generally speaking, there are two main possibilities for conservation: (1) the conservation of individual endangered species, and (2) the conservation of whole ecosystems in which endangered species are living.

The conservation of individual endangered species is primarily performed outside their original, natural habitat, i.e. conservation of species without (or disconnected from) the conservation of their natural ecosystems. This type of conservation varies from conservation under fully-captive (artificial) con-ditions, to conservation in semi-reserves (populated with animals that were bred in captivity and subsequently placed in protected areas outside the original distribution area of the species), and to replacement of wild-caught animals (sometimes of entire populations) directly into areas where they are less threatened. In other words, this type of conservation does not take away the initial human threat imposed on the species concerned and its natural environ-ment, but rather withdraws the species (and only the one species) from the threat by protecting it under various degrees of captivity, leaving its original home range, including the remaining species belonging to the same ecosystem, to be destroyed.

There are important fundamental objections to the isolated conservation of individual species which will be dealt with later on. Apart from these, some objections can also be raised against conservation under captive conditions that directly relate to the quality of this type of conservation for the species itself. Some of these objections, applicable to varying degrees depending on how much the captive conditions differ from the original natural circumstances, are given below:

1. Conservation in captivity often disregards the existence of geographical intraspecific differences, and animals from different geographic origins are often mixed. This practice is based on the species concept, in which all individuals belonging to a given species may be crossed without risks of reduced fertility in the offspring. However, in the majority of cases the species descriptions are based on morphological and anatomical studies

344

of museum material, while other characteristics that may distinguish between geographical forms, such as behaviour, chromosomes, isozymes, etc. have not been studied so far in many species. Even in those cases where geographically determined differences in external appearance have been reported, these differences are often neglected. The institutions maintaining endangered animals care about subspecific differences only in a handful of species, and these are mainly the 'impressive' species that of old received more than just superficial attention in the great animal kingdom encyclopaedia used as standard reference works in zoological gardens. In tigers, giraffes and gorillas, for instance, emphasis is laid on breeding pure subspecies, but in most of the lesser primates, as another example, the clear indications of the existence of great intraspecific geographic differences are simply ignored. Unnatural mixing of animals from different populations, in the first place, leads to the creation of a mixed gene pool that does not exist in the wild. As such, this is an undesirable practice, since conservation should aim at conserving natural populations, or at least groups of individuals that reflect the natural state as closely as possible. In the second place, uncontrolled mixing of animals from different geographic origins may lead to considerable problems of compatibility. Modern methods have demonstrated that at least in some cases the currently used species descriptions are not correct in that the differences between subspecies or geographic populations are more extensive than was believed previously on the basis of morphological and anatomical studies. The existence of wider variation between geographical forms may result in incompatibility when pairing or grouping animals from different localities, or reduction of fertility in their hybrid offspring.

2. If captive populations are not properly managed genetically they are liable to risks of inbreeding and loss of genetic variability. The risk of inbreeding increases with the number of units into which a captive population is divided. Especially in zoo populations the numbers of animals per collection are very low, resulting in high risks of inbreeding. Only good cooperation between the participating institutions can lead to a reduction of this risk.

Loss of genetic variability in small captive populations is a risk at least as important when we are speaking of conservation. Conservation should aim at conserving gene pools including all polymorphic systems and rare variants that were present in the population at the time conservation was started. Such a gene pool, moulded by ages of natural selection, forms the basis of the species on the diversity of which it has to rely in the struggle for existence under natural circumstances. Thus, considerable reduction in variability in captivity may close the way back to the wild for the population, or at least considerably reduce the chance of survival under unprotected conditions. As with the inbreeding problem, the limited size of the captive population and its subdivision into numerous small groups, determines the risk of loss of genetic variability, and again only good cooperation between the institutions maintaining animals of the species may reduce the risk. However, the limited size of the captive population always involves a certain loss of variability over the generations due to genetic drift. Breeding according to optimal schemes only slows down the process.

3. Under any kind of captive condition the selective forces are different from those acting in the wild. Theoretically, a total absence of natural selective forces would not do much harm to the captive population. Although the frequencies of deleterious genes would rise since they are would no longer be selected against, a new equilibrium would establish itself and all gene frequencies would remain constant. None of the genes present in the captive population at the time it was established would be lost. However, this would apply only for a very large population; a small captive population without selection genetic drift will have a much greater influence than in the wild, since there is no selection against possibly increasing frequencies of genes that would be disadvantageous under natural conditions. Thus, the possible increase in frequency of disadvantageous genes, accompanied by a decrease of the corresponding advantageous genes could render the population less fit for reintroduction at a later stage.

More dangerous still is the fact that complete absence of selective forces in captivity does not occur; there are many selective pressures, some of which may act in the opposite direction of those acting in the wild. This may lead to a considerable shift in the original gene frequencies and after a given number of generations, to the total disappearance of certain genes that in the wild may be of crucial importance for survival. In part, such unnatural selective forces are imposed by the captive conditions themselves (housing, group composition, diet, climate, etc.). Also, they may be partly imposed by man, who often tends to select animals for breeding according to personal ideas of which individuals meet the requirements he sets for the species regarding physical appearance, character, etc. As such man is easily tempted to act as a breeder of domestic races and selects only those individuals for breeding which fulfil his own demands. In fact, all unnatural selective forces may result in features of domestication, considerably changing the population genotypically and phenotypically, and reducing the chance of survival of such a population under natural conditions. How strong the unnatural forces in captivity are depends on the species concerned, the places where the animals are housed and the attitude of the keepers. The total absence of unnatural selection, however, seems an unattainable ambition.

4. The unnatural conditions of captivity may lead within a few generations to the loss of non-genetic types of behaviour (behaviour patterns which each new generation learns from the preceding one, as opposed to genetically determined, instinctive behaviour). Such behaviour may be crucial for survival under natural conditions. The relative importance of learned behaviour to instinctive behaviour depends very much on the species concerned. Thus, the risk of losing essential behavioural capabilities varies from one species to the other. It goes without saying that genetic-based types of behaviour are subject to the influence of genetic

drift and unnatural selective forces in captivity, and may be lost as well. On the other hand, new behaviours of any kind may be developed in captivity which could be disadvantageous in the wild.

Summarising, most if not all objections to captive conservation of endangered species involve the loss of species-specific characteristics. In fact, captive conservation has a high risk of preserving only part of the total of characteristics that constitute the biological species. As such, captive conservation has some similarity to the conservation of material in natural history museums, which only preserve morphological and anatomical characteristics of a species, representing only a small part of the total number of traits. Captive conservation, therefore, may easily turn out to be a way of satisfying man's desire to rescue species, rather than being in the interest of the species and nature themselves.*

Only when the risks of captive conservation mentioned above are minimised is there some chance of a real, unselfish contribution to the prolonged existence of the true biological species.

The conservation of endangered species in and together with their natural ecosystems has a completely different starting point. It aims at the preservation of a complete natural system, not only the endangered species living in it. In fact, the entire system, a complex assembly of a vast number of biological species, is considered as being endangered, while in practice often only a small number of the constituent species would receive this qualification if considered separately. The representatives of each of these species are left to live their natural lives in their original environment and in the circumstances to which they are adapted through innumerable generations of natural selective forces. In its ideal form this type of conservation approaches most nearly a true conservation of nature on behalf of life on earth, not merely on behalf of man alone.

Ideally, those ecosystems to be preserved would be protected against any human influence and the species living in them would be allowed to develop under completely uncontrolled conditions. However, in many cases it is not enough just to protect the area against habitat destruction, hunting and other human activities. The ecosystem to be protected may already be partly damaged, certain species may have already vanished, while on the other hand the system may have been invaded by animals that fled from other threatened regions. Such invaders may include new predators, new competitors, conspecifics (possibly introducing new diseases) or conspecific races. Even when no such changes have occurred, the area still occupied by the ecosystem is usually much smaller than it originally was. In addition, the relict area may be split into several isolated parts, prohibiting natural exchanges between populations or subpopulations. The various species of the ecosystem may react differently to these kinds of changes in their own habitat. The one species may reach critical

---

* It should be noted that in itself there is nothing against conserving species purely in the interest of mankind. Most of us would have been very happy if someone before us had conserved a species of dinosaur, even if the remaining characteristics would show us only a glimpse of what a living dinosaur looked like, just as we are glad to know that a mammoth was covered with long hair thanks to the specimens preserved in permafrost soils, and to have a film of the last thylacines living in a Tasmanian zoo. The problem, however, is that we should not feign to act on behalf of the species when we only act on behalf of ourselves.

population size much more rapidly than the other (and thus may require a much larger area to survive); the animals of some species may withdraw with the boundary of the vanishing habitat and eventually be concentrated (or even be overcrowded) in the remaining area (with the possible risk of mixing different geographical forms that invaded each other's area), while others may not; isolation of subareas may be much more effective for some species than for others, etc. A certain degree of human control and management, therefore, cannot always be avoided. In fact, this control has already started with the choice of the areas to be protected as the best locations for reserves. These areas then need to be monitored intensively, especially the endangered species in them. Above all, extensive research is necessary to obtain an insight into the interactions within the ecosystem to enable determination of the demands of the individual species and detection of undesirable developments.

It goes without saying, that this latter type of conservation, i.e. conservation of whole ecosystems, is preferable to the preservation of individual, isolated species, as it ensures that a maximum of the species' characteristics are conserved. However, whenever it is decided to choose this type of conservation, there should be good indications of success. It should be certain that the area(s) concerned can be sufficiently protected against any kind of harmful human activity, that the habitat is sufficiently undisturbed and that the populations of endangered species are large enough to survive. Extensive knowledge and research therefore seems crucial before final decisions can be taken.

In cases where the chances of success for this mode of conservation are considered too small, captive conservation of isolated species may be the only alternative. In fact there are some species that have been saved solely by the capture of the last living specimens for captive breeding, e.g. Père David deer, Przewalski horse, Hawaiian goose, though it should be noted that these early examples are also accompanied by failures such as the quagga and the marsupial wolf. Most probably, the number of species that are becoming extinct in the wild, while the last specimens are surviving in captivity will increase considerably in the near future. Apart from such more or less 'accidental' cases, in which zoos happen to be in the possession of animals whose conspecifics in the wild became extinct, during the past few years there seems to be a more purposeful movement towards the establishment of captive populations of highly endangered species, such populations being meant to act as 'genetic reserves', a kind of security in the event that conservation in the wild should fail.

In all cases in which the maintenance of captive populations is accepted as a means of conserving species, the ultimate goal is (or should be) the reintroduction of captive-bred animals to the wild sometime in the future. The possible chance of reaching this goal depends on various factors, such as:

1. Will there be sufficient suitable habitat in the future? (This depends on the type of habitat required and whether or not it will be possible to extend the available area.)
2. Will proper protection of habitat be possible in the future? (Future protection depends on changes of philosophy of local governments, local inhabitants as well as in the world in general.)

3. Is the species suitable for reintroduction? (This depends on the complexity of interactions with the habitat, the complexity of intraspecific relationships and the relative importance of non-instinctive learned behaviour.)
4. Did substantial loss of crucial characteristics and genetic variability occur during the period of captive propagation? (This is determined by the genetic management during this period, the number of animals and the number of generations.)
5. Is there a serious risk of disturbing the habitat by the reintroduced specimens? (This includes such risks as mixing conspecifics from different geographic origins and introducing new diseases by reintroductants to wild conspecifics or other species.)

Thus, reintroduction programmes may have a much greater chance of success in one species than in another. There are species for which there does not seem to be a realistic possibility of reintroduction (these necessarily can only be preserved as 'museum species' in captivity), while for others the chance may be fairly high. In both cases, however, optimal circumstances should be created in order to maintain the captive populations as *biological populations*.

The orang utans

Turning to the orang utan, it should be mentioned first of all, that when speaking of conservation there are two orang utans, the Bornean and the Sumatran one. It is not the place here to discuss in detail the academic question as to whether or not these two forms should be treated as subspecies or as separate species. Recent studies, however, have made it clear that the differences between the two are great enough for us to consider them as distinct biological forms which deserve to be conserved independently. They should in no way be considered as complementary; the loss of one of them cannot be compensated for by the other, either in the wild, or in captivity. When discussing orang utan conservation we should in fact always pluralise and speak about the 'orang utans'.

Since the early 1960s when the critical situation of the orang utans in the wild was made clear to the world, a considerable amount of research has been done on wild living orang utans. These studies have rendered a great quantity of data on the social organisation, behaviour, reproduction, food choice, migrations, etc. Information has been obtained on all the various kinds of interactions between orang utans and their natural surroundings, giving us a much greater insight into their habitat demands and population densities. Compared with the situation of 1960 it is now possible to make much better estimates of the chances of successfully conserving the orang utans in their natural habitats. The current situations of Bornean and Sumatran orang utans and their ecosystems have been reviewed in depth by Rijksen (1982, this volume), and if adequate protection measures can be taken and if proper monitoring is made possible, wild conservation seems to have a realistic chance of success. In addition, there certainly is no better way of preserving the orang utans. These animals are so highly specialised to the life in their natural habitat, that their uniqueness is only fully expressed in their natural surroundings. The surroundings (i.e. ecosystems), on the other hand, are so unique themselves that they deserve conservation too, not only for the sake of the orang utans, but also for that of the extremely complex assembly of many thousands of species which constitute a unique biological entity in which the orang utan is only a small, albeit not unimportant, link. On the one hand, the complexity of the ecosystem leaves no hope of regaining such a habitat in the future once it has been destroyed. On the other hand, it reduces the chance of successfully reintroducing captive-bred orang utans, at least if the reintroductants are meant to become truly wild animals showing all natural interactions with their surroundings, which are needed by the species as well as by the ecosystem. Thus, currently there can be no doubt about the validity of wild conservation: as the chances seem to be still realistic, every effort should be made to reach this goal.

Nevertheless, not all problems have been solved yet. Important gaps still exist in our knowledge, which should be filled in order to ensure proper conditions for wild conservation. In particular, the situation in Borneo does not seem to be clear enough. The exact distribution and population densities of orang utans there are still unknown, while the possible differences between geographical forms in Borneo have not been studied in enough detail. Much research will have to be done to solve this and many other problems. In addition, the small relict populations, more vulnerable as they are to hazards of any kind, need constant and careful monitoring.

In captivity we are facing a situation with three populations, Sumatran orang utans, Bornean orang utans and hybrids (constituting approximately 350, 250 and 100 living specimens, respectively). Orang utans have been held in captivity for some 200 years, but only during the past few decades have improvements been made in captive maintenance. Bornean and Sumatran orang utans can now be distinguished unambiguously and the mixing of the two forms, though still the case in practice, may belong to the past within a few years. Breeding success has increased drastically, thanks to better housing and dietary conditions, improved medical care and extensive studies on behaviour, endocrinology, genetics, etc. As a result, there is good hope that the Sumatran and Bornean captive populations will prove to be self-sustaining for a prolonged period, and that no new imports from the wild will be necessary. As such, the captive population no longer forms a threat to the wild populations as was the case previously. An excess of births is even to be expected soon, leading to a surplus of captive animals. This may result in more natural birth rates with longer birth intervals (resulting in more natural social relationships) once the zoological gardens and related institutions no longer feel the necessity to produce orang utans, but on the contrary are forced to reduce breeding.

Nevertheless, here too great problems remain to be solved. Apart from distinguishing between Bornean and Sumatran orang utans hardly anything is done as yet to manage the captive populations genetically. In spite of the theoretical knowledge presently available, and in spite of the existence of a well-organized studbook for the orang utan presenting all basic data, practical difficulties in transferring animals from one collection to another and lack of understanding as to the necessity of managing the captive stock as a biological population leads to the continuation of the old situation. Much has to be improved in this context if the orang utans in captivity are indeed to be maintained as healthy populations for more than just a few generations. Much has to be done also with regard to housing and the prerequisites for optimal social and behavioural conditions. Even then, if maximum attention were given to all such aspects, a captive orang utan will only be a poor reflection of the orang utan in its own habitat.

Zoological gardens sometimes do not hesitate in giving the world the impression that they are going to save the orang utans from extinction by suggesting that wild conservation has little chance and that the zoo population is of more importance than the 'rapidly vanishing' wild populations. Zoos are easily tempted to do so because they believe they need the publicity and to reduce the negative approach towards zoos from certain sectors of the public. This, however, does not benefit conservation itself. It is as if the zoo world and the 'fieldworkers' were competing as to who is allowed to conserve the orang utans. This has a rather negative effect, especially on wild conservation, since the impression is given that captive conservation would be a good alternative to wild conservation and that the latter *a priori* has no chance of success. Neither of these impressions is true: in the case of the orang utan there is no reasonable alternative to wild conservation; as long as there are good reasons to believe in the possible success of wild conservation nothing should be done to distract from it. On the contrary, every effort should be made to focus attention on it.

This, however, does not mean that there would be no role for captive orang utan populations in conservation. The captive populations exist for historical reasons, and they should be used optimally. They can play important roles for conservation in two main ways:

1. *Education.* It is often believed that acquainting the public in large parts of the world with exotic animals intensifies their love of nature in general, and generates the feeling that nature (in which such unique animals live) should indeed be conserved. This may not be true *per se*, but if the exhibition of endangered species in zoo collections is accompanied by good information on the species, on its distribution, habits and habitat, and especially on the causes of threat to the species, there may well be a positive effect. In the case of the orang utans it would be of great importance to make clear to the visitors that the main cause of threat to this species lies in themselves (i.e. the peoples of developed countries, where most zoological gardens are situated), since most of the commercial logging activities threatening the orang utan's habitat, are initiated by themselves by their own demands for luxury (wood). If additional information of this kind (also indicating the uniqueness of the particular ecosystem and its role in the biosphere, and the great difference in free ranging and captive orang utans) is supplied with the exhibits, the captive orang utans may make an essential contribution towards changing world opinion in the interest of conservation.

2. *Research.* It has already been stressed several times that the success of conservation in the wild (necessarily under partly human controlled conditions) depends at least in part on a better knowledge of the species

to be conserved. Research in the wild, however, is limited, for practical reasons, to a few fields of interest. Additional information, especially when this requires experimental research, is often much more easily obtained from studies on captive animals. Fortunately, apart from the work that has been done and is still being done in the great primate centres, there is a tendency to increase research activities in zoos; some of the larger zoos have developed research centres, while more and more others have appointed zoologists (not as managing directors, but at least partly as researchers). It is to be expected and hoped, that this trend will be continued and that the amount of basic data produced by it will grow considerably, not only in the interest of captive maintenance, but also in the interest of the species to be conserved in the wild. Some of the contributions in the present volume indicate that zoo research can indeed contribute in an important way.

As indicated above, the third possible function of the captive orang utan populations with regard to conservation, that of 'genetic reserve', is rather debatable. As far as the orang utans are concerned there is currently no need for such reserves, and if by some misfortune wild conservation should fail, there is no realistic hope of regaining truly natural populations with the aid of captive specimens. In fact, however, the question as to whether or not the captive orang utan populations could act as genetic reserves hardly matters, since the best way to maintain and manage these populations is to treat them as if they were genetic reserves. This way of maintenance and management ensures best the possibility for the continued existence of populations with a maximum of genetic variability and a minimum of risks of inbreeding depression. Thus, those in charge of the captive orang utans should unite forces and consider their orang utans as true populations, not as personal properties.

Although actual reintroduction of captive orang utans seems difficult, and at present it certainly seems undesirable because of the risks of contamination (in various ways) of the vulnerable wild populations, the possibility of a surplus of captive specimens in the near future (even after slowing down birth rates to natural frequency) requires research into a possible solution for this developing problem. Plans should be made and experiments carried out to investigate the possibilities of semi-reserves or parks in which larger groups of orang utans could live. It should be stressed that this has nothing to do with direct conservation of the species, but solely with the fact that the captive stock should not be limited in its propagation because of lack of expensive housing and intensive care. It should be noted that the orang utan, although an animal highly specialised to a very particular way of life in a very particular ecosystem, as a higher primate must be a highly adaptable animal too. As such it may well be able to adapt to the rather unnatural circumstances of semi-reserves or parks. This certainly will not result in a truly wild population, but in any case would not be worse than maintenance in purely artificial conditions. In addition it would offer a possibility of increasing the carrying capacity of captive orang utans and thus limit the influence of genetic drift. Moreover, from the point of view of conservation education, such parks could be more effective than traditional zoo exhibits.

Conclusions

To summarise, the following conclusions may be drawn as to the conservation of the orang utans:

1. There still seem to be realistic chances of rescuing the natural populations of orang utans from extinction, and thus, all our attention should be focussed on the conservation of these populations in their original habitat as wild conservation is the only possible way to conserve all characteristics of this species.
2. Every possible effort should be made to convince all parties involved of the necessity for taking appropriate measures for the optimal protection of the wild populations.
3. The wild populations should be studied and monitored carefully in order to obtain further insight into the species' demands and to detect possible ill developments in time.
4. Conservation education in Indonesia, the home of the orang utans, should continue and be extended at all levels of society.
5. Conservation education outside Indonesia with special emphasis on orang utans and the rain forest ecosystem should be increased. All zoological gardens maintaining orang utans should focus attention on the conservation of this species in the information supplied with the exhibits and clearly indicate the causes of threat to the species.
6. The research potential of the captive orang utan stock should be used optimally and the type of research carried out should contribute to the knowledge necessary for conservation in the wild.
7. Apart from its importance for education and research the captive orang utan stock currently does not and can not play a direct role in the conservation of the species. However, for the sake of its own continued existence, it should be treated, maintained and managed in such a way that it resembles as far as possible the natural populations, which involves the acceptance of modern ideas on population genetics and biological species.
8. Thought should be given to the forthcoming problem of surplus orang utans in captivity, including the possibilities of reducing birth rates in a natural way, and investigating the possibilities of establishing semi-reserves or parks.

Above all, it should be kept in mind that the conservation of the orang utans is a very precarious matter. Success stands or falls with prompt and adequate action of all parties involved.